Treatise on Energy Quantum Theory (EQT), Volume 7

Mind–Matter Unity

Consciousness and Subjectivity in EQT Process Philosophy of Nature

Kaisheng Li and Longji Li

2026

This work is a **pre-publication academic manuscript**. The authors reserve the right to revise, expand, or withdraw any part of the text in future editions.

May 29, 2026

Acknowledgements

This book represents the culmination of a long intellectual pursuit and is the collective outcome of our many years of deep collaboration.

It is dedicated to our families.

Having been granted the privilege of life in this world, "life" itself gradually became our most persistent enigma in the years that followed. In seeking to understand why life arises and why it is finite, this quest led us—often unexpectedly—from the mysteries of biological existence into the foundations of physics, and eventually toward a critical examination of the deepest structural assumptions underlying contemporary physical theory. Through our joint efforts, this long and unplanned journey has finally converged into the theoretical framework presented in this book. This exploration has occupied much of our adult lives; that we could follow this path to its conclusion is owed not to professional obligation, but to a sustained and uncompromising fascination with physics itself.

We wish to offer our most special thanks to our respective wives.

Finally, this book is offered to those readers who retain a serious and undiminished curiosity about the nature of life and the structure of the universe. Whether or not definitive answers can ever be reached, the act of sustained questioning remains, in itself, a meaningful response to the limits of a finite existence.

Kaisheng Li & Longji Li
May 2026

Contents

Part I.

Introduction: The Hard Problem, the Isomorphism Program, and Ontological Positioning

1. The Hard Problem as a Problem of Isomorphism

1.1. A Problem Repeatedly Restated Yet Still Unresolved

Consciousness studies face a fundamental obstacle that has persisted for more than a century. Over the past half century, neuroscience has described the material correlates of consciousness with increasing precision: which cortical regions are activated in wakefulness, which rhythms recede in deep sleep and anesthesia, what integrity of thalamo–cortical loops is necessary for awareness to be maintained, and which pathways transform light signals on the retina into an instance of "seeing". These achievements are real, and they continue to accumulate. Yet even after such mechanisms have been fully clarified, one question remains suspended as before: why is this entire set of physical processes accompanied by subjective experience? Why do neuronal firing, transmembrane ionic flow, and synchronized oscillations of neural ensembles correspond to what it is like to "see red" or what it is like to "feel pain"—that is, to a qualitative character given from within, rather than operating silently in the dark without any qualitative character at all?

David Chalmers named this question the "hard problem" of consciousness, distinguishing it from a group of "easy problems". The easy problems—the allocation of attention, the integration of information, the report of states, and the discrimination of stimuli—ask what functions a cognitive system performs. Although such problems are technically very difficult, they are in principle open to mechanistic explanation and can eventually be answered in the language of function and dynamics. The hard problem asks about something else: why is the execution of these functions accompanied by first-person given-ness? This question does not arise within the functional chain; it remains valid even after the functional chain has been completely specified. Even if a system is behaviorally and informationally indistinguishable from a conscious subject, the question "does it really have experience?" can still be raised independently in logical terms. This is precisely where the "hardness" of the hard problem lies.

The difficulty of the hard problem does not arise from its complexity, but from its structure. Any physical theory ultimately explains structure, function, and dynamics: how one physical quantity varies with another, how a configuration maintains stability, and how a process propagates and evolves. The entire vocabulary of physical explanation is a vocabulary of "what operates and how". The hard problem, however, arises precisely after "what operates and how" has been fully described, and asks further why this operation is experienced. Within the vocabulary of structure and function, there appears to be no concept capable of receiving this question. This is what Joseph Levine called the "explanatory gap": even if a physical account is functionally complete, there remains a gap between it and "why there is this experience", a gap not logically entailed by the former.

For this reason, theories that claim to offer a direct and comprehensive solution to the hard problem usually fall into one of two difficulties, from which this book must distinguish itself at the outset. The first is topic substitution: replacing the hard problem implicitly with one of the easy problems—for example, by directly defining consciousness as the "broadcasting of global information" or as a certain "measure of integrated information"—and then declaring the hard problem solved once that easy problem has been solved. This is a substitution of the topic, not a solution to it. The second is overcommitment: using a single term ("emergence", "identity", or "unity") to cover the explanatory gap between physical vocabulary and the first person, making the gap appear closed while in fact it has merely been obscured by words and remains intact. This book rejects both paths and adopts the following discipline: a theory that disguises a problem it cannot solve as one already solved is farther from the truth than a theory that clearly marks the limits of its explanation.

The strategy of this book, therefore, is not to attack the entire problem head-on, but first to ask: does the hard problem appear as an indivisible whole precisely because it mixes, within a single formulation, several links that can in fact be separated and treated individually? This book answers in the affirmative. The following chapters will show that, once an appropriate decomposition is applied, most of the hard problem can be treated mechanistically: the entire structure of experience can be carried by a set of physical isomorphism relations, and the given-ness of experience can be characterized by a class of physical co-evolutionary relations. What is genuinely irreducible will be compressed into a single residual point with a definite location. This book does not pretend to cross that residual point; it marks it precisely and argues that it has been compressed to the smallest possible scale.

1.2. The First-Principle Thesis of This Book: The Isomorphism between Experience and Gradient Configurations

This book proposes and argues for the following first-principle thesis:

First-person experience is isomorphic to a certain class of gradient configurations of the frequency-resolved energy quantum density field ρ_f, and this book maintains, in the strongest sense, that this isomorphism is identity—experience is what a specific class of ρ_f gradient processes presents from within.

This thesis requires clarification word by word, because each qualification carries argumentative weight.

The "frequency-resolved energy quantum density field ρ_f" is the unique fundamental object stipulated in the foundational volume of EQT (its postulates, master equation, and resonance kernel will be cited in Chapter 2 and will not be rederived here). Its essential point is that the foundation of the universe is not particles, nor fields in a spacetime background, but an energy density distribution partitioned by frequency families and its gradient structure; the various entities and interactions in the phenomenal world are all manifestations of ρ_f under different frequencies, gradient configurations, and stabilization conditions. This book inherits that ontology and extends it into the domain of consciousness.

A "gradient configuration" refers to the overall pattern of gradients exhibited by ρ_f within a given spatiotemporal region and across several frequency families—its directions, magnitudes, topology, singularities, resonance modes, and temporal evolution. This book will argue that it is precisely the rich structure of this pattern that corresponds to the rich structure of experience.

"Isomorphism" in mathematics means a structure-preserving mapping: there is a one-to-one correspondence between two structures, and the correspondence preserves the relations internal to each structure. This book maintains that such a mapping exists between the structural space of experience and the structural space of ρ_f gradients—each structural feature of experience has its counterpart on the gradient side, and conversely.

"Maintains, in the strongest sense, that this isomorphism is identity" is the key decision that distinguishes this book from all weaker approaches, and it must be made explicit here. In the face of structural correspondence between mind and matter, there are in principle three claims of decreasing strength. The strongest is identity: experience is what a certain gradient configuration is from within; mind and matter are two aspects of the same process. They are not causally correlated, nor merely corresponding in a model, but are the same reality. The intermediate claim is structural isomorphism: the structure of experience corresponds one-to-one to the structure of gradients, but it does not assert that experience is the gradient itself; the two may still be conceived as two kinds of things that are only structurally parallel. The weakest claim is modeling correspondence: the ρ_f gradient is the best physical model of experience, and isomorphism holds in the modeling sense; this has in effect retreated to correlation and nearly gives up the claim of "unity". This book adopts the strongest identity claim. The reason will be discussed in detail in Chapter 5; here its core may be stated first: only the identity claim can fulfill the actual meaning of the title "mind–matter unity"; and only the identity claim can reduce the explanatory burden within the hard problem to the minimum—because identity does not require an explanation of a "relation". There is no correspondence relation between A and A awaiting explanation. Thus the question left by the intermediate claim, namely "why does this isomorphic relation hold?", does not arise from the outset under the identity claim.

Yet precisely because the identity claim is strong, it carries a burden that must not be avoided, and the success or failure of this book depends on whether it can bear that burden. This burden may be called the problem of structural sufficiency. Experience is extremely rich: it has continuously varying qualia, mutually heterogeneous qualitative types, a flowing stream of consciousness, a present field that is both unified and differentiated, memory layers of different durations, and temporal experience with thickness. If experience is claimed to be a certain gradient configuration, then the gradient side must itself possess sufficient structural richness—rich enough to establish one-to-one correspondences with all structural features of experience. If there is even one structural feature of experience for which the gradient side cannot provide a counterpart, then "is" immediately descends into "approximately models", and the identity claim collapses back into the weakest modeling claim. Almost all historical identity theories that claimed "consciousness is some physical quantity" failed precisely here: they identified consciousness with a single scalar—some quantity of information, field intensity, or frequency. The structure of a single scalar is too poor to support the richness of experience, so such theories can only claim correlation and cannot establish identity.

The fundamental difference between this book and those approaches lies here: what this book identifies with experience is not a scalar, but a multidimensional gradient-configuration space of extremely rich structure. The next section will show that the richness of this space has five definite sources, and that these five sources respectively carry five classes of structural features of experience. The reason the problem of structural sufficiency can be addressed directly within this framework depends entirely on this point.

1.3. How Identity Is Possible: Five Sources of Gradient Richness

This book maintains that the structural richness of the ρ_f gradient-configuration space is sufficient to be isomorphic to the structure of experience. This richness comes from five sources that are mutually independent yet mutually cooperative. This section first lists the five as the argumentative program of the whole book; their individual development and falsifiable comparison constitute the main body of Part II (Chapters 5 to 11).

First, the richness of single-frequency gradient configurations. Within a single frequency family, the gradient of ρ_f is already a high-dimensional structure: it has direction, magnitude, topology of spatial distribution, singularities, and critical points. This layer of structure carries continuous variation within a single qualitative kind of experience—for example, continuous gradations from dark to light and from warm to cool within the color-experience space of "red", or differences in intensity and bodily localization within "pain". This is the finest texture of the isomorphism: how experience changes continuously within one qualitative kind corresponds to how a single-frequency gradient configuration changes continuously.

Second, the richness of the energy quantum frequency dimension. ρ_f is partitioned by frequency families, and frequency itself is an independent structural dimension. This dimension carries the type differences among distinct qualia in experience—why red and pain, sound and taste are different kinds of experience rather than merely different quantities of the same experience corresponds to their belonging to different frequency families. This source is crucial, because the problem of qualitative kinds in qualia—why experience has mutually irreducible "kinds of quality" such as red, pain, and bitterness, rather than only more or less of a quantity—has always been one of the hardest obstacles for identity theories. Without the frequency dimension, a theory can at most explain "more and less", but cannot explain "the difference between red and pain"; with the frequency dimension, qualitative type differences have a counterpart on the gradient side.

Third, the richness of evolution constrained by the master equation. The evolution of ρ_f is constrained by the unified EQT Master Equation, which specifies all admissible forms by which the gradient field may change over time. This source carries the dynamic structure of experience—the flow of the stream of consciousness, the shift of the attentional focus, and how one thought turns into the next. Experience is by no means a series of static snapshots; it has an internal law of temporal evolution. The counterpart of this law on the physical side is precisely the gradient evolution constrained by the master equation.

Fourth, the richness of same-frequency and cross-frequency resonance modes. This is the heaviest-bearing source, because it carries the two most central structural features of experience: unity and binding. Same-frequency resonance corresponds to coherence within a single subject—why various experiences belong to the same "I" rather than remaining mutually isolated. Cross-frequency resonance corresponds to feature binding—why the color, shape, and motion of an object are experienced as attributes of the same object rather than as three disconnected segments of experience. Chapter 9 will argue that it is precisely this richness of resonance modes that allows the "I" to be understood as a single coherent resonance rather than as the sum of many microscopic subjects. The combination problem (how many interiors become one) will also be dissolved here; see Section 1.5 and Chapter 9.

Fifth, the richness of the gradient transient timescale spectrum. The gradient configurations of ρ_f are not permanently resident; they appear as transients, that is, stable but not eternal forms of existence. Transients in different frequency families have different durations of stability—high-frequency transients are short, while low-frequency transients are long. This source carries the temporal thickness and memory layers of experience: the present is not an extensionless point; it has retention (the resonance of what has just passed) and protention (the anticipation of what is to come). Memory, likewise, has different timescales from working memory to long-term memory. High-frequency short-duration transients correspond to short-term retention and working memory, low-frequency long-duration transients correspond to long-term memory, and forgetting corresponds to the decay of transients. The temporal dimension of experience therefore also has a counterpart on the gradient side, and is not limited to the spatial dimension.

Taken together, the five sources have a unified function: they jointly show that the ρ_f gradient-configuration space has sufficiently high dimensionality and sufficiently rich structure to establish feature-by-feature correspondence with the structural space of experience—variation within a qualitative kind, qualitative types, the stream of consciousness, unity and binding, temporal thickness and memory. Once this correspondence is fulfilled for every structural feature of experience, the identity thesis that "experience is what this gradient configuration is from within" is transformed from a declaration into an argument supported by structural sufficiency. This is precisely what distinguishes this book from all single-scalar identity theories: others identify consciousness with a number; this book makes a fivefold-rich gradient structural space isomorphic to consciousness.

An honest qualification must be added immediately, and it will itself be formulated as a falsifiable proposition (see Chapters 5 and 11): this book maintains that these five sources exhaust the structural dimensions of experience. This is a strong claim, and it is also a claim that can be refuted by counterexample. If a structural feature of experience is found in the future—for example, the sense of self, the sense of agency, or affective tone—for which no counterpart can be found in these five sources, then the identity thesis will have a gap, and the completeness claim of this book will fail. This book prefers to state the claim in this falsifiable form rather than evade testing through the vague phrase “broadly covers”. (It must be indicated in advance that the three items most likely to be cited as counterexamples—the sense of self, the sense of agency, and affective tone—will be argued in Chapter 11 to belong to endogenous perturbation types within the “perturbation axis”, rather than to a sixth class of structure outside the fivefold structural sources; completeness is thereby maintained.)

1.4. Given-ness: As Internal–External Co-Evolution, Not an Intrinsic Property of the Interior

Up to this point, what the fivefold richness carries is the structure of experience—what experience is like, how unified it is, what qualitative kind it belongs to, how it flows, and how much temporal thickness it has. But one matter has not yet been touched, and it is precisely the sharpest core of the hard problem: even if a gradient configuration is structurally fully isomorphic to experience, why is it given (experienced), rather than silently operating with the same structure in a non-qualitative darkness?

The answer of this book is a key shift: given-ness is not an intrinsic property of an internal gradient configuration, but a relation—more specifically, a class of ρ_f detection relations. This shift draws on the mechanism of “observation as participation” from the EQT volume on detection (introduced in detail in Chapter 13). Its central point is that observation is never a non-perturbative reading of the fixed properties of a closed object by an external subject. The detector and the probed system belong to the same ρ_f field. Observation is a co-evolutionary process of local coupling, controlled perturbation, nonlinear feedback, and transient freezing between the two within this field. Its result is a stabilized output within the detection relation, not an absolute property detached from the relation.

When this mechanism is applied to consciousness, the given-ness of experience is not the intrinsic flavor of a coherent resonance inside the brain (if it were stipulated in this way, we would fall back into the dead point of "why does the interior have flavor?"), but the continuous internal–external co-evolution between this internal configuration and what is external to it within the same ρ_f field. Given-ness is a relational event: an internally coherent configuration "manifests" to that interior because it is always situated within co-evolutionary coupling with the exterior. An internal configuration fully isolated and entirely decoupled from the outside, however structurally rich it may be, would merely operate in the dark. It is given precisely because it is situated within an internal–external co-evolutionary detection relation.

Here it is necessary to clarify what is meant by "internal" and "external" in order to avoid misunderstanding. By "external", this book does not primarily mean the environment outside the body, but a functional distinction within the detection relation: the interior is the coherent configuration in which the given content is located, while the exterior is the remaining configuration that co-evolves with it and enables it to be given. In most cases, during normal wakefulness, this exterior is connected to the real environment through the sensorimotor channels of the body. Yet in purely inward thinking, meditation with closed eyes, or vivid dreams, internal–external co-evolution occurs among the brain's own ρ_f structures and does not require connection to the environment. This precisely explains why, even under substantial decoupling from the environment, given-ness still exists and may even become more vivid. What given-ness requires is the co-evolutionary relation between interior and exterior itself, not that the exterior must be the body or the world.

However, if given-ness is tied only to "internal–external co-evolution", an objection immediately arises: coupling exists everywhere in the field. A gravitational-wave detector and its environment also continuously co-evolve and also freeze recordable outputs. Why is the internal–external co-evolution of living systems alone given, while the co-evolution of a detector is merely a third-person, non-interior physical process? In other words, "internal–external co-evolution" seems too broad; it would generalize given-ness to all coupled systems and slide into a form of panexperientialism. The answer of this book to this objection constitutes the next section and also anchors its anti-panpsychist position.

1.5. Boundary as Gate: Given-ness Anchored in Life

The resolution of the above objection lies not in frequency, nor in threshold, but in boundary.

The key is this: the establishment of given-ness presupposes the existence of an "interior"; and the existence of an "interior" presupposes a boundary. For a system to have the "interior" in "internal–external co-evolution", it must first possess a boundary that genuinely distinguishes itself from the environment—a topological "inside". Most matter in the universe does not have such a boundary. Chapter 3 will discuss this in detail: from the Big Bang to galaxy clusters, aggregation driven by the four fundamental interactions operates through attraction and condensation. It can produce structures of greater density and larger scale, but it does not produce an inside–outside distinction—neither celestial bodies nor molecules have an "inside". The first thing to produce a topological interior is a different kind of aggregation, namely the enclosure of lipid membranes (called in this book the second axis of cosmic aggregation). Through enclosure rather than attraction, it first closes off a region with an inside–outside distinction, capable of self-maintenance and selective exchange. This event is precisely the boundary between life and non-life.

The objection is thereby dissolved: a gravitational-wave detector and its environment do indeed co-evolve and do indeed produce freezing, but the detector does not have an interior. It does not possess a boundary that distinguishes itself from the environment. It is merely an interaction between two regions of the field, and there is no "inside" to which that co-evolution could be given. Living systems are different: they have membranes and therefore a genuine interior. The same internal–external co-evolutionary freezing occurs within the framework of "for an interior", and therefore this freezing is given to that interior. Thus given-ness is firmly anchored in the boundedness of life—where there is boundary, there is inside and outside; internal–external co-evolution is given to the interior. Where there is no boundary, there is no inside and outside; co-evolution is merely field interaction, and there is no sense in which it is given. The gate is thereby established: it is not frequency matching, not a perturbation threshold, but the presence or absence of boundary, that is, the distinction between life and non-life. The anti-panpsychist position of this book is secured by this gate—interiority is not a background color diffused through all things, but an achievement with a definite origin, beginning with the first membrane.

It is worth noting that this gate falls precisely on the nature of the membrane and is not an externally imposed constraint. A membrane never seals the interior completely. It produces a boundary of selective exchange: it both demarcates inside and outside (allowing an interior to exist) and maintains continuous coupling between inside and outside (allowing this interior to be given rather than dark). The two sides of enclosure—demarcation and exchange—correspond exactly to two aspects of experience: there is an interior (so that experience has structure and perspective), and this interior is awake (so that experience is given). The anchoring of given-ness in boundary is therefore not a temporary patch, but a conclusion read directly from the physical nature of life.

1.6. The Final Irreducible Cell and the Honest Boundary of This Book

Through the preceding arrangement, most of the hard problem has been received mechanistically: the entire structure of experience is carried isomorphically by the fivefold richness of gradients (Part II), and the given-ness of experience is characterized by internal–external co-evolutionary freezing of a bounded interior and anchored in the boundary of life (Part III). Through boundary, the question "why is it given?" is reduced to "why is there a bounded interior?", and the latter is answered by the fifth mechanism.

Yet this book does not claim that the hard problem has been eliminated. It claims only that it has been reduced and compressed into a final irreducible cell. This final cell is the following: even if the entire structure of a bounded interior—boundary, inside and outside, co-evolution, and freezing—is fully given, the fact that "this structure is experienced from the first-person interior" is still not logically entailed by these structural facts. From "there is an interior, and inside and outside undergo co-evolutionary freezing" to "this interior feels something", the language of structure has been extended to the final cell, but between "being experienced" and "possessing all structures required for being experienced", there remains a gap not entailed by structure. The fifth mechanism and internal–external co-evolution have narrowed this gap to the utmost—so narrow that only the single-cell difference remains between "the structure of a bounded interior" and "the experience of that structure"—but they cannot eliminate it, because this cell is precisely the irreducible core of the hard problem itself.

It must be stated in advance that this final cell will undergo two compressions in the book, so as to prepare the later argument. The treatment proceeds in two steps. The first part of Part III (up to Chapter 15) completes the first compression, locating the residual point as "why the given-ness of a bounded interior is experienced", which this book calls the given-ness residual point. The later part of Part III (Chapters 17 to 20) then shows that the given-ness of human experience is further a given-ness that is cognized and bounded by subjectivity as "mine". Accordingly, Chapter 20 completes the second compression, further compressing and locating the residual point from the general given-ness residual point to the subject-attribution residual point (why the activity of being cognized and bounded as "mine" is experienced). It must be emphasized that the second compression does not eliminate the residual point, but makes it smaller and more precisely located. After two compressions, the gap is forced to its most precise position, yet it remains irreducible. The soluble domain advances twice; the irreducible cell is twice reduced and located, while remaining honestly in place.

This book adopts a clear and non-negotiable attitude toward this final cell: it precisely marks it, acknowledges that it cannot be crossed, and never disguises any identity thesis or mechanistic statement as its solution. One prohibition must be declared in particular, and it runs through the entire book: even though this book maintains that "given-ness is internal–external co-evolutionary freezing of a bounded interior", this mechanistic statement must never be used by itself as the answer to the final cell. "What physical process does given-ness correspond to?" (answer: internal–external co-evolutionary freezing of a bounded interior) is a measurable, falsifiable proposition belonging to the objective third-person side. But "why is this process uniquely experienced from the first-person interior?" is the final cell and belongs to the irreducible residual point. The former has been answered; the latter is honestly handed over. To conflate the two—to use "it just is that process" to deflect "why is it experienced?"—would repeat the error of topic substitution that this book rejects from the beginning. This prohibition also applies to the second compression: even though Chapters 17 to 20 clarify the mechanisms of cognization and demarcation, it is never permissible to use "experience is the activity of being cognized and demarcated as mine" to masquerade as an answer to "why is this activity experienced?" Interface, demarcation, and mechanistic statement all belong to the objective side of the gap; none may usurp the role of solving the final cell (see Chapters 20 and 27).

It must be emphasized that acknowledging this final cell does not weaken the achievement of this book. On the contrary, it precisely defines that achievement. What this book accomplishes is not to claim that it has solved the entire mystery of consciousness, but to condense the fog spread across consciousness studies into a single point with a definite location; to reduce "why is it given?" to "why is there a bounded interior?", and to answer the latter through the fifth mechanism; and finally to show that what remains irreducible is only this single cell (after two compressions, located at the most precise position of the subject-attribution residual point). A theory that can so precisely indicate "where my explanation stops, where what I cannot explain exactly lies, and why only this cell remains" is far more credible and more robust under scrutiny than any claim to have "fully solved the hard problem". The reason why several earlier explorations by the author were unsatisfactory lies precisely here: they either used a word to obscure the final cell and thereby seemed to cheat, or they maintained discipline so strictly that they set aside the soluble parts together with the irreducible cell and thereby seemed to concede defeat. The posture of this book is a third one: solve all that can be solved, mark precisely what cannot be solved, and keep the boundary between the two clear, each in its proper place.

1.7. The Place of This Chapter in the Book and the Division of Labor among Chapters

This chapter completes only one task: it establishes the entry point and overall program of the book. It proposes the first-principle identity thesis (experience is what a ρ_f gradient configuration is from within), indicates that the success or failure of this thesis depends on the richness of gradient structure, lists the five sources of richness as the argumentative program, repositions given-ness as internal–external co-evolution and anchors it in the boundary of life, and finally compresses and locates the irreducible remainder as the final cell. This chapter does not provide the full argument for any of these components; that is the task of the subsequent parts.

On this basis, the book is divided into five parts and twenty-eight chapters. Part I (including this chapter and Chapters 2 to 4) is the introduction: Chapter 2 cites the ρ_f ontology and methodological discipline; Chapter 3 discusses in detail boundary and life as the preconditions of given-ness; Chapter 4 locates this book in relation to dualism, physicalism, functionalism, panpsychism, and existing theories of consciousness. Part II (Chapters 5 to 11) forms the center of the argument, unfolding the fivefold richness one by one, fulfilling the isomorphism between experiential structure and gradient structure, dissolving the combination problem in Chapter 9, and in Chapter 11 placing the sense of self, the sense of agency, and affective tone within the "perturbation axis" while declaring completeness. Part III (Chapters 12 to 20) treats given-ness and subjectivity, in two stages: the first stage (Chapters 12 to 16) introduces the co-evolutionary mechanism from the detection volume, anchors given-ness in boundary, situates the "I" as the boundary of the dominant coherent resonance, and completes the first compression of the final cell in Chapter 15 (the given-ness residual point); the second stage (Chapters 17 to 20) advances to the level of "being cognized and attributed" that is specific to human experience—Chapter 17 discusses knowledge as the higher-order organization of memory-like gradient transients, Chapter 18 discusses the experiential knowledge base and conscious activity (the comparison mechanism of cognized perception), Chapter 19 discusses the subjective boundary and the $\mathscr{B}$ demarcation mechanism of "mine", and Chapter 20 completes the second compression of the final cell (the subject-attribution residual point). Part IV (Chapters 21 to 25) turns to the measurable side, repositioning the state tensor Ψ as the observable projection of the isomorphic structural space and giving measurement interfaces and falsifiable predictions (including two ρ_f-specific predictions). Part V (Chapters 26 to 28) provides the philosophical closure: Chapter 26 directly states the fulfillment of mind–matter unity; Chapter 27 analyzes the nature of the final cell (a boundary rather than an object, and one not dissolved by identity, interface, or demarcation); Chapter 28 is the conclusion of the book.

The three previous works must be clearly positioned here. *The Information World of the Cell: Pooling Networks, Continuous-State Computation, and the Architecture of Life Information Processing*, *The Second Axis of Cosmic Aggregation: Membranes, Life, and Consciousness*, and *The Origin of Cellular Proto-Consciousness: Bi-*

ological Foundations and Operational Criteria were all early works prior to the formulation of EQT and belong to the level of phenomenological observation: they saw the right phenomena—the complex information processing within the cell (pooling networks), the special status of life relative to non-life (enclosure), and the embryonic form of proto-consciousness—but had not yet reached their underlying mechanism. The formulation of EQT arose from reflection on the four interactions: the four forces (together with the fifth mechanism added in early form) are all phenomenal descriptions, answering "how force manifests", but not "what the mechanism of force is"; the ρ_f gradient answers the latter. Therefore, the main line running through this book is the ρ_f gradient principle stipulated in the foundational volume. The three previous works are incorporated under it as provisional phenomenological evidence, and the phenomena they observed are supplied in this book with the ρ_f gradient mechanism. This book is the culminating volume of the series: with a unified mechanism, it repositions previously scattered phenomenological insights and advances them to first-person given-ness and the unity of the "I", and further to the cognization and attribution of experience.

After reading this chapter, the reader should be able to ask of any claim that "the mystery of consciousness has been solved": what have you answered—the structure of experience (what it is like), the givenness of experience (why it is experienced), or the irreducible final cell (why the structure of a bounded interior uniquely has an inner side)? Have you solved the final cell, or merely concealed it with a phrase? If this question is pursued consistently, many ambiguities in consciousness theories will reveal their argumentative gaps. The rest of this book is a direct answer to the first two questions and an honest marking of the third.

1.8. Chapter Summary

This chapter establishes the overall program of the book. First, it proposes the first-principle identity thesis: first-person experience is isomorphic to a certain class of gradient configurations of ρ_f, and this isomorphism is identity in the strongest sense. Second, it indicates that the success or failure of the identity thesis depends on the sufficient richness of gradient structure, and lists five sources of richness—single-frequency gradient configuration, the frequency dimension, master-equation evolution, resonance modes, and the transient timescale spectrum—as the argumentative program of Part II. These respectively carry variation within qualitative kinds, qualitative types, the stream of consciousness, unity and binding, temporal thickness and memory. Third, it repositions given-ness as internal–external co-evolutionary freezing of a bounded interior (drawing on the detection volume), and uses boundary—the boundary between life and non-life, established by the fifth mechanism—as the gate that anchors given-ness in life and secures the anti-panpsychist position. Fourth, it compresses and locates the irreducible remainder of the hard problem as the final cell: "why is the structure of a bounded interior experienced?", and anticipates that it will undergo two compressions (the given-ness residual point in Chapter 15 and the subject-attribution residual point in Chapter 20), being located at its most precise point while remaining irreducible. The book acknowledges that this cell cannot be crossed and never disguises mechanistic statements, interface, or identity as its solution. Fifth, it positions the three previous works as pre-EQT phenomenological evidence, establishes the ρ_f gradient as the main line running through the book, and states that the whole book consists of five parts and twenty-eight chapters.

The failure conditions of this chapter are as follows (only with respect to the programmatic claims newly introduced in this chapter; the failure conditions of each specific argument will be given in the corresponding chapters and summarized in Appendix E).

- F1.1 The identity level cannot be maintained. If the gradient structural space is shown in principle to be insufficient for establishing feature-by-feature correspondence with experiential structure (that is, if structural sufficiency cannot be satisfied), then the identity level cannot be maintained, and the first-principle thesis can at most retreat to the structural-isomorphism level or the modeling level.

- F1.2 The five sources are incomplete. If there exists a structural feature of experience for which no counterpart can be found in any of the five sources of richness (and which cannot be incorporated into the perturbation axis of Chapter 11), then the completeness claim that "the five sources exhaust the structural dimensions of experience" fails, and a gap appears in the isomorphism program.

- F1.3 The relational characterization of given-ness fails. If given-ness is shown to be an intrinsic property of the internal configuration and to have no systematic relation to internal–external co-evolution (for example, if given-ness is unaffected when internal–external co-evolution is severed), then the characterization of given-ness as internal–external co-evolution fails.

- F1.4 The boundary gate fails. If there is decisive evidence that an unbounded (non-living) system possesses first-person given-ness, then boundary as gate fails, and the anchoring of given-ness and the anti-panpsychist position collapse together.

- F1.5 The final cell is counterfeited. Any account that uses a mechanistic statement such as "given-ness is a certain physical process", or an upgraded interface, subjective demarcation, or an identity thesis, by itself as an answer to "why is this process experienced?" violates the core prohibition of this book and constitutes topic substitution.

Table 1.1.. The Boundary between What This Chapter Establishes and What It Cannot Derive

What this chapter establishes	What this chapter cannot derive
The first-principle identity thesis and the choice among its three levels of strength (adopting the identity level).	It cannot prove the structural sufficiency of gradients in this chapter (to be fulfilled chapter by chapter in Part II).

Continued on next page...

Continued table

What this chapter establishes	What this chapter cannot derive
The five sources of richness as an isomorphism program, and their correspondences with five classes of experiential structure.	It cannot develop the specific isomorphism argument for any one source in this chapter (to be left to Chapters 5 to 11).
Given-ness as internal–external co-evolution, anchored in the boundary of life.	It cannot provide the complete mechanism and measurable predictions of given-ness in this chapter (to be left to Parts III and IV).
The compression and location of the irreducible remainder of the hard problem (the final cell, to be located through two compressions as the subject-attribution residual point).	It cannot solve the final cell (this is explicitly handed over to the philosophical boundary in Chapter 27).
The positioning of the three previous works and the establishment of the ρ_f main line; the whole book consists of five parts and twenty-eight chapters.	Mechanistic statements, interface, or identity must not be used to masquerade as answers to the final cell (a prohibition throughout the book).

2. ρ_f Ontology and Methodological Citation

2.1. The Nature of This Chapter: Transferring the Foundation, Not Rebuilding It

Chapter 1 established the overall program of the book: experience is isomorphic to ρ_f gradient configurations (at the identity level), given-ness is internal–external co-evolutionary freezing of a bounded interior, and the final cell is handed over to philosophy. Every link in this overall program depends on something that has not yet been formally introduced in this book—the ρ_f ontology and its mechanisms. The task of this chapter is to transfer this foundation formally into place.

The nature of this chapter must be stated at the outset, because it determines what this chapter should and should not do. This book is one applied volume of EQT (Energy Quantum Theory)—it applies the ontology and mechanisms established in the foundational volume of EQT to the specific domain of consciousness. The foundational volume has its own complete system: starting from several postulates, it derives the unified master equation, reinterprets the four fundamental interactions, general relativity, quantum mechanics, and related structures as manifestations of ρ_f under different frequency and gradient conditions, and offers a series of falsifiable predictions in the domain of physics. All of this belongs to the foundational volume. This book cites it without re-proving it. This chapter does not restate the derivations of the foundational volume, does not defend its postulates, and does not evaluate its success or failure in physics—those belong to the argumentative scope of the foundational volume itself. This chapter does only two things: first, it introduces the ontological and mechanistic items that will be repeatedly used in the remainder of this book, and for each item explains what it means for the consciousness volume and how this book will use it (Sections 2.2 to 2.6); second, it establishes the methodological discipline of the whole book (Section 2.7).

This stance of "citing without re-proving" must be correctly understood. It does not treat the foundational volume as an unquestionable doctrine—whether the foundational volume holds depends on its own arguments and tests, and this book does not prejudge that issue. Rather, it means that the work of this book is conditional: if the ρ_f ontology holds, then experience can be understood in the manner argued here. The credibility of this book is therefore partly dependent on the foundational volume. This is normal for any applied work (the credibility of an application of mechanics depends on mechanics itself). What this book can independently guarantee is the rigor of the step "from ρ_f ontology to consciousness"; the warrant for the ρ_f ontology itself lies in the foundational volume. This conditionality is openly acknowledged and is explicitly listed among the failure conditions in Section 2.7.

2.2. ρ_f: The Frequency-Resolved Energy Quantum Density Field (A1)

The first and most fundamental item cited by this book is the basic ontology of EQT (Postulate A1 of the foundational volume):

The unique fundamental object of the universe is an energy quantum density field ρ_f partitioned by frequency families—energy is distributed in units of "energy quanta", each energy quantum has its own frequency, the entire field is organized by frequency families, and the distribution of its density across spacetime and the gradients of that distribution constitute its basic structure.

The significance of this postulate for the consciousness volume must be fully grasped, because the first-principle thesis of this book (experience is the inner side of a ρ_f gradient configuration) rests entirely upon it. It provides three points on which this book repeatedly relies.

First, unique ontology. ρ_f is the unique fundamental object—there are not two entities, "matter" and "mind", standing alongside it. This uniqueness is the ontological premise for this book's adoption of the "identity level" (Chapter 5) rather than dualism: if mind and matter are to be "one", there must first be a unique reality of which both are manifestations. ρ_f is precisely that.

Second, frequency-family organization. ρ_f is organized by frequency families—frequency is an intrinsic dimension of this ontology and is independent of the magnitude of density. This frequency dimension is the sole resource by which this book carries qualitative type differences (Chapter 7), and it is also the basis for distinguishing matter and mind (Chapter 4: same origin, different frequency bands). An ontology without a frequency dimension, such as a single scalar field, would be unable to carry either of these; the fact that ρ_f carries a frequency dimension intrinsically is the key that allows many of the arguments in this book to hold.

Third, gradient as basic structure. The basic structure of ρ_f is not merely the magnitude of density, but the gradient of density—the rate and direction of density variation across spacetime, together with its topology. The isomorphism proposed in this book is not "experience is isomorphic to the magnitude of density" (which would fall back into a single scalar), but "experience is isomorphic to a gradient configuration". Precisely because gradient configurations are high-dimensional and rich (directions, magnitudes, topology, singularities; Chapter 6), they are able to carry the rich structure of experience. The word "gradient" is the source from which the structural sufficiency of this book (Chapter 5) becomes possible.

2.3. Frequency as an Energetic Spectral Quantity (A2)

The second item cited is the physical meaning of frequency (Postulate A2 of the foundational volume):

> The frequency of an energy quantum is related to its energy through $E = h\nu$—frequency is the spectral quantity of energy, and different frequencies correspond to different energy scales.

This postulate establishes "frequency" not as a mere label, but as a quantity with definite physical meaning: frequency is the spectrum of energy. Its significance for the consciousness volume is that the "frequency dimension" invoked by this book (which carries qualitative types in Chapter 7 and distinguishes matter and mind in Chapter 4) is not a suspicious parameter introduced ad hoc for consciousness, but a quantity grounded in the firm physical meaning of $E = h\nu$. When this book says that "different qualitative types correspond to different frequency families", "frequency" here is the same quantity as frequency in the physics of the foundational volume and has the same meaning as an energetic spectrum. This ensures the continuity between the frequency dimension of the consciousness volume and the physics of the whole series, rather than smuggling in a homonymous concept with a different meaning at the point of consciousness. Whenever the term "frequency" is used in the following chapters, it is used in this A2 sense.

2.4. The Unified Master Equation (A3)

The third item cited is the unified master equation that governs the evolution of ρ_f (Postulate A3 of the foundational volume):

> The spatiotemporal evolution of ρ_f is constrained by a unified master equation. The equation contains nonlinear terms that describe self-sustaining and structure formation, as well as dissipative terms that describe decay and irreversibility. The formation, maintenance,

and decay of all ρ_f structures are solutions of this equation under the corresponding conditions.

This book does not restate the specific form of the master equation (its form and derivation belong to the foundational volume; Appendix B of this book will cite the parts relevant to consciousness). The significance of this postulate for the consciousness volume is concentrated in two classes of terms, on which the following chapters repeatedly rely.

The nonlinear terms describe self-sustaining and structure formation. This book uses them to carry two classes of experiential structure: first, the formation and maintenance of the attentional focus in the stream of consciousness (Chapter 8: a locally enhanced configuration produced by nonlinear amplification); second, the self-sustaining of coherent resonance (Chapter 9: the coherent modes on which unity and binding depend, whose maintenance requires nonlinear self-sustaining to overcome dissipation). Without nonlinear terms, there would be no self-sustaining structure, and therefore no maintainable coherence or focus.

The dissipative terms describe decay and irreversibility. This book uses them to carry two structures: first, the temporal directionality of the stream of consciousness (Chapter 8: the arrow of experiential time shares its origin with the irreversible direction set by dissipation); second, the decay of transients and forgetting (Chapter 10: the decay of transients during their duration and the gradual forgetting of memory correspond to dissipation). Without dissipative terms, there would be no arrow of time, no finite duration of transients, and no forgetting.

The competition between nonlinearity and dissipation also carries the mode of existence of transients themselves: whether a structure can form and be maintained as a transient depends on whether nonlinear self-sustaining can overcome the flattening effect of dissipation. The balance between the two determines the stable duration of the transient, and this stable duration differs by frequency (short at high frequency, long at low frequency), which is precisely the physical basis of the timescale spectrum in Chapter 10. The master equation is therefore not a distant background, but the direct basis for the carrying work performed in Chapters 8, 9, and 10.

2.5. The Lorentzian Resonance Kernel (A4)

The fourth item cited is the resonance kernel that governs cross-frequency coupling (Postulate A4 of the foundational volume):

> The coupling between different frequency families is constrained by a Lorentzian resonance kernel $g(f, f_0)$—coupling strength is strongest under frequency matching (when f is close to f_0) and decays in a Lorentzian form as frequency detuning increases.

This postulate is the core mechanistic basis of Part III (given-ness) and Chapter 9 (binding). Its significance has three aspects.

First, it makes coupling selective. The resonance kernel means that not any two frequency configurations are strongly coupled—only those whose frequencies match are strongly coupled, while detuned configurations are weakly coupled. This selectivity is crucial for multiple arguments in the book: in Chapter 7, synesthesia is interpreted as "abnormal cross-frequency coupling between families that should have been separated"; in Chapter 9, feature binding is interpreted as "the connection of configurations located in different frequencies through cross-frequency resonance"; and in Chapters 13 and 14, the condition of "frequency matching" in detection/given-ness is mechanistically grounded in this resonance kernel.

Second, it prevents the uniqueness of the ontology from sliding into "pan-relationalism". ρ_f is the unique ontology, and all things belong to the same field (A1). This might have led to a kind of "everything couples with everything and is visible to everything" pan-relationalism. The resonance kernel $g(f, f_0)$ blocks precisely this: belonging to the same field does not mean that everything is strongly coupled; only frequency-matched configurations are strongly coupled. This selective gate is supplied physically by the resonance kernel (the detection volume uses it to distinguish "visible and invisible"), and this book cites it. But it must be noted that the gate of this book's anti-panpsychism is something else—boundary (Chapter 3), not the resonance kernel (see Chapter 14 for the distinction between the two). The resonance kernel governs "whether coupling can be established"; boundary governs "whether given-ness can be attributed". The two must not be confused.

Third, it provides two resonance modes: same-frequency and cross-frequency resonance. The resonance kernel describes cross-frequency coupling (where f and f_0 differ but match), and it also covers same-frequency coherence (where f and f_0 are the same). Chapter 9 of this book argues precisely on the basis that "same-frequency resonance carries unity, while cross-frequency resonance carries binding". Its mechanism is this A4 resonance kernel.

2.6. Meta-Norm M: The Requirement of Falsifiability

The fifth item cited is not an ontological postulate, but a meta-norm from the foundational volume (denoted as M):

Every core module of EQT must retain predictions that are falsifiable in the near term—that is, it must be able to specify under what observable conditions the module would be overturned.

This meta-norm has a special and subtle significance for this book, and it must be explained here separately, because this book does not simply copy it wholesale, but observes it in two levels.

The final cell of the hard problem (Chapters 1 and 15)—"why the structure of a bounded interior is experienced"—is, by its nature, not falsifiable through near-term observable predictions. This is precisely why it is "hard". If M were applied indiscriminately to the entire book, it would repeatedly classify this final cell as "non-compliant" and demand its deletion. The book would then be forced to abandon the very core it seeks to touch, and would degenerate into a theory that deals only with observable states. This is one of the reasons why several earlier explorations of this project were unsatisfactory.

Therefore, this book adopts two-level compliance with M (discussed in detail in Chapters 5 and 15; the rule is first established here). First, all contents belonging to the measurable side—the observable neural proxies of isomorphism, variations of given-ness under different conditions, transitions among conscious states, and the Ψ state space (Part IV)—strictly comply with M: they must provide near-term falsifiable predictions and accept support or refutation by data. Second, all contents belonging to the final cell—the irreducible residual point of being-experienced—are not governed by M, because by their nature they cannot be falsified by observation. They are governed by another level of standard: "internal coherence and phenomenological fit". Their destination is the philosophical boundary (Chapter 27), not observable adjudication. This book makes this two-level structure explicit precisely in order both to preserve scientific discipline (the measurable side is never exempt from falsification) and to prevent that discipline from wrongly eliminating the final cell, which is by nature unfalsifiable yet genuinely present.

It must be emphasized that this two-level structure is not a violation of M, but an appropriate restriction of M: M requires that "core modules retain falsifiable predictions", whereas the final cell, strictly speaking, is not a "module" (it makes no mechanistic assertion and only marks an irreducible boundary). Excluding a boundary that makes no assertion from "modules that must be falsifiable" does not violate the spirit of M. Every genuine module on the measurable side of this book strictly retains its falsifiable predictions, without exception.

2.7. Methodological Discipline

After citing the ontology and mechanisms, this section establishes the methodological discipline that will run consistently throughout the book. These disciplines are the rules by which the book constrains itself, and also the standards by which the reader may examine it.

First, cite without re-proving. Whenever an ontology, mechanism, or conclusion has already been established in the foundational volume or in other EQT volumes (such as the detection volume, Chapter 13), this book cites it, does not derive it again, and clearly marks its source and its status as "citation". The new work of this book is clearly marked as "new". The boundary between citation and new contribution must remain clear throughout, so that the reader does not mistake what is cited for what this book proves, or misattribute what this book proves to what has been cited. The independent contribution of this book lies only in the step "from ρ_f to consciousness"; the ontological warrant outside that step belongs to what is cited.

Second, whatever enters formulas must undergo dimensional checking. Whenever this book introduces or uses a physical quantity, its dimension must be made clear, and the dimensions on both sides of an equation must be consistent. In particular, this book distinguishes two quantities of consciousness intensity: one is a dimensional physical power (denoted A, with units of power), and the other is a dimensionless relative quantity obtained through normalization (denoted $\tilde{A}$); the two must not be confused (see Chapters 22 and 22). In addition, the cross-frequency coupling coefficient χ is an interface coefficient (characterizing coupling strength at a specific interface), not a universal constant—it varies with the interface and must not be used as if it were a universal quantity like Planck's constant. These distinctions concerning dimensions and properties will be stated one by one when the corresponding quantities are introduced.

Third, every claim must identify its proposition type. The claims of this book fall into three types, and every core claim must indicate which type it belongs to: definition (stipulating the meaning of a symbol or concept, not true or false, but assessed by usefulness), model (asserting a certain structure or mechanism, testable through its implications), and empirical proposition (asserting an observable fact, adjudicated by observation). Confusing these three types is a source of disorder—to treat a definition as an empirical proposition to be "confirmed", or to treat a model as a definition and thereby exempt it from testing, are both common errors. This book seeks to assign every claim to its proper type.

Fourth, two-level evaluative standards. As established in Section 2.6: the measurable side follows the standard of falsifiability, while the final cell follows the standard of "coherence and phenomenological fit". The boundary between the two remains clear.

Fifth, the system of failure conditions. At the end of each chapter, this book lists the failure conditions of the core claims of that chapter—that is, under what circumstances those claims would be overturned. The failure conditions of the whole book are summarized in Appendix E and divided according to the two levels (failure conditions at the falsifiable level, and failure conditions at the coherence level). This system is the most concrete form of the book's self-discipline: a theory unwilling to say when it would be wrong is not a responsible theory. This book would rather mark in advance every possible point of error than seek a superficial irrefutability through vagueness.

As a demonstration that this chapter itself follows this system, the failure conditions of this chapter (as a citation chapter) mainly concern the validity of the cited ontology and the fidelity of the citation. They are listed at the end of the chapter.

2.8. Conclusion of This Chapter

This chapter has formally transferred the ρ_f ontology and mechanisms into place as the foundation of the whole book, and has established its methodological discipline. The five cited items and their uses for the consciousness volume are as follows: the ρ_f frequency-resolved energy quantum density field (A1: unique ontology, frequency dimension, and gradient as basic structure; the ontological premise of the identity level, qualitative types, and structural sufficiency); frequency as an energetic spectral quantity (A2: ensuring the continuity between the frequency dimension of the consciousness volume and the physics of the whole series); the unified master equation (A3: its nonlinear terms carry focus and coherent self-sustaining, while its dissipative terms carry the arrow of time and forgetting; it is the basis of Chapters 8, 9, and 10); the Lorentzian resonance kernel (A4: carrying frequency matching in binding and given-ness and preventing ontological uniqueness from sliding into pan-relationalism; it is the mechanism of Chapters 9, 13, and 14); and Meta-Norm M (two-level compliance: the measurable side follows falsifiability, while the final cell follows coherence and phenomenological fit). The five methodological disciplines are: cite without re-proving; dimensional checking (A and $\tilde{A}$, and χ as an interface coefficient); identification of three proposition types; two-level evaluation; and the system of failure conditions. From this point onward, the book stands on this foundation. The following chapters will not re-prove it, but will only cite it.

2.9. Chapter Summary

This chapter is the citation and methodological chapter of the book, and it strictly follows "cite without re-proving". First, it states that this book is the consciousness-application volume of EQT; its work is conditional (dependent on the validity of the foundational volume), and its independent contribution lies only in the step "from ρ_f to consciousness". Second, it cites five ontological and mechanistic items and explains the use of each for the consciousness volume: A1 (ρ_f: unique ontology, frequency dimension, gradient structure), A2 (frequency as an energetic spectral quantity), A3 (master equation: nonlinear and dissipative terms, carrying focus/coherence/the arrow of time/forgetting), A4 (Lorentzian resonance kernel: carrying frequency matching in binding and given-ness, and preventing pan-relationalism), and M (Meta-Norm). Third, it establishes two-level compliance with M: the measurable side follows falsifiability, while the final cell follows coherence and phenomenological fit, and argues that this is an appropriate restriction of M rather than a violation (the final cell makes no mechanistic assertion and is not a "module" that must be falsifiable). Fourth, it establishes five methodological disciplines: cite without re-proving; dimensional checking (the distinction between $A/\tilde{A}$, and χ as an interface coefficient rather than a universal constant); identification of three proposition types (definition/model/empirical proposition); two-level evaluative standards; and the system of failure conditions.

The failure conditions of this chapter are as follows (this chapter is a citation chapter, and its failures mostly concern the cited ontology and the fidelity of citation; they are summarized in Appendix E).

- F2.1 The cited ontology does not hold. If the ρ_f ontology (A1–A4) itself is falsified, then the foundation on which the entire book depends fails, and this book, as its application, fails accordingly (this is the direct consequence of the book's conditionality).

- F2.2 The citation is unfaithful. If this book's citation of A1–A4 or M is shown to deviate from the original meaning of the foundational volume (for example, by assigning to the frequency dimension or the resonance kernel properties that the foundational volume does not assign to them), then the citation is unfaithful, and the arguments based on that citation must be re-examined.

- F2.3 Two-level compliance is in fact an exemption from falsification. If this book's "two-level compliance" with M is shown to provide a back door for measurable modules to avoid falsification (that is, if a module that should be falsifiable is exempted by being placed under the pretext of the "final cell"), then the two-level structure degenerates into an evasion and violates M.

- F2.4 Dimensions or proposition types are confused. If later chapters confuse A and $\tilde{A}$, mistakenly treat the interface coefficient χ as a universal constant, or confuse the three proposition types of definition/model/empirical proposition, then they violate the discipline established in this chapter, and the relevant arguments must be corrected.

Table 2.1.. The Boundary between What This Chapter Establishes and What It Cannot Derive

What this chapter establishes	What this chapter cannot derive
The citation of the ρ_f ontology and mechanisms (A1–A4, M), and the use of each item for the consciousness volume.	It cannot, and does not, re-prove the ρ_f ontology itself (its argument and warrant belong to the foundational volume).
The conditionality of this book's work: it depends on the foundational volume, and its independent contribution lies only in the step "from ρ_f to consciousness".	It cannot guarantee the validity of the foundational volume (this book does not prejudge its success or failure in physics).
Two-level compliance with Meta-Norm M, and the argument that this is an appropriate restriction rather than a violation.	It cannot use the two-level structure as an excuse to exempt measurable modules from falsification (no measurable-side module is exempt).
Five methodological disciplines (citation, dimensional checking, three proposition types, two levels, and failure conditions).	It cannot later confuse $A/\tilde{A}$, misuse χ, or conflate proposition types (any violation must be corrected).

3. Boundary and Interior: Life as the Precondition of Given-ness

3.1. The Carrying Task of This Chapter, and an Order That Must First Be Clarified

Chapter 1 ultimately tied the establishment of given-ness to one precondition—the existence of an "interior"; and the existence of an "interior", in turn, depends on boundary. The task of this chapter is to secure this precondition: to argue why the "interior" is an achievement with a definite physical origin, namely the boundary of life (the membrane), and why the boundary between life and non-life is precisely the boundary of whether given-ness can occur, and therefore the point on which this book's anti-panpsychist position depends.

An order of argument must first be clarified, so that the position of this chapter is not misunderstood. This chapter does not explain given-ness itself—"what process given-ness is" belongs to Chapter 14 (internal–external co-evolutionary freezing of a bounded interior). This chapter explains only the precondition of given-ness: for "given-ness" to occur, there must first be an "interior" to which it can be attributed; this chapter argues where this "interior" comes from. The precondition and given-ness itself are two different matters: this chapter concerns "why there is an interior", while Chapter 14 concerns "why the internal–external co-evolution of this interior is given". Only by distinguishing the two does this chapter avoid overstepping its boundary. Likewise, this chapter does not touch the final cell (the irreducible residual point of being-experienced, Chapter 15). Everything discussed in this chapter remains at the articulable physical and organizational level: boundary, interior, and mode of aggregation. Not a single statement concerns "why the interior is experienced".

This chapter is also the formal source of several claims to which the whole book will repeatedly refer: boundary as the gate of givenness (Chapter 14), the boundary of the "I" nested within the boundary of life (Chapter 16), and matter as a stable transient (Chapter 10 on long-term memory, and Chapter 4 on matter and mind). These later claims are rooted here.

3.2. The Four Fundamental Interactions: Passive Aggregation, Unable to Produce an "Interior"

To see enclosure as a wholly new form of aggregation, one must first understand the nature of the forms of aggregation by which the universe increased its complexity before enclosure, and why they cannot produce an "interior".

From the Big Bang to galaxy clusters, the formation and complexification of cosmic structure have been driven by the four fundamental interactions: the strong interaction, the weak interaction, electromagnetic interaction, and gravity. According to EQT (Chapter 2), all four are manifestations of ρ_f under different frequency and gradient conditions (their reduction belongs to the foundational volume, which this book cites). For the purpose of this chapter, however, what matters is not their respective physical details, but one property they share as mechanisms of aggregation: they are all passive aggregation—using attraction and condensation as their mechanism, they gather matter into regions of higher density and structures of larger scale.

Gravity draws diffuse matter into nebulae, stars, and galaxies; electromagnetic interaction binds charged particles into atoms and connects atoms into molecules and crystals; strong and weak interactions bind quarks into hadrons and gather nucleons into atomic nuclei. These forms of aggregation can undoubtedly produce structures of great grandeur and exquisite complexity—from the lattice of a crystal to the spiral arms of a galaxy. Yet no matter how grand or intricate these structures are, they share one fundamental lack: they do not have an "interior".

This "lack of an interior" must be stated precisely, because it is the pivot of the whole chapter. A crystal, a star, or a molecule, of course, occupies space and has a "region inside it" (atoms inside the crystal, plasma in the stellar core). But this "region inside" is only a "middle part" in a spatial sense, not an interior that genuinely distinguishes itself from the environment and maintains itself. Between the inside and outside of a crystal, there is no actively maintained interface that distinguishes "belonging to this crystal" from "not belonging"; between a star and the interstellar medium, there is a continuous transition of density rather than a self-maintained boundary. What passive aggregation produces is density contrast—more matter here, less matter there—and density contrast, however extreme, does not constitute an "inside–outside distinction". The difference between a high-density region and its surrounding low-density region is a difference of quantity (more and less), not a difference of kind (inside and outside). Passive aggregation can increase density contrast and structural intricacy without limit, but it can never cross the threshold from "contrast" to "inside and outside", because the nature of attraction and condensation is to draw matter together, not to enclose a region and make it a bounded, self-maintaining "inside".

This lack is principled, not a matter of degree: it is not that the interior produced by passive aggregation is "not yet good enough", but that passive aggregation does not produce this thing called an interior at all. No matter how much time and matter are given to the four interactions, what they produce are ever more complex distributions of density contrast, not the spontaneous emergence of a boundary that separates inside from outside and maintains itself. To cross this threshold, a different kind of aggregation is required—one whose mechanism is not attraction and condensation, but closure and enclosure.

3.3. The Second Axis of Aggregation: Enclosure, the First Production of a Topological "Interior"

What crosses this threshold is a form of aggregation that this book calls the "second axis of aggregation"—enclosing aggregation, whose material instance is the lipid membrane.

The formation of a lipid membrane is mechanistically different from the four interactions. To be sure, the physical basis of the membrane still depends on electromagnetic interaction (the amphiphilicity of lipid molecules and the hydrophobic effect), so the fifth mechanism is still built upon the four forces and is not a fifth "force" standing alongside them. This point must be made explicit to avoid misunderstanding (this book calls it the "second axis of aggregation" rather than a "fifth force" precisely for this reason). Yet as a mechanism of aggregation, what the membrane accomplishes is something that the passive aggregation of the four forces cannot accomplish: it does not pull matter into a denser cluster, but encloses a region of space—the lipid bilayer spontaneously closes into a vesicle, producing for the first time an "interior" in the topological sense.

This "topological interior" must be clarified, because it is precisely what passive aggregation can never produce, and what enclosure produces at once. A region enclosed by a membrane is essentially different from the "middle part in space" found inside a crystal: it has a self-maintaining interface (the membrane) that divides "inside the membrane" and "outside the membrane" into two topologically distinct sides. The inside of the membrane is a bounded and closed region, while the outside of the membrane is its environment. Between the two there is not a continuous transition of density, but a definite and self-maintaining boundary. More importantly, this interface is selectively permeable: it does not seal the interior completely (which would prevent the interior from persisting), but permits controlled exchange of matter and energy between inside and outside—some things pass through, while others are blocked. What the membrane produces, then, is an interior that is at once separated from the environment (with an inside–outside distinction), connected to the environment (through selective exchange), and able thereby to maintain itself.

This is the physical origin of the "interior": not density contrast produced by passive aggregation, but a topologically closed region with a self-maintaining interface first produced by enclosure. At this point in cosmic evolution, there is for the first time an "inside"—not the spatial middle, but an interior belonging to a self-maintaining system and distinguished from its environment. This book maintains that this "having an interior" is precisely the structural precondition of the first-person perspective. The first person is first-personal because it has an "inside" and an "outside", a point at which a perspective can be anchored; and this "interior", physically speaking, begins with the topological closure produced by enclosure. In a world of passive aggregation, there is no perspective, because there is no "inside" in which a perspective could reside. Once enclosure produces an interior, perspective for the first time has a physically possible place.

The exact weight of this chapter's claim must be made clear: this chapter claims that the interior produced by enclosure is the physical precondition of given-ness (and therefore of perspective and the first person)—without this interior, there can be no given-ness. This is a claim about precondition, a model proposition (asserting a physical necessary condition). This chapter does not claim that "once there is an interior, there is given-ness" (that would be sufficiency, which requires internal–external co-evolution, Chapter 14), still less does it claim to explain "why the interior is experienced" (that is the final cell, Chapter 15). Enclosure produces an interior; this is a necessary condition for given-ness, not a sufficient condition, and still less an explanation of the being-experienced of given-ness.

3.4. The Boundary between Life and Non-Life Is the Presence or Absence of Boundary

The interior produced by enclosure marks a crucial boundary in the universe: the boundary between life and non-life.

The appearance of the membrane is the beginning of life—all known life has membrane-enclosed cells as its basic unit. Without a membrane, there is no cell; without a cell, there is no interior that is separated from, connected to, and self-maintained in relation to the environment; and without such an interior, there is no life. Non-living things—rocks, stars, crystals, and molecules—are products of passive aggregation. They have density contrast, but they have no enclosing boundary and therefore no interior. Thus:

The boundary between life and non-life is the presence or absence of boundary (the interior of enclosure): life is a system with boundary and interior, while non-life is density contrast without boundary and without interior.

This boundary is decisive for this book, because it is precisely the gate of its anti-panpsychist position (anticipated in Chapter 1 and fully used in Chapter 14). This book maintains that interiority (the interior, perspective, and the precondition of given-ness) has a definite physical origin—it begins with life, with the first membrane. Non-life has no interior and therefore no interiority. This position is secured by this boundary:

It distinguishes this book from panpsychism. Panpsychism maintains that interiority requires no origin and exists universally (everything has a trace of experience). This book rejects that view: interiority is not a background color diffused through all things, but an achievement produced by enclosure and possessing a definite origin. A rock has no interior not because its interior is "too weak", but because it has no boundary and no such thing as an interior at all.

Interiority in it is not merely low in degree; it simply does not apply.

It also distinguishes this book from functionalism. Functionalism maintains that a sufficiently complex functional organization is enough to generate consciousness, regardless of its physical realization. This book rejects that view: the criterion lies not in the complexity of function, but in physical enclosure. A system that is functionally complex but lacks an enclosing boundary (such as a computer or an unmembraned chemical reaction network) has no interior and therefore lacks the precondition of given-ness. Complex functions can be realized in an unbounded system, but the precondition of given-ness (the interior) cannot—it must be produced by the specific physical organization of enclosure.

The exact position of the gate established in this chapter must again be emphasized, in order to distinguish it from the resonance kernel cited in Chapter 2 (Chapter 14 will distinguish them in detail). This book has two gates that must be kept distinct: the first is the resonance kernel $g(f, f_0)$ (A4, Chapter 2), which governs "whether coupling can be established" and prevents ontological uniqueness from sliding into pan-relationalism (whether everything is visible to everything else); the second is boundary (this chapter), which governs "whether given-ness can be attributed" and prevents interiority from diffusing into panpsychism (whether everything has an interior). These two gates govern two different matters, stand in two different places, and must not be conflated. The anti-panpsychist gate of this book is the latter—boundary, not the former—the resonance kernel.

3.5. Membrane as Matter, That Is, Stable Transient: Joining with the Ontology of the Whole Book

This chapter must connect "membrane", "boundary", and "interior" back to the ρ_f ontology of Chapter 2; otherwise, the fifth mechanism would hang outside the ontology of the whole book and become an ungrounded assertion. This connection also provides the source of the claim, repeatedly recalled throughout the book, that "matter is stable transient".

According to Chapter 2, all structures of ρ_f exist in the form of transients—stable but not eternal, differing only in their duration of stability. The membrane, as a material structure, is therefore a highly stable transient of ρ_f: it is constituted by a ρ_f configuration under specific frequency and gradient conditions (lipid molecules and lipid bilayers are both stable transients of ρ_f in the material band), stable enough to maintain the boundary of a cell throughout its life, yet still transient—membranes age and rupture, cells die, and boundaries eventually dissipate. "Membrane is matter, and matter is stable transient" are therefore not two isolated assertions, but the same point: the membrane is a highly stable gradient transient of ρ_f in the material band.

This connection has two meanings. First, it prevents the fifth mechanism from hanging outside the ontology: enclosure is not a mysterious new force external to ρ_f, but a specific organizational form realized by a material transient of ρ_f (the lipid membrane)—an organizational form that seals an interior through a closed interface. Enclosure can produce an interior that the four forces cannot produce not because it appeals to something outside ρ_f, but because it is a specific, closed spatial organization of a ρ_f transient, whereas passive aggregation is another (non-closed) organization. The difference lies in organizational form, not in ontological kind—life and non-life are both ρ_f transients, the same in substance but different in organization (this is the same line of thought as Chapter 4's "matter and mind are both ρ_f, differing in frequency band and mode", and Chapter 13's "detector and object belong to the same field, differing in function": sameness of substance with difference of organization/function).

Second, it lays the basis for Chapter 10's account of "long-term memory as ultra-low-frequency long-duration transient". Since matter (including membranes and neural structures) is a stable transient of ρ_f, the material basis of memory (such as synaptic structure) is also a stable transient. This allows Chapter 10's understanding of long-term memory as a low-frequency long-duration transient to remain continuous with this chapter's claim that "matter is stable transient", rather than being an ad hoc assertion.

3.6. Positioning the Three Previous Works: Pre-EQT Phenomenological Observations, Mechanistically Completed by ρ_f

Since this chapter formally introduces the fifth mechanism, it must also clarify the relation between this book and the three previous works—because concepts such as the "fifth mechanism" and "cellular interior" first appeared in those three pre-EQT works, and their position must be accurately specified. Otherwise, the reader might mistakenly think that this book is their direct continuation, or that the fifth mechanism is the foundation of the whole book.

The three previous works—*The Information World of the Cell: Pooling Networks, Continuous-State Computation, and the Architecture of Life Information Processing* (the earliest work, centered on complex information processing inside the cell, namely "pooling networks"), *The Second Axis of Cosmic Aggregation: Membranes, Life, and Consciousness* (which, on the basis of the former, proposed enclosure as the distinction between life and non-life), and *The Origin of Cellular Proto-Consciousness: Biological Foundations and Operational Criteria* (which explicitly understood cellular information processing as proto-consciousness)—were all written before EQT was proposed, and belong to the level of phenomenological observation. Their insights are real and important: they saw the right phenomena—there is indeed complex information processing within the cell; life does have a special status relative to non-life marked by enclosure; and proto-consciousness does indeed have the cell as its minimal unit. Yet they had not yet reached the underlying mechanisms of these phenomena. They described "what it is" (enclosure, interior, information processing), but did not answer "what the mechanism is" (why these are possible and what their physical basis is).

The proposal of EQT is precisely a deepening from phenomenon to mechanism: the four interactions (together with the fifth mechanism added in its early form) are all phenomenological descriptions—they characterize "how aggregation manifests", but do not answer "what the mechanism of aggregation is"; the ρ_f gradient answers the latter (the foundational volume). Therefore, in this book, the phenomenological observations of the three previous works are incorporated as supporting evidence, and their mechanisms are supplied by ρ_f: the mechanism of enclosure (the insight of *The Second Axis of Cosmic Aggregation: Membranes, Life, and Consciousness*) is the closed organization of material transients of ρ_f (Section 3.5); the mechanism of information processing inside the cell (the insight of *The Information World of the Cell: Pooling Networks, Continuous-State Computation, and the Architecture of Life Information Processing*) is the evolution of ρ_f gradient configurations within that interior (Chapter 8); proto-consciousness (the insight of *The Origin of Cellular Proto-Consciousness: Biological Foundations and Operational Criteria*) is incorporated in this book into the framework that "all bounded life has a minimal degree of given-ness" (Chapters 6 and 16).

It must therefore be made clear where the main line of this book lies: the main line running through this book is the ρ_f gradient principle (the foundational volume), not the fifth mechanism. The role of the fifth mechanism in this book is not that of the foundation, but that of "the boundary gate that provides the precondition (interior) for given-ness"—it is a key link under the ρ_f main line (the presence or absence of boundary), not the main line itself. This positioning must be maintained: if the fifth mechanism were treated as the foundation of the whole book, the book would fall back to the pre-EQT level of phenomenological description, whereas it is precisely the ρ_f mechanism that goes beyond that level. This book is the culminating volume of the series—with the mechanism of ρ_f, it repositions the scattered phenomenological insights of the previous works and advances them to given-ness and the "I".

3.7. Conclusion of This Chapter

This chapter has established the precondition of given-ness and has placed the fifth mechanism in its proper position under the ρ_f main line. The four fundamental interactions are passive aggregation: through attraction and condensation they produce density contrast, but they cannot produce the "inside–outside distinction"—this is a principled lack. Enclosure (the fifth mechanism, whose material instance is the lipid membrane), by contrast, first produces a topological, self-maintaining, selectively permeable "interior" through a closed interface; and this "having an interior" is precisely the physical precondition of given-ness (perspective, the first person). The boundary between life and non-life is the presence or absence of boundary (interior); this boundary is the anti-panpsychist gate of this book—interiority is not a background color diffused through all things, but an achievement beginning with the first membrane. Therefore non-life (without boundary) has no interior and lacks the precondition of given-ness, and the same applies to systems that are sufficiently complex yet lack enclosure. The membrane, as matter, is a highly stable transient of ρ_f; the fifth mechanism is thereby connected back to the ontology of the whole book (same substance, different organization), and it lays the root for Chapter 10's account of long-term memory as long-duration transient. The three previous works are pre-EQT phenomenological observations and are incorporated in this book as supporting evidence; their mechanisms are completed by ρ_f. The main line of this book is the ρ_f gradient, while the fifth mechanism is one link under it, the "boundary gate", not the foundation.

This chapter strictly observes its boundary: what it establishes is the precondition of given-ness (the physical origin of the interior), not given-ness itself (Chapter 14), and still less the final cell of being-experienced (Chapter 15). Everything discussed in this chapter remains at the articulable physical and organizational level; not a single sentence concerns "why the interior is experienced".

3.8. Chapter Summary

This chapter is the heaviest-bearing chapter of Part I. It establishes the precondition of given-ness and formally introduces the fifth mechanism. First, it clarifies the order: this chapter concerns the precondition of given-ness (why there is an interior), not given-ness itself (Chapter 14), and not the final cell (Chapter 15). Second, it argues that the four interactions are passive aggregation: through attraction and condensation they produce density contrast, but in principle they cannot produce an "inside–outside distinction" (density contrast is a quantitative difference, while inside and outside are a difference of kind). Third, it argues that enclosure (the fifth mechanism, whose material instance is the lipid membrane, and which is still built upon the four forces rather than being a fifth force) first produces a topological, self-maintaining, selectively permeable interior; this "having an interior" is the physical precondition of given-ness/perspective/the first person (a model proposition, claiming a necessary condition rather than sufficiency, and still less an explanation of being-experienced). Fourth, it establishes that the boundary between life and non-life is the presence or absence of boundary, as the anti-panpsychist gate; and it distinguishes this gate (boundary, governing the attribution of given-ness) from the resonance-kernel gate (A4, governing the establishment of coupling and preventing pan-relationalism). Fifth, it joins the argument to the ontology: the membrane is a highly stable transient of ρ_f, and the fifth mechanism is the closed organization of a ρ_f transient (same substance, different organization), thereby laying the root for Chapter 10's account of long-term memory. Sixth, it positions the three previous works as pre-EQT phenomenological observations whose mechanisms are supplied by ρ_f; the main line is ρ_f, while the fifth mechanism is one link under it and not the foundation.

The failure conditions of this chapter are as follows (summarized in Appendix E).

- F3.1 Passive aggregation can produce an interior. If the passive aggregation of the four interactions (without enclosure) is shown to be able to produce an "interior" with a self-maintaining interface and an inside–outside distinction, then the claim that "the interior must be produced by enclosure" fails, and the premise of this chapter collapses.

- F3.2 Interior is not a necessary precondition of given-ness. If "given-ness/perspective" is shown to be able to arise in a system without any topological interior (without boundary), then the claim that "interior is the precondition of given-ness" fails, and the anti-panpsychist gate fails with it (linked to F14.5 in Chapter 14).

- F3.3 The boundary gate and the resonance-kernel gate are confused. If this book actually conflates the "boundary gate" (anti-panpsychism) with the "resonance-kernel gate" (anti-pan-relationalism)—using the resonance kernel as the anti-panpsychist gate, or conversely—then the distinction between the two gates fails, and the argument becomes confused.

- F3.4 The membrane cannot be incorporated as a ρ_f transient. If the membrane (and matter in general) is shown in principle to be impossible to understand as a stable transient of ρ_f, then the connection that joins the fifth mechanism back to the ontology fails, and the fifth mechanism hangs outside the ontology.

- F3.5 The positioning of the previous works is mistaken. If the fifth mechanism is in fact the indispensable foundation of the whole book (rather than one link under the ρ_f main line), that is, if the book's argument actually depends on the fifth mechanism rather than the ρ_f gradient, then the positioning that "the main line is ρ_f" is mistaken, and the entire structure of the book must be reassessed.

Table 3.1.. The Boundary between What This Chapter Establishes and What It Cannot Derive

What this chapter establishes	What this chapter cannot derive
The four forces are passive aggregation and cannot produce the inside–outside distinction (a principled lack).	It cannot explain given-ness itself (internal–external co-evolutionary freezing of an interior belongs to Chapter 14).
Enclosure (the fifth mechanism) first produces a topological interior, serving as the physical precondition of given-ness (necessary, not sufficient).	It cannot derive "having an interior already means being given" (sufficiency requires co-evolution, Chapter 14).
The boundary between life and non-life is the presence or absence of boundary, serving as the anti-panpsychist gate (distinct from the resonance-kernel gate).	It cannot, and does not, address "why the interior is experienced" (the final cell, Chapter 15).
The membrane is a stable transient of ρ_f, and the fifth mechanism is connected back to the ontology (same substance, different organization).	It cannot establish the fifth mechanism as the foundation of the whole book (it is one link under the ρ_f main line).
The three previous works are pre-EQT phenomenological observations, whose mechanisms are completed by ρ_f.	It cannot provide the observable proxies of boundary in this chapter (these belong to Part IV, Chapters 16 and 23).

4. Relations to Existing Theories of Consciousness and Positions on the Mind–Matter Problem

4.1. The Task of This Chapter: Placing This Book on the Intellectual Map

The first three chapters have established the overall program of this book (Chapter 1), its ontology and methodology (Chapter 2), and the precondition of given-ness (Chapter 3). As the conclusion of Part I, this chapter performs a work of positioning: it places this book on the existing intellectual map of consciousness studies, clarifying one by one where it stands in relation to various positions—where it converges with them, where it departs from them, and which argument established in the preceding chapters explains that departure.

This positioning is not a ceremonial survey; it has substantive necessity. First, a theory clarifies what it is only by clarifying what it is not. The weight of many claims in this book (the identity level, the frequency dimension, and the boundary gate) becomes visible precisely through contrast with neighboring positions. Second, many of this book's arguments are developed in response to difficulties in existing positions (for example, Chapter 5 addresses single-scalar identity theories, while Chapter 3 addresses functionalism and panpsychism). Concentrating these contrasts in one chapter allows the reader to see the exact coordinates of this book on the intellectual map.

For each position, this chapter seeks first to state its strongest point fairly—a positioning that attacks only weakness and fails to recognize strength is not credible—and then to explain where and why this book takes another path. One classificatory tool that will be used repeatedly must be stated in advance: the threefold distinction established in Chapter 1 (structure, given-ness, and the final cell). This book will show that many disagreements and difficulties among existing positions often arise from failing to separate these three, and that the position of this book is largely defined by this threefold distinction.

This chapter first discusses five philosophical positions (Sections 4.2 to 4.6), then three scientific theories (Section 4.7), and finally gives the overall coordinates of this book (Section 4.8).

4.2. Dualism: This Book Rejects Its Separation, While Preserving Its Respect for Given-ness

Dualism maintains that mind and matter are two different kinds of entities or properties, and that there is a mental reality in addition to the physical world. Its strongest point lies in its respect for given-ness: it squarely recognizes the fact that first-person experience cannot be exhausted by third-person physical description, and it refuses to reduce experience hastily to physical process. The hard problem becomes a problem partly because of this dualist intuition: experience indeed has something that is not entailed by structure and function.

This book rejects the separation of dualism, while preserving its respect for given-ness; the two are not in conflict within this book. The reason for rejecting separation lies in the ontology of Chapter 2: ρ_f is the unique ontology, and there is no second kind of entity standing alongside it. Mind and matter are both manifestations of ρ_f (Section 4.6 will make clear that this book is a monism). Yet this book preserves the truth that dualism respects—given-ness is indeed not entailed by structure—and places it in a precise position within a monist framework: it is not a second entity, but the irreducible final cell (Chapter 15). This book may therefore be described as "accommodating the truth seen by dualism within a monist ontology": it acknowledges the irreducibility of given-ness (what dualism sees), but does not establish a second entity on that basis (rejecting the separation of dualism), and instead locates this irreducible remainder as an irreducible boundary within a single ontology. Dualism misreads "given-ness is irreducible" as "there must be a second entity". This book shows instead that what is irreducible may be only a boundary, not an entity.

4.3. Physicalism: This Book Shares Its Monism, but Differs in Its Treatment of Given-ness

Physicalism (here meaning reductive or eliminative physicalism) maintains that everything is physical, and that experience can either be reduced to physical processes or is fundamentally a kind of illusion that should be eliminated. Its strongest point lies in ontological parsimony and unity: it does not multiply mysterious entities and insists that the world is fundamentally a unified physical world. This parsimony and unity are fully accepted by this book (the unique ontology of Chapter 2 embodies the same spirit).

This book shares physicalism's commitment to monism and parsimony, but differs in its treatment of given-ness. Reductive physicalism maintains that given-ness can be reduced to functional or informational processes. This book rejects that view, for the reason given in Chapter 15: given-ness (the final cell) is not entailed in principle by structure and process, and any "reduction" will fall into topic substitution (replacing the hard problem with an easy problem). Eliminative physicalism maintains that there is in fact no phenomenal consciousness to be explained and that given-ness is an illusion. This book also rejects that view, for the reasons given in Chapters 12 and

15: given-ness is the starting point and the most certain datum of the entire discussion; to eliminate it is to saw off the branch on which one sits. Moreover, "why this stubborn illusion exists" itself becomes a problem, and no explanatory labor has been saved.

This book is therefore a non-reductive and non-eliminative monism: it shares monism and parsimony with physicalism, but does not force given-ness into a reductive framework, nor does it dispose of it by elimination. Instead, it preserves for given-ness a position within a single ontology that is irreducible (yet neither mysterious nor a second entity). This is the position physicalism misses when it rushes to reduce or eliminate.

4.4. Functionalism: This Book Rejects Its Substrate Independence; the Criterion Lies in Enclosure, Not Function

Functionalism maintains that mental states are defined by their functional roles, and that consciousness depends on functional organization regardless of its physical realization—the same functional organization, whether realized by neurons, silicon chips, or something else, has the same mind. Its strongest point lies in its recognition of the importance of functional organization and the resulting intuition of multiple realizability: mind does indeed seem related to a certain organizational pattern rather than to a particular material. This book does not deny that functional organization shapes the structure of experience (in Part II, experiential structure is carried by gradient configurations, and gradient configurations have functional realizations at the neural level; Chapter 6 makes clear that neural coding belongs to the level of implementation).

But this book rejects the core claim of functionalism—substrate-independence (that given-ness depends only on function and not on physical realization), for the reason given in Chapter 3. This book maintains that the precondition of given-ness is interior, and that interior is produced by enclosure (a physical boundary). The criterion therefore lies in physical enclosure, not in functional complexity. A system that is functionally complex but lacks an enclosing boundary (an unmembraned chemical network or a purely computational device) has no interior and therefore lacks the precondition of given-ness, no matter how faithfully its functional organization reproduces that of a conscious subject. The error of functionalism lies in exaggerating "the structure of experience is related to functional organization" (which this book partly accepts) into "given-ness depends only on function" (which this book rejects): structure can be realized by functional organization, but the precondition of given-ness (interior) cannot be produced by function. It must be produced by the specific physical organization of enclosure.

At this point one sharp implication of this book's position must be faced directly, because evading it would be dishonest: according to this book, an artificial system without an enclosing boundary, however complex its functions and however convincing its behavior, does not possess the physical precondition of first-person given-ness. This is a pointed and, in principle, contestable assertion. This book does not avoid it, and marks it as an implication open to discussion and, in an appropriate sense, testing. It follows directly from "boundary as gate" (Chapter 3), and is the natural result of this book's anti-functionalist position, not an additional dogma. Its exact scope must be noted: this book asserts that what lacks enclosure lacks the precondition of given-ness; it does not assert that what lacks enclosure cannot have any functional "information integration". Function can exist without enclosure; the precondition of given-ness cannot.

4.5. Panpsychism: This Book Rejects Its Universality; the Disagreement Concerns Whether Interiority Has an Origin

Panpsychism maintains that experience (or its prototype) is a universal property of matter—everything has a trace of interiority, and the experience of complex systems is composed of micro-experiences. Its strongest point lies in its direct response to the hard problem: it does not evade the question "where does given-ness come from?", but gives a radical answer (given-ness need not arise from nothing, because it is already universally present), thereby avoiding the problem in emergentism of "how experience emerges from non-experiential matter".

The disagreement between this book and panpsychism is fundamental, and this disagreement is precisely the central point of Chapter 3. This book rejects the universality of interiority: interiority is not a background color diffused through all things, but an achievement with a definite physical origin—it begins with life, with the interior produced by the first membrane. Rocks and electrons have no interior not because their interiority is "too weak", but because they simply have no boundary and no such thing as an interior. Interiority in them is not low in degree; it does not apply at all (Section 3.4). This rejection has two reasons. First (epistemological): panpsychism makes "where there is an interior" empirically empty. If everything has a trace of interiority, then "whether there is an interior" no longer marks any observable difference; by contrast, the boundary criterion of this book (the presence or absence of enclosure) is sharp and nearly empirically decidable. Second (theoretical consequences): panpsychism makes the combination problem exceptionally difficult. Since interiority is universally distributed as micro-subjects, combination becomes a genuine deadlock of "adding subjects". This book, by contrast, maintains that the "I" is a single coherent resonance (not the sum of micro-subjects), and the combination problem is dissolved (Chapter 9). This dissolution is possible precisely because this book does not accept the panpsychist premise that "micro-subjects are everywhere".

The difference between this book and panpsychism must be made precise, because both at first sight appeal to "interiority". The difference is this: panpsychism places interiority at the basic physical level (in everything, everywhere in ρ_f), whereas this book places it at a specific organizational level (life with an enclosing boundary). At the basic level of ρ_f, this book does not claim that there is any experience or interiority—ρ_f everywhere is not given. Only when ρ_f organizes an interior through enclosure does the precondition of given-ness arise. Therefore this book is not "panpsychism at the cellular level", but "the emergence of interiority beginning with enclosure". Yet this emergence is not a mysterious emergence of experience from non-experience (the difficulty of emergentism), but an origin triggered by a definite physical event (enclosure producing an interior), with a definite physical precondition. This book therefore stands between panpsychism and emergentism: unlike panpsychism, it does not make interiority universal without origin; unlike emergentism, it does not make interiority emerge from nothing, but specifies a definite physical precondition for interiority (the interior produced by enclosure), to which internal–external co-evolution (Chapter 14) must be added for given-ness to occur.

4.6. Russellian Monism: This Book's Closest Neighbor, and Its Two Decisive Advances

Among the positions considered here, the closest to this book is Russellian monism. It maintains that physical science describes only the relational–structural properties of reality (how things interact and how they are measured), but does not reach the intrinsic properties of reality (what things are in themselves). Mind (experience) may precisely be the intrinsic property of physical reality—the same reality, seen from the outside, is the relational structure described by physical science, while seen from the inside, it is experience. Its strongest point lies in its ability, within a monist ontology (without adding a second entity), to accommodate both the objectivity of the physical and the interiority of experience. This is precisely what Sections 4.2 and 4.3 seek. Thus, in spirit, this book is closest to it: this book's claim that "mind and matter are two aspects of the same ρ_f" is a Russellian posture.

Yet relative to Russellian monism, this book makes two decisive advances. These two advances are precisely what make this book more than a philosophical posture, and turn it into a theory with specific content and testability.

First, by using a frequency-resolved gradient structure, this book fulfills structural sufficiency and thereby prevents "identity" from falling back into the difficulty of a single scalar. Russellian monism is a general philosophical framework: it maintains that experience is the intrinsic side of the physical, but does not specify what structure the physical side must possess in order to match the richness of experiential structure. This lack of specification creates a hidden risk: if the intrinsic side of the physical is only a poor "primitive feeling", then it repeats the combination problem of panpsychism; if it appeals to some single intrinsic quantity, then it repeats the structural insufficiency of single-scalar identity theories (Chapter 5). The advance of this book is to take the frequency-resolved gradient configuration of ρ_f as the structure on the physical side—a fivefold-rich structural space (Chapters 1 and 5), sufficient to be isomorphic to the entire structure of experience relation by relation (Part II). This turns "experience is the intrinsic side of the physical" from a general assertion into an isomorphic claim supported by structural sufficiency and testable relation by relation. Russellian monism says that "experience is the intrinsic side"; this book says further that "the intrinsic side is this fivefold-rich gradient structure, and it is isomorphic to experience relation by relation". This is an advance from posture to content.

Second, by using boundary and internal–external co-evolution, this book characterizes given-ness and thereby defines the range of the intrinsic side. Russellian monism faces a well-known difficulty: if experience is the intrinsic side of the physical, where does this intrinsic side exist? Does everything have it (sliding into panpsychism and its combination problem), or do only certain systems have it (which requires a criterion)? Russellian monism itself does not provide this criterion. The advance of this book is to define the range of the intrinsic side by boundary (Chapter 3)—only systems whose interiors are produced by enclosure (life) have an interior to which given-ness can be attributed, and given-ness itself is characterized by internal–external co-evolution (Chapter 14). This makes the "intrinsic side" no longer a metaphysical stipulation of unclear range, but something with a definite boundary (life) and a definite mechanism (internal–external co-evolutionary transient freezing). Russellian monism says that "experience is the intrinsic side"; this book says further that "this intrinsic side exists only where there is an enclosing boundary, and its given-ness is the freezing of internal–external co-evolution". This, again, is an advance from posture to content.

Taken together, these two points position this book as a Russellian monism enriched by EQT: it inherits the Russellian monist posture (mind and matter as two aspects of the same reality), but enriches the structure of the "intrinsic side" with frequency-resolved gradient structure (solving structural sufficiency), and defines the range and mechanism of the "intrinsic side" through boundary and co-evolution (solving range and given-ness). What remains after these two advances, and what Russellian monism and this book jointly face as irreducible—the being-experienced of the intrinsic side itself—is the final cell (Chapter 15). This book shares this final cell with Russellian monism, but compresses it to the smallest scale (only the single cell of "why the structure of a bounded interior is experienced"). This compression is precisely the result of the two advances described above.

4.7. Three Scientific Theories: IIT, GNW, and Predictive Processing

Turning to the three major frameworks in contemporary consciousness science, the relation between this book and them can be clarified uniformly through the threefold distinction of "structure/givenness/final cell". Their insights and limitations often depend on which of these three they address and which they conflate.

Integrated Information Theory (IIT). IIT maintains that consciousness corresponds to a system's quantity of integrated information Φ, and it characterizes its informational structure. Its strongest point lies in its serious treatment of the structure of experience: it is not satisfied with "more or less consciousness", but attempts to characterize the informational geometry of experience. This book moves in the same direction in its emphasis on experiential structure, but there is a fundamental disagreement: IIT centers on the quantity Φ, while Chapter 5 will argue that any identity theory centered on a single scalar cannot carry qualitative types (the difference between red and pain), because it lacks structural sufficiency (a quantity has only the one dimension of "magnitude"). This book replaces the single Φ with a fivefold-rich gradient structure precisely in order to remedy this structural insufficiency. In addition, the criterion of IIT is information integration (a functional–informational property), and therefore in principle it does not exclude consciousness in systems without enclosure. This book, by contrast, uses boundary as gate (Chapter 3), and thus diverges from IIT. One may say that this book shares IIT's ambition to take structure seriously, but supplements the insufficiency of its single scalar with frequency-resolved structure and supplies a boundary criterion where IIT lacks one.

Global Neuronal Workspace Theory (GNW). GNW maintains that consciousness corresponds to information being "broadcast" to a global workspace and made available to many subsystems. Its strongest point lies in its precise characterization of one class of functional features of consciousness—accessibility, reportability, and global availability—and in its strong neural evidence. This book holds that what GNW characterizes largely belongs to the easy problems in the threefold distinction of Chapter 1 (function and access), rather than to given-ness itself: that information is globally broadcast is a functional fact; why it is experienced is not answered by GNW (nor does GNW claim to solve the hard problem). This book and GNW therefore do not so much conflict as divide labor: GNW excellently characterizes how conscious contents are broadcast and accessed (in this book's framework, this corresponds to how the dominant coherent resonance integrates and accesses configurations across the system, Chapters 9 and 16), while the nature of given-ness is carried by this book's account of internal–external co-evolution. GNW concerns "how content becomes globally accessible"; this book concerns "why this access is given".

Predictive processing. Predictive processing maintains that the brain is a prediction machine, and that experience is the continuous tuning between the brain's best predictions (generative model) of sensory input and the errors of actual input. Its strongest point lies in its unified characterization of the dynamics of perception, action, and learning, and its explanatory scope is broad. This book is closely connected with it at the level of dynamics: the "generative-model drive and error tuning" of predictive processing can, within this book's framework, be understood as endogenous perturbations (generative models, predictions) and exogenous perturbations (sensory input) jointly driving the evolution of the master equation and becoming given through internal–external co-evolution (Chapters 8 and 14). Prediction and error are precisely one organization of endogenous and exogenous perturbations. But predictive processing likewise does not answer given-ness: why the operation of a system performing prediction–error tuning is experienced is not answered by predictive processing. The division of labor is therefore as follows: predictive processing excellently characterizes how experiential contents are generated and tuned (how perturbations drive evolution), while this book undertakes the account of what given-ness is.

Taken together, the three theories exhibit a common pattern: IIT, GNW, and predictive processing each handle one aspect of the structure or functional dynamics of experience with considerable success, but none has (and most do not claim to have) directly answered given-ness or the final cell. The relation between this book and them is therefore mostly not one of competition, but of division of labor and supplementation: this book accepts their characterizations of structure and function (repositioning them within the ρ_f framework), while adding what they do not reach—given-ness (internal–external co-evolution) and the honest marking of the final cell. The real disagreements between this book and them concentrate on two points: the criterion (this book uses boundary as gate and rejects a purely functional/informational criterion) and structural sufficiency (this book uses frequency-resolved structure and rejects the single scalar).

4.8. The Overall Coordinates of This Book

Synthesizing the positioning developed in this chapter, the coordinates of this book on the intellectual map may be stated in one sentence: this book is a Russellian monism enriched by EQT, neither reductive nor eliminative; it solves structural sufficiency through frequency-resolved gradient structure, solves the range and mechanism of given-ness through enclosing boundary and internal–external co-evolution, and honestly compresses and locates the irreducible remainder of the hard problem as a single final cell.

From these coordinates, this book's selections and rejections in relation to the various positions can be summarized together: it takes from dualism its respect for given-ness, but rejects its separation (given-ness is a boundary, not a second entity); it takes from physicalism its monism and parsimony, but rejects its reduction and elimination (given-ness is irreducible and cannot be eliminated); it takes from functionalism its emphasis on functional organization (for the structure of experience), but rejects its substrate-independence (the criterion lies in enclosure, not function); it takes from panpsychism its direct attitude toward the hard problem, but rejects its universality (interiority has an origin and is not diffused through all things); it is closest to Russellian monism, but advances it through frequency structure and boundary co-evolution, transforming it from a posture into a theory with content and testability. As for IIT, GNW, and predictive processing, this book is mostly complementary to them in division of labor (accepting their characterizations of structure and function), while disagreeing with them on two points: criterion (boundary) and structural sufficiency (frequency resolution).

With this chapter, Part I is complete. Chapter 1 established the overall program; Chapter 2 cited the ontology and discipline; Chapter 3 established the precondition of given-ness (boundary); Chapter 4 has located this book on the intellectual map. The following parts will then fulfill the isomorphism of experiential structure, characterize given-ness, turn to measurable projection, and provide philosophical closure. With the coordinates of the book now marked, the reader may enter the detailed arguments of the following parts with this map in hand.

4.9. Chapter Summary

This chapter concludes Part I by positioning this book on the intellectual map of consciousness studies, using the threefold distinction of "structure/given-ness/final cell" as its classificatory tool. First, dualism: this book rejects its separation, but takes its respect for given-ness, placing given-ness as a boundary within a monist ontology (the final cell) rather than as a second entity. Second, physicalism: this book shares its monism and parsimony, but differs in its treatment of given-ness—it neither reduces it (Chapter 15: given-ness is not entailed by structure) nor eliminates it (given-ness is the most certain datum). Third, functionalism: this book accepts its emphasis on functional organization (for structure), but rejects its substrate-independence (the criterion lies in enclosure, Chapter 3), and directly acknowledges the sharp implication that an artificial system without enclosure lacks the precondition of given-ness. Fourth, panpsychism: this book takes its direct attitude toward the hard problem, but rejects its universality (interiority has an origin and is not diffused; it does not apply at the basic physical level, and appears only at the level of enclosing organization), thereby avoiding both the difficulty of emergentism and the combination deadlock of panpsychism. Fifth, Russellian monism: it is the closest neighbor of this book, and this book advances it decisively in two ways—by using frequency-resolved gradient structure to fulfill structural sufficiency (avoiding retreat into a single scalar), and by using boundary and internal–external co-evolution to define the range and mechanism of the intrinsic side. This book is therefore "Russellian monism enriched by EQT", sharing the final cell with it while compressing that cell to the smallest possible scale. Sixth, IIT/GNW/predictive processing: they respectively handle structure (IIT), functional accessibility (GNW), and dynamics (predictive processing) with success; this book is mostly complementary to them in division of labor, with disagreements centered on criterion (boundary versus function/information) and structural sufficiency (frequency resolution versus single scalar). Seventh, the chapter gives the overall coordinates of the book.

The failure conditions of this chapter are as follows (summarized in Appendix E; this chapter is a positioning chapter, and its failures mostly concern the accuracy and fairness of positioning).

- F4.1 Unfair positioning. If this chapter's account of a given position is shown to distort the strongest form of that position (attacking a straw man rather than its strongest form), then the positioning at that point is unfair and must be restated.

- F4.2 The difference from Russellian monism does not hold. If the two advances that this book claims over Russellian monism (frequency structure and boundary co-evolution) are shown not to provide content absent from Russellian monism (that is, if this book has no substantive difference from Russellian monism), then the independent positioning of this book fails.

- F4.3 The claimed "division of labor" with scientific theories is in fact avoidance. If the claimed "division of labor and complementarity" with IIT/GNW/predictive processing is shown to be an evasion of genuine competition (that is, if this book uses "division of labor" to avoid points of direct conflict with them), then the positioning by division of labor is inaccurate.

- F4.4 The anti-functionalist implication is too strong or inconsistent. If the implication that "an artificial system without enclosure lacks the precondition of given-ness" is shown to be inconsistent with claims elsewhere in this book (or if its criterion of "enclosure" cannot be applied consistently to artificial systems), then this implication must be revised.

Table 4.1.. The Boundary between What This Chapter Establishes and What It Cannot Derive

What this chapter establishes	What this chapter cannot derive
This book's selections and rejections relative to dualism, physicalism, functionalism, and panpsychism, and the reasons for them.	It cannot re-prove those reasons in this chapter (their arguments are in Chapters 3, 5, 9, and 15).
This book as "Russellian monism enriched by EQT", and its two advances relative to Russellian monism.	It cannot prove in this chapter that the two advances hold (they are fulfilled in Part II and Chapter 14).
This book's relation to IIT/GNW/predictive processing is mostly one of division of labor and complementarity, with disagreements on criterion and structural sufficiency.	It cannot adjudicate in this chapter the empirical success or failure of those scientific theories (that belongs to their respective empirical domains).
The overall coordinates of this book, and the unified selection and rejection of various positions based on those coordinates.	It cannot provide any new substantive argument in this chapter (this chapter is positioning; the arguments are in other chapters).

Part II.

Mind–Matter Isomorphism: The First-Principle Thesis and Its Fivefold Richness

5. The Precise Formulation of the Isomorphism Thesis and Its Sufficiency Conditions

5.1. From Program to Argument: The Task of This Chapter

Chapter 1 proposed the first-principle thesis of this book—first-person experience is isomorphic to a certain class of gradient configurations of ρ_f, and this isomorphism is identity in the strongest sense—and tied its success or failure to the sufficient richness of gradient structure. There, however, terms such as "isomorphism", "identity", and "structural richness" still remained at the programmatic level: they indicated a direction, but were not yet claims that could be tested. If a thesis is to serve as the axis of the whole book, direction alone is not enough; it must be made precise enough for one to state clearly under what conditions it holds and under what conditions it fails. The task of this chapter is precisely to forge the programmatic identity thesis of Chapter 1 into a claim with clear structure, clear criteria, and clear failure conditions.

This requires four steps in sequence. First, one must define what is meant by "the structure of experience" and "the structure of gradients"—isomorphism is a relation between two structures, and if the structures at both ends have not been clarified, "isomorphism" is empty talk (Section 5.2). Second, one must give the strict meaning of "isomorphism"—it is not a vague "similarity" or "sense of correspondence", but a mathematically defined, structure-preserving mapping (Section 5.3). Third, one must choose among the three levels of "identity", "structural isomorphism", and "modeling", and justify that choice, especially by directly assuming the burden carried by the identity level (Sections 5.4 to 5.6). Fourth, one must state the completeness claim that "the five sources exhaust the structural dimensions of experience" as a proposition that can be overturned by counterexample, and explain how it is to be tested (Section 5.7). This chapter does not develop the isomorphism argument for each of the five sources; that is the task of Chapters 6 to 11. What this chapter establishes is the common framework and criterion on which those arguments depend.

5.2. Two Structural Spaces: The Experiential Side and the Gradient Side

Isomorphism is a relation between structures. To speak of the isomorphism between experience and gradients, one must first explain what structure exists on the experiential side and what structure exists on the gradient side.

Consider first the experiential side. To say that experience "has structure" means that experience is not an undifferentiated, unanalyzable whole, but an organization in which elements stand in definite relations to one another. These relations include at least the following classes, which this book collectively calls the structural features of experience. The first is the continuous order within a quality: within the same qualitative kind, experience exhibits continuous gradation, and this gradation is ordered—scarlet and crimson are closer to each other than scarlet and sky blue, and this "closeness" and "distance" constitute a metric order internal to the color-experience space; the same is true of pain as it ranges from mild to severe, or from diffuse to sharp. The second is the type distinction between qualities: red, pain, bitterness, and humming belong to mutually irreducible qualitative kinds. The differences among them are not distances along the same dimension, but differences in dimension itself. The third is the temporal order of evolution: experience flows; the transition from one state to the next follows certain constraints and is not an arbitrary jump—the stream of consciousness has its direction and coherence. The fourth is synchronic unity and differentiation: at any given moment, diverse experiences are unified as experiences of "the same I", while internally differentiated into discernible objects and features. Some features are bound as aspects of the same object (this object's redness, roundness, and motion), while others belong to different objects. The fifth is temporal thickness and hierarchy: the present is not an extensionless point; it contains retention of what has just passed and protention of what is about to come; memory, in turn, is layered according to its duration of retention. These five classes of relations constitute what this book calls the "structural space of experience"—each "point" in it is a possible experiential state, and the relations among these points are the structure of that space.

It must be emphasized that what is listed here is the structure of experience, not the given-ness of experience. What structure an experiential state has (what it is like and how it relates to other states), and the fact that this state is experienced (that it manifests to someone), are two questions that must be separated. This chapter and the whole of Part II treat only the former—structure. The latter—given-ness—is the subject of Part III, and its characterization follows a different path from the characterization of structure (internal–external co-evolution, not internal isomorphism). Separating the two is central to the methodology of this book: the structure of experience can be exhaustively characterized by isomorphism with the structure of gradients, whereas given-ness is not contained within this isomorphism and must instead be taken up through relational co-evolution. Confusing these two is the source of much ambiguity in theories of consciousness.

Now consider the gradient side. As an energy density field partitioned by frequency families, ρ_f is not an unstructured mass either; its gradient configurations are high-dimensional organizations with definite relations. This book maintains that the structure on the gradient side can likewise be analyzed into several dimensions: within a single frequency family, the direction, magnitude, spatial topology, and distribution of singularities of the gradient constitute a continuous order; the distinctions among frequency families constitute an independent dimension; the constraints that the master equation imposes on gradient evolution constitute a set of temporal evolutionary relations; the resonance modes of same-frequency and cross-frequency relations constitute relations of coherence and coupling; and the frequency-dependent timescales of gradient transients constitute a hierarchy of temporal retention. These five groups of structures correspond precisely to the five sources of richness listed in Chapter 1, and constitute what this book calls the "structural space of gradients".

The isomorphism thesis may therefore be stated preliminarily as follows: there exists a one-to-one correspondence between the structural space of experience and the structural space of gradients, and this correspondence preserves the structural relations on both sides. The continuous order within qualitative kinds in experience corresponds to the continuous order of single-frequency configurations in gradients; the type distinctions between qualitative kinds in experience correspond to the distinctions among frequency families in gradients; and so on. The next section will explain what it means to "preserve structural relations", and why this requirement is far more stringent than "appearing similar".

5.3. The Strict Meaning of Isomorphism: What It Means to "Preserve Structure"

In ordinary language, words such as "correspondence", "similarity", and "parallel" are loose; they may refer to any degree of resemblance. If what this book calls isomorphism were merely such loose resemblance, then it would be almost unfalsifiable—between any two sufficiently complex systems, one can always find some kind of "similarity". This book adopts the strict mathematical concept of isomorphism, which requires far more than resemblance.

In mathematics, an isomorphism between two sets equipped with structure is a one-to-one mapping that preserves structure: any structural relation that holds on one side also holds on the other side after mapping, and conversely. A simple example may clarify this. Suppose one side consists of a group of cities and the road distances among them, and the other side consists of a set of points on a scale map and the distances among those points on the map. If there is a one-to-one correspondence such that the actual distance between any two cities is proportional to the map distance between the corresponding two points, then the map is isomorphic to the group of cities—the map "preserves" the distance structure among the cities. Note the stringency here: it is not enough for the map to resemble the cities roughly; the distance for every pair of cities must be proportional to the corresponding distance on the map. If even one pair of cities has a map distance that is not proportional to its actual distance, then the map is not a faithful isomorphism of that group of cities—it "distorts" the structure at that point.

When this stringency is applied to experience and gradients, we obtain the true weight of this book's isomorphism thesis. It does not say that experience and gradients are "similar in some respects", but that every structural relation in the structural space of experience has a corresponding structural relation in the structural space of gradients that matches it precisely, and that this matching is bidirectional and without omission. More specifically, at least the following three requirements must be satisfied.

First, the correspondence must be without omission. Every structural feature of experience must find a counterpart on the gradient side; there must be no structural relation in experience for which the gradient side cannot provide a corresponding structure. This is the "coverage" requirement of isomorphism with respect to the experiential side, and it leads directly to the completeness claim in Section 5.7.

Second, the correspondence must preserve relations. It is not only that "points" of experience correspond to "points" of gradients; the relations among experiential points (adjacent, distant, different in kind, temporally prior, bound as one) must correspond to the relevant relations among gradient points. If scarlet and crimson are adjacent in experience, then the two corresponding gradient configurations must also be adjacent in gradient space; if red and pain belong to different kinds in experience, then the two corresponding gradient configurations must also occupy different structural dimensions (according to this book, different frequency families). A correspondence that does not preserve relations is not an isomorphism.

Third, the correspondence must be bidirectional. Not only must the structures of experience have counterparts in gradients, but the relevant structures of gradients should also have counterparts in experience—at least within the class of gradient configurations that "carry consciousness". This bidirectionality gives the isomorphism thesis predictive force: a structural relation in the gradient structural space that has not yet been noticed should predict the existence of a corresponding, as-yet-unnoticed structural relation in experience, and conversely (its empirical consequences will become falsifiable predictions in Part IV).

Precisely because isomorphism is strict here rather than loose, the isomorphism thesis becomes a substantive and fallible thesis. Its modes of failure are clear: if one finds a structural feature of experience that has no counterpart on the gradient side (violating the first requirement), or finds a relation in experience whose corresponding gradient relation does not hold (violating the second requirement), then the isomorphism breaks at that point. This book willingly states the thesis with this stringency precisely so that it can withstand testing, rather than secure a superficial success through looseness.

5.4. Three Levels of Strength: Identity, Structural Isomorphism, and Modeling

After establishing the strict meaning of isomorphism, a higher-level choice must be faced: even if there is a strict isomorphism between experiential structure and gradient structure, what ontological strength should be assigned to this isomorphism? Faced with the same fact of structural correspondence, there are in principle three claims of decreasing strength. This book must choose one explicitly and justify the choice.

The strongest level is identity. It maintains that experience is what that class of gradient configurations presents from within; mind and matter are not a correspondence between two kinds of things, but two aspects of the same reality. At this level, "experience" and "the inner side of that gradient configuration" are not two things connected by a bridge, but two names for the same thing—just as "morning star" and "evening star" are two names for the same Venus. Structural isomorphism is interpreted here as follows: the reason the structures on the two sides correspond one-to-one is that they are already the same structure, accessed from two directions—from the outside, in the third person, it appears as a gradient configuration; from the inside, in the first person, it appears as experience.

The intermediate level is structural isomorphism (without claiming identity). It maintains that experiential structure and gradient structure correspond strictly one-to-one, but that experience and gradients remain two kinds of things that happen to be structurally parallel. At this level, mind and matter are two different realities, and there is an "isomorphic bridge" between them such that every structure on one side has a mirror image on the other. This level acknowledges correspondence, but does not reduce that correspondence to identity.

The weakest level is modeling correspondence. It maintains that ρ_f gradient configurations are the best physical model of experience, and that isomorphism is a correspondence in the sense of modeling—just as a differential equation is a model of planetary motion. At this level, the relation between gradient configuration and experience is the relation between model and modeled object; the model may be extremely accurate, but a model is not the modeled thing itself. This level has in effect retreated to a kind of correlationism: it describes experience and predicts experience, but does not claim that experience is ontologically identical to gradients.

The differences among these three levels are by no means merely verbal; they are differences in ontological commitment, and they directly determine the explanatory burdens each must bear. The following two sections discuss them in turn.

5.5. This Book Adopts the Identity Level: Reasons

This book adopts the strongest identity level. This choice has three reasons, which must be stated one by one, because it is the heaviest ontological commitment of the whole book.

First reason: only the identity level fulfills the real meaning of "mind–matter unity". The title of this book is *Mind–Matter Unity*; if "unity" is taken in its strict sense, it means that mind and matter are the same reality—not two realities closely related to each other, nor one reality well modeled by the other. The structural-isomorphism level allows mind and matter to remain ultimately two kinds of things; that is a picture of "dualism with parallelism", far from the meaning of "unity". The modeling level does not even claim parallel realism and has even less to do with "unity". Only the identity level makes "mind–matter unity" a thesis about the structure of reality, rather than a thesis about the accuracy of description. Since this book takes "unity" as its title, it cannot stop ontologically at correspondence or modeling.

Second reason—and the more argumentative one: only the identity level minimizes the explanatory burden. At first glance this seems counterintuitive—how can the strongest claim carry the smallest burden? The key is that identity requires no explanation of a "relation". At the structural-isomorphism level, mind and matter are two kinds of things, and a heavy question immediately arises: why does the strict correspondence relation between these two kinds of things hold? What makes every structure on the physical side happen to have a mirror image on the mental side? This kind of "pre-established harmony" itself requires an explanation, and any explanation would introduce new, often more difficult assumptions (where are the piers of the bridge connecting two heterogeneous domains to stand?). At the identity level, this question does not arise from the beginning: there is no "relation" between A and A waiting to be explained, because there is only one A. The structures correspond one-to-one not because two things happen to be parallel, but because they are already the same structure. Thus the difficulty borne by the intermediate level—"why does the correspondence relation hold?"—is removed at once under the identity level. The strongest ontological claim thereby clears away one of the most difficult explanatory debts. This is precisely the argumentative economy of the identity level.

Third reason: the identity level compresses the irreducible residual point of the hard problem to a single cell. As stated in Chapter 1 and to be discussed in detail in Section 5.6 and Part III, this book does not claim to eliminate the hard problem, but claims to compress it into an irreducible final cell. The identity level is necessary for this compression. At the modeling or structural-isomorphism level, in addition to the difficulty of "why it is given", there are extra explanatory debts such as "why is the model so accurate?" and "why are the two domains so parallel?". The residual points are diffuse and multiple. Only at the identity level are these additional debts eliminated, and the irreducible remainder is gathered into a single cell—"why the same reality, from its inner side, is experienced". The identity level is therefore not only the strongest claim, but also the claim that makes the residual point smallest and most precise.

It must be admitted that the identity level also has its cost, and this book does not avoid it. It asks one to accept a thesis that is not intuitively easy: what appears from the outside as fluctuations in the gradient of energy density is, from the inside, living, qualitative experience itself. This thesis cannot be "proved" by something more basic—it is an ontological stipulation. Its defense does not lie in deductive proof, but in whether it can uniformly, coherently, and with the smallest residual point accommodate all relevant phenomena. The remaining parts of this book develop precisely this defense.

5.6. The Burden of the Identity Level: Structural Sufficiency and Its Criterion

To adopt the identity level is to assume directly its heaviest burden. Chapter 1 named this burden the problem of structural sufficiency; this section turns it into an operational criterion.

The burden arises as follows. If one merely claims that experience is produced by gradients (as in some form of emergentism), then the gradient side need not be as structurally rich as experience—the producer may be simpler than the produced, just as simple rules can produce complex patterns. But this book claims identity: experience is the inner side of gradient configurations. Identity is symmetrical—if A is B, then B must possess all structural features of A, and conversely. Therefore every structural feature of experience must also be possessed by the gradient configuration. If there is even one structure in experience for which the gradient side cannot provide a counterpart, then "A is B" breaks at that point—because identical things cannot differ in structure. This is the unique and non-negotiable stringency of the identity level: it does not allow the gradient side to lack any degree of structural richness.

It is precisely this stringency that requires this book to distinguish itself from all single-scalar identity theories, and explains why the latter are bound to fail. Suppose a theory claims that "consciousness is the quantity of integrated information Φ" or that "consciousness is the intensity of a certain field". Φ is a scalar, a number; the structure of a number is extremely poor—it has only one dimension, "greater or smaller". But the structure of experience is far more than "greater or smaller": it has continuous order within qualities (already multidimensional), type distinctions among qualities (differences of dimension), temporal evolution, unity and binding, and temporal thickness. To identify a scalar that has only the one dimension of "magnitude" with a multidimensionally rich experiential structure is to fail structural sufficiency at the first step. No matter how that scalar is adjusted, it cannot provide the type dimension required for "the difference between red and pain". The reason single-scalar identity theories ultimately can only retreat to correlationism ("the higher Φ is, the stronger consciousness is") is precisely that they are structurally insufficient from the start: they can match "strength", but they cannot match "kind", "binding", or "temporal thickness". Their failure is not an accidental technical difficulty, but the structural verdict of the identity level upon a single scalar.

From this follows the structural sufficiency criterion of this book, which will serve as the unified testing standard for the chapters of Part II:

For every class of structural features of experience, one must identify a specific structural source on the gradient side, such that the structural dimension provided by that source is sufficient to carry all the order relations of that class of experiential structure; and this carrying must be relation-by-relation—every structural relation that holds in experience must have a corresponding relation that holds in the identified source.

According to this criterion, the task of this book is decomposed into a series of fulfillments: the continuous order within qualities is carried by the continuous structure of single-frequency gradient configurations (Chapter 6); the type distinction among qualities is carried by the frequency-family dimension (Chapter 7); the temporal order of evolution is carried by evolution constrained by the master equation (Chapter 8); unity and binding are carried by resonance modes (Chapter 9); temporal thickness and memory hierarchy are carried by the transient timescale spectrum (Chapter 10). The success or failure of each chapter is measured by this criterion: it must identify the relevant structural source and show relation-by-relation that the carrying holds, rather than merely provide a vague "sense of correspondence". The stringency of this criterion embodies this book's willingness to argue in a fallible form—if any chapter cannot fulfill its carrying relation by relation, the isomorphism breaks at that point, and the identity level has a gap there.

5.7. The Completeness Claim: A Falsifiable Proposition

The structural sufficiency criterion requires that every class of structural features of experience be carried by a gradient source. But "every class" presupposes that we already know how many classes of structural features experience has. Chapter 1 and Section 5.2 of this chapter listed five classes, and correspondingly invoked five gradient sources. Here lies one of the claims that most needs to be stated in a falsifiable form throughout the book:

Completeness claim: the structural features of experience are exhausted by five classes—the continuous order within qualities, the type distinction among qualities, the temporal order of evolution, synchronic unity and binding, and temporal thickness and hierarchy; and these five classes are carried relation-by-relation by the five gradient structural sources, respectively.

This is a strong claim. This book does not soften it with the vague phrase "roughly covers", but prefers to state it in the form most vulnerable to attack, for two reasons. First, only when stated in falsifiable form is it a scientific–philosophical proposition rather than a rhetorical posture; a "completeness" that cannot be touched by counterexample is not completeness, but emptiness. Second, the falsifiability of this claim indicates the most valuable direction for later research—to seek that "sixth class".

The mode of failure of the completeness claim is clear: if one can identify a structural feature of experience that neither belongs to any of the five classes listed above nor can be carried by any of the five gradient sources, then the completeness claim fails and the isomorphism program has a gap. This book here identifies in advance several candidates most likely to become counterexamples. They will be examined directly in Chapter 11, not avoided:

The first is the sense of self—experience seems always to be accompanied by a sense of ownership, "this is my experience". Is this sense of ownership entailed by the five structural classes listed above, or is it an independent sixth structural class? The second is the sense of agency—the sense accompanying the initiation of an action or thought, "I am doing this", differs structurally from the sense of passive reception. Can this difference be incorporated into the five listed classes? The third is affective tone—experience is often suffused with a pervasive tone (anxious, peaceful). Is this tone a kind of quality (and thus included in the dimension of type), or is it an independent structural layer spanning multiple qualities? The position of this book is that all three can be reduced to configurations of the five classes of structure at a higher organizational level, rather than constituting an independent sixth class (see Chapter 11 for the argument). But this position is stated as a claim that can be overturned, not as an unquestionable premise. If the reduction in Chapter 11 is shown not to hold—if one of the sense of self, the sense of agency, or affective tone cannot be carried by the five sources—then the completeness claim fails at that point, and the isomorphism program of this book must either add a further source or retreat to a weaker level.

The failure consequences of the completeness claim must be distinguished from those of the identity-level claim; they are not the same. If the isomorphic carrying of a particular item fails (for example, if the frequency dimension is shown unable to carry qualitative types), then the identity level breaks at that point, and the problem is serious enough to shake the first-principle thesis. But if the completeness claim fails (a sixth class of structure is discovered), this does not necessarily shake the five isomorphisms already fulfilled. It indicates instead that the program is incomplete and must add a sixth gradient source—an expansion of the program, not necessarily a collapse of its foundation. This book lists these two failures separately in the failure conditions (see the end of this chapter and Appendix E) in order not to confuse their severity.

5.8. Conclusion of This Chapter and the Development of Part II

This chapter has forged the programmatic identity thesis of Chapter 1 into a claim with precise structure, explicit criteria, and a falsifiable form. Its main points may be summarized as follows.

The experiential side and the gradient side each have their own structural space, and isomorphism is a strict, structure-preserving, bidirectional and non-omissive one-to-one correspondence between the two—not a loose similarity. Faced with this isomorphism, this book chooses the strongest identity level among identity, structural isomorphism, and modeling, for three reasons: only the identity level fulfills the real meaning of "mind–matter unity"; only the identity level minimizes the explanatory burden by eliminating the question "why does the correspondence relation hold?"; and only the identity level gathers the residual point of the hard problem into a single final cell. The cost of the identity level is that it must bear the non-negotiable burden of structural sufficiency: every structural feature of experience must be carried relation-by-relation on the gradient side. If there is a lack at one point, identity breaks at that point. The failure of single-scalar identity theories is precisely evidence of this burden. From this follows the structural sufficiency criterion, which serves as the unified testing standard for the chapters of Part II. Finally, this book states "the five classes of experiential structure are exhausted by the five gradient sources" as a falsifiable completeness claim, and identifies in advance the sense of self, the sense of agency, and affective tone as the most likely candidate counterexamples, to be examined directly in Chapter 11.

The remaining chapters of Part II (Chapters 6 to 10) will, according to the structural sufficiency criterion, fulfill one by one the carrying of the five classes of experiential structure by the five sources. Chapter 11 will then test the overall coverage and directly address the candidate counterexamples to completeness. The framework and criterion established in this chapter are the common standard on which those chapters depend: every chapter will be asked the same question—does it fulfill, relation-by-relation and without omission, the class of carrying for which it is responsible?

5.9. Chapter Summary

This chapter completes the precision of the isomorphism thesis. First, it defines the structural space of experience (five classes of relations: intra-qualitative order, qualitative types, temporal evolution, unity and binding, and temporal thickness) and the structural space of gradients (the corresponding five groups of gradient structures), and explicitly separates the structure of experience from the given-ness of experience—the former belongs to the isomorphism of this part, while the latter belongs to the co-evolution of Part III. Second, it establishes the strict meaning of isomorphism: a structure-preserving, bidirectional and non-omissive one-to-one correspondence, not loose similarity. Third, among identity, structural isomorphism, and modeling, it adopts the identity level and justifies it by three reasons: fulfilling the real meaning of unity, reducing the explanatory burden, and gathering the residual point into a single cell. Fourth, it proposes the structural sufficiency criterion as the unified standard of Part II, and uses it to explain the inevitable failure of single-scalar identity theories. Fifth, it states the completeness claim as a falsifiable proposition and identifies candidate counterexamples in advance.

The failure conditions of this chapter are as follows (summarized in Appendix E).

- F5.1 Isomorphism degenerates into loose similarity. If the isomorphism used in this book actually amounts only to "loose similarity" rather than strict structure-preserving correspondence (that is, if there exists a structural relation that holds in experience but not in gradients, or conversely, and yet this is not counted as a rupture of isomorphism), then the isomorphism thesis loses falsifiability and degenerates into rhetoric.

- F5.2 The argumentative reasons for the identity level do not hold. If "eliminating the explanatory burden of the correspondence relation" is shown to be a false advantage (for example, if the identity level in fact introduces an equal or heavier burden elsewhere), then the core argumentative reason for adopting the identity level fails, and this book must reassess the choice among the three levels.

- F5.3 The structural sufficiency criterion cannot be satisfied. If the gradient structural space is shown in principle to be unable to carry relation-by-relation a certain class of experiential structure (not merely that it has not yet done so, but that it cannot do so in principle), then structural sufficiency cannot be satisfied, the identity level breaks, and the first-principle thesis retreats to a weaker level.

- F5.4 The completeness claim is overturned by counterexample. If there exists a structural feature of experience that neither falls into the five listed classes nor can be carried by the five gradient sources, then the completeness claim fails, and the isomorphism program must either add a source or retreat to a weaker level (this is an expansion debt of the program, whose severity differs from F5.3).

Table 5.1.. The Boundary between What This Chapter Establishes and What It Cannot Derive

What this chapter establishes	What this chapter cannot derive
The definition of the structural space of experience and the structural space of gradients, and the separation of structure from given-ness.	It cannot prove in this chapter the specific carrying relation of any source (to be fulfilled chapter by chapter in Chapters 6 to 10).
The strict meaning of isomorphism (a structure-preserving, bidirectional, non-omissive one-to-one correspondence).	It cannot characterize given-ness in this chapter (that belongs to Part III and follows co-evolution rather than isomorphism).
The choice among three levels of strength (adopting the identity level) and its three argumentative reasons.	It cannot deductively prove the identity level to be true (its defense lies in the unified accommodation achieved by the whole book, not in deduction).
The structural sufficiency criterion as the unified testing standard of Part II.	It cannot determine in this chapter that the completeness claim is true (its truth or falsity depends on whether a sixth class exists, to be examined in Chapter 11).
The falsifiable formulation of the completeness claim and the advance identification of candidate counterexamples.	It cannot guarantee that the five sources are already complete (completeness is a claim accepted for testing, not a presupposition).

6. The Richness of Single-Frequency Gradient Configurations: The Continuous Order within Qualities

6.1. The Carrying Task of This Chapter, and a Demonstrative Responsibility

Chapter 5 established the structural sufficiency criterion and decomposed the work of Part II into five carrying tasks. This chapter undertakes the first of these tasks, and also the finest one: to argue that the continuous order exhibited by experience within a single quality is carried by the gradient configurations of ρ_f within a single frequency family.

The scope of this task must first be delimited carefully, because it is easily confused with Chapter 7. This chapter does not deal with differences between different qualities—the difference between red and pain, or between sound and taste. That is the problem of qualitative type and belongs to the frequency-family dimension; it is left to Chapter 7. This chapter deals only with continuous variation within the same quality: within the quality of "red", the continuous transition from vermilion to crimson to pink; within the quality of "pain", the continuous variation from dull pain to severe pain, from diffuse to sharp, and from one bodily location to another. In one sentence: Chapter 7 concerns "why red and pain are two kinds of things", while this chapter concerns "why there are such continuous layers within red". This chapter maintains that the latter is carried by single-frequency gradient configurations—because the same quality corresponds to the same frequency family, and the continuous layers within that quality correspond to the continuous structure of gradient configurations within that family.

This chapter also bears a responsibility beyond its own content: as the first chapter in Part II to fulfill a carrying relation, it must demonstrate "how a carrying relation is to be fulfilled". What Chapter 5 requires is not a vague "sense of correspondence", but a relation-by-relation, testable isomorphism—every structural relation on the experiential side must have a corresponding relation on the gradient side. This chapter will make this "relation-by-relation" standard as concrete as possible, so that Chapters 7 to 10 can follow it as a model: first characterize the structure to be carried on the experiential side (Section 6.2), then characterize the structure on the gradient side that will carry it (Section 6.3), then establish and test the correspondence relation by relation (Sections 6.4 and 6.5), and finally address the most likely objections directly (Section 6.6).

6.2. What Is to Be Carried: What Kind of Structure Is the Continuous Order within Qualities?

To carry a structure, one must first characterize that structure itself clearly—otherwise "carrying" has no target. This section characterizes what kind of structure the "continuous order within qualities" is on the experiential side.

Take color vision as an example, since it is the most readily grasped instance of intra-qualitative order. If "red" is taken as a quality, it is by no means a point without interior: within "red", there are many hues that differ from one another and yet are still all "red"—vermilion, crimson, magenta, pink, dark red, and so on. Among these hues, there are definite order relations. This book summarizes them into the following classes, and these are the objects that this chapter must carry by gradients.

First, the order of similarity. Between any two hues there is a difference between "similarity" and "dissimilarity", and this similarity is ordered: vermilion is more similar to crimson than to pink, and still more similar than to a purplish red. Hues therefore form a space with "nearness" and "distance"—similar ones are adjacent, dissimilar ones are farther apart. This "nearness" is not an arbitrary subjective impression, but a stable order that can be systematically measured by psychophysical methods, such as hue-discrimination thresholds and similarity ratings.

Second, dimensionality. Variation within a quality is often not along a single dimension, but unfolds along several dimensions. Within color vision, there are at least three dimensions: hue (the cycle of red, yellow, green, and blue), saturation (from gray to pure), and lightness (from dark to bright). Any concrete color is a point in this three-dimensional space. Intra-qualitative order is therefore a multidimensional structure, not a line.

Third, the topology and boundary of this space. This multidimensional qualitative space has a specific topological form, rather than being a featureless block. The hue dimension is cyclic (red–yellow–green–blue–purple–back to red, with the two ends joined into a ring); the saturation dimension is bounded (one end is the singular point of colorless gray, the other end is maximal purity of saturation); the lightness dimension also has black and white as endpoints. There are also special points in the space: pure gray is a singular point at which all hues converge when saturation is zero (at that point hue loses meaning). These topological features—where there is a ring, where there is a boundary, and where there is a singularity—are part of the structure of intra-qualitative order, and are a relatively rigid part that can be consistently described.

It must be emphasized that this chapter characterizes the structure of experience, not the given-ness of experience (the separation introduced in Chapter 5): we characterize "what shape the internal order of the quality-space of red has", not "that this red is experienced". Likewise, this chapter characterizes the intra-qualitative structure, not the perturbation that triggers a particular color experience (the perturbation axis of Chapter 11): the shape of hue space belongs to the structural axis, while which red is seen at this moment belongs to the perturbation axis. This chapter concerns only the former—the shape of that space.

Beyond color vision, other qualities likewise have their own internal structures of order and can be included in the same way. Within pain there are dimensions of intensity (from dull pain to severe pain), character (dull, stabbing, burning, cramping), and localization (where in the body), forming the qualitative space of pain; within pitch there is a nearly one-dimensional structure that nevertheless carries a cyclic feature (the pitch helix: similarity across octaves). What they share is this: every quality is not internally a point, but a qualitative space with an order of similarity, dimensions, and specific topology and boundaries. The structure of this space is what this chapter must carry by single-frequency gradient configurations.

6.3. What Carries It: What Kind of Structure Is a Single-Frequency Gradient Configuration?

We now turn to the gradient side and characterize the structure that is to carry the qualitative space described above: the gradient configuration of ρ_f within a single frequency family.

According to the EQT ontology (cited in Chapter 2), ρ_f is partitioned by frequency families. This chapter focuses on a single family and temporarily sets aside relations among different families (the subject of Chapter 7). Within a given frequency family, the gradient of ρ_f—that is, the rate of density variation across space—constitutes a vector field. This vector field is far from being a number; it is itself a high-dimensional object with definite structure. Its structure includes at least the following classes, which can correspond to the structures listed on the experiential side in Section 6.2.

First, continuous variation and metric of gradients. Within a single frequency family, gradient configurations can vary continuously—their directions can rotate, and their magnitudes can increase or decrease. Between any two gradient configurations, one can define a "distance" by their difference (how far apart the configurations are). Thus single-frequency gradient configurations themselves constitute a space with "nearness" and "distance"—similar configurations have small distance, different configurations have large distance. This is a continuous space with a metric.

Second, the dimensionality of this configuration space. A gradient configuration is characterized by several independent quantities (directional components, magnitude, and, under a more refined description, higher-order spatial structures such as curvature), and these independent quantities constitute the dimensions of the configuration space. The space of single-frequency gradient configurations is therefore multidimensional, with its dimensionality determined by the independently variable degrees of freedom of gradients within that family.

Third, the topology and singularities of this space. The gradient-configuration space has its topology: some degrees of freedom are cyclic (such as the direction angle of a gradient, which returns to itself after a full rotation and thereby forms a ring), while others are bounded (such as gradient magnitude, whose lower bound is zero). Singularities are especially important: when the gradient magnitude approaches zero, the direction of the gradient loses definition (the zero vector has no direction). This is a structural singularity in the configuration space, where the dimension of "direction" collapses. In addition, a vector field in space may contain topological defects (such as vortices, sources, and sinks), which are singular structures of gradient configurations. These topological features—where a ring is formed, where there is a boundary, and where direction collapses into a singularity—are rigid parts of the structure of single-frequency gradient-configuration space.

Placing Sections 6.2 and 6.3 side by side already reveals a structural resonance: both sides are continuous spaces with an order of similarity, multiple dimensions, and specific topology and singularities. But Chapter 5 repeatedly warned that the structural sufficiency criterion does not require "apparent resonance", but relation-by-relation correspondence—every structural relation on the experiential side must have a corresponding relation on the gradient side. The next section makes this relation-by-relation correspondence concrete and subjects it to testing.

6.4. Relation-by-Relation Isomorphism: Making the Correspondence Concrete

This section follows the criterion of Chapter 5 and establishes the isomorphism between intra-qualitative order and single-frequency gradient configurations relation by relation. By relation-by-relation, it means that we do not stop at a general resemblance such as "both sides have dimensions and topology", but require that every structural relation on the experiential side have a corresponding structural relation on the gradient side, and that the correspondence preserve the relation.

Correspondence of the order of similarity. On the experiential side: the similarity among hues forms an ordered nearness and distance (vermilion is close to crimson and farther from a purplish red). On the gradient side: the distance defined by differences among single-frequency gradient configurations forms an ordered nearness and distance. The isomorphism here requires that two hues with high experiential similarity correspond to two configurations with small distance in gradient-configuration space; hues with low similarity correspond to configurations with large distance. This correspondence is order-preserving—it requires the entire order relation of experiential similarity to be preserved one-to-one by the order relation of distances among gradient configurations, rather than matching only a few individual points. This is a strong requirement, and it is also testable: if there are three hues A, B, and C such that experientially A is more similar to B than to C, while their corresponding gradient configurations place A closer to C than to B, then order preservation fails and the isomorphism breaks at that point.

Correspondence of dimensionality. On the experiential side: the qualitative space of color vision has three dimensions—hue, saturation, and lightness (pain has the three dimensions of intensity, character, and localization). On the gradient side: the independent degrees of freedom of single-frequency gradient-configuration space must correspond to them in number and organization. However many dimensions the experiential qualitative space has, the gradient-configuration space carrying it must have the corresponding number of independent degrees of freedom. This correspondence is not merely numerical equality; it also requires that each experiential dimension correspond to a definite gradient degree of freedom, and that relations among experiential dimensions (such as mutual independence among the three dimensions, or coupling between two of them) correspond to the relevant relations among gradient degrees of freedom.

Correspondence of topology and singularities. This is the most rigid and therefore the most testable part. On the experiential side: the hue dimension forms a cycle, the saturation dimension is bounded, and when saturation is zero, all hues converge into the singularity of gray. On the gradient side: the gradient direction angle forms a cycle, gradient magnitude is bounded, and at zero magnitude, direction collapses into a singularity. The isomorphism here requires one-to-one correspondence between the topological features on both sides: the ring of experiential hue corresponds to the ring of the gradient direction angle; the singularity in which experiential saturation approaches zero and converges into gray corresponds to the singularity in which gradient magnitude approaches zero and direction collapses. This correspondence is the most testable because topology is rigid—a ring is a ring and cannot be continuously turned into a line; a singularity is a singularity and cannot be smoothed away. If a topological feature of experiential qualitative space (such as the cyclicity of hue) has no corresponding topological feature in gradient-configuration space, or if the correspondence is misplaced (for example, an experiential ring corresponds to a gradient line), then the isomorphism breaks at the topological level. A break at the topological level is the most serious break of isomorphism, because it cannot be remedied by parameter adjustment.

If these three correspondences—order-preserving similarity, corresponding dimensions, and one-to-one topology and singularities—hold one by one, then according to the criterion of Chapter 5, "single-frequency gradient configurations" are sufficient to carry the experiential structure of "the continuous order within qualities". It must be reiterated: what is required here is relation-by-relation validity, not overall resemblance. It is precisely this strictness of relation-by-relation mapping that makes the carrying claim of this chapter a fallible claim, rather than an unfalsifiable analogy.

6.5. A Verificatory Detail: Why the Correspondence of Singularities Is Strong Evidence

Among the relation-by-relation correspondences, the correspondence of singularities deserves separate discussion, because it provides a test that goes beyond "general similarity" and is almost adjudicative—just as the "error modes of binding" will provide a test in Chapter 9. This section takes gray in color vision as an example to show the weight of this evidence.

Consider a phenomenon in color vision that appears ordinary but is in fact profound: when saturation approaches zero, all hues converge into the same colorless gray, and the differences among hues disappear at that point. In pure gray, the question "is this a reddish gray or a bluish gray?" loses its meaning, because the hue dimension collapses when saturation is zero. This is a definite singularity in experiential qualitative space: a high-dimensional structure (the hue ring) collapses into a dimensionless structure (a single gray point) under a certain condition (saturation approaching zero).

Now consider the gradient side: when the magnitude of a vector field approaches zero, direction loses definition—the zero vector has no direction, and therefore the dimension of "direction" collapses into a singularity at zero magnitude. This is a singular structure determined purely by the mathematical nature of a vector field, and it is unavoidable.

The singularity structures on the two sides correspond precisely: in experience, "saturation $\to$ zero, hue dimension collapses, convergence into gray" corresponds to, in gradients, "magnitude $\to$ zero, direction dimension collapses, convergence into the zero vector". This correspondence is strong evidence because it is not artificially arranged to match. The singularity of hue converging into gray on the experiential side is a fact independently established in color phenomenology; the singularity of direction converging into zero on the gradient side is a fact independently established in the mathematics of vector fields. Two singular structures independently valid in their own domains correspond precisely under the isomorphism mapping. This kind of "unarranged coincidence" supports the truth of the isomorphism far more strongly than a "similarity" that can be obtained by adjusting parameters. If an isomorphism precisely matches independently established singularities on both sides, it is unlikely to be merely a coincidental analogy.

The limit of this evidence must be stated honestly: the correspondence of a single singularity is a strong indication, but strict carrying requires the entire structure of the qualitative space (the order of similarity, all dimensions, and all topological features) to correspond relation by relation, not only the singularity. The correspondence of singularities is a strong support for the carrying claim of this chapter, but it is not the whole of that claim. The full test must rest on the three correspondences in Section 6.4 as a whole, and must ultimately accept the test of the observable proxies in Part IV (for example, comparing the geometry of the qualitative space measured psychophysically with the geometry of the single-frequency gradient-configuration space inferred from neural measurements).

6.6. Objections That Must Be Addressed Directly

The carrying claim of this chapter will face several strong objections, which must be addressed directly, because a carrying relation that avoids objections is not credible.

Objection 1: the geometry of qualitative space is conventional, not intrinsic to experience. One might argue that the three-dimensional space of color vision and the ring of hue are artificial conventions of colorimetry (such as the construction of the CIE chromaticity diagram), and do not necessarily reflect the structure of experience itself. To use gradients to carry a convention would then have no deep significance. Response: what this chapter carries is not any particular conventional coordinate system of colorimetry, but the coordinate-independent structures in experiential qualitative space—the order of similarity, the number of dimensions, and the topological features (rings, boundaries, singularities). These are coordinate-independent: no matter what coordinates are used to describe color space, it remains true that "the hue dimension forms a ring" and that "saturation approaching zero converges into the singularity of gray", just as no matter what coordinates are used to describe the Earth's surface, it remains true that "it is a closed surface and has two poles". This chapter carries these coordinate-independent structures, and they are precisely intrinsic to experience and can be jointly confirmed by phenomenology and psychophysics, rather than being artificial conventions.

Objection 2: the order of similarity can be explained by neural coding and does not require gradient configurations. One might argue that hue similarity can be explained by the response curves of cone cells and the coding of opponent channels; this is mature neuroscience, so why invoke ρ_f gradients? Response: this chapter does not compete with explanations by neural coding, but maintains that the two occupy different levels and are compatible. Neural coding characterizes the implementation level—how the structure of qualitative space is implemented in neural organization (cones and opponent channels). The isomorphism claimed by this book lies at the more fundamental ρ_f level—these neural codings themselves, according to EQT, are specific organizations of single-frequency gradient configurations of ρ_f. Neural coding and ρ_f gradients are not two competing explanations, but descriptions of the same structure at two levels: neural coding is the implementation of that single-frequency gradient configuration in biological organization. What this chapter claims is that the structure of qualitative space is isomorphic to the structure of single-frequency gradient configurations; this isomorphism has its counterpart at the neural implementation level (the geometry of cones and opponent channels), rather than being replaced by the neural implementation level.

Objection 3: the topological correspondence may be forced interpretation. One might argue that "the ring of hue corresponds to the ring of the direction angle" and "the singularity of gray corresponds to the singularity of the zero vector" sound orderly, but how do we know this is not a retrospective imposition? Response: the difference between forced interpretation and isomorphism lies in whether the claim is relation-by-relation and falsifiable. The correspondence in this chapter does not list only those items that match; it requires all coordinate-independent structures of experiential qualitative space to correspond one by one with gradient-configuration space, and it explicitly gives failure conditions (at the end of Section 6.4 and in Section 6.7): if a topological feature has no counterpart, if the counterpart is misplaced, or if the order of similarity is not preserved, then the isomorphism breaks. A claim that states clear failure conditions and requires the entire structure to correspond one by one is not forced interpretation. The mark of forced interpretation is precisely that it selects only what matches, avoids what does not, and offers no falsifiability. This chapter states openly that mismatches would make the claim fail, in order to distinguish itself from forced interpretation.

6.7. Conclusion of This Chapter

This chapter has fulfilled the first carrying task of Part II: the continuous order within qualities is carried relation by relation by the richness of single-frequency gradient configurations. Its path is the model that the other chapters of Part II will follow: first characterize the structure to be carried on the experiential side (the order of similarity, dimensionality, topology, and singularities of qualitative space), then characterize the structure on the gradient side that carries it (the distance order, degrees of freedom, topology, and singularities of single-frequency gradient-configuration space), then establish the order-preserving, dimension-corresponding, and topology-matching correspondences relation by relation, using the correspondence of singularities as a strong support of unarranged coincidence, and finally address directly the three objections of conventionality, neural coding, and forced interpretation. This chapter strictly observes two separations: what is carried is the structure of qualities (not givenness, which belongs to Part III), and it is the qualitative space on the structural axis (not what is seen at this moment on the perturbation axis, which belongs to Chapter 11).

The carrying relation established in this chapter will be advanced one level in Chapter 7: this chapter concerns the continuous order "within a single quality", while Chapter 7 concerns the type distinctions "between different qualities". Together, they carry the full structure of experiential qualities—both the continuous gradation within qualities and the type differences among qualities.

6.8. Chapter Summary

This chapter is the first chapter in Part II to fulfill an isomorphic carrying relation, and it demonstrates "how carrying is to be fulfilled". First, it delimits the scope: this chapter carries the continuous order within qualities (such as gradations in hue space, and the intensity and localization of pain), and does not concern the type differences between qualities (which belong to Chapter 7). Second, it characterizes the structure to be carried on the experiential side: the order of similarity, dimensions (hue/saturation/lightness in color vision), topology and singularities of qualitative space (the cyclicity of hue, and the singularity in which saturation approaching zero converges into gray). Third, it characterizes the carrying structure on the gradient side: the distance order, degrees of freedom, topology and singularities of single-frequency gradient-configuration space (the cyclicity of the direction angle, and the singularity in which direction collapses as magnitude approaches zero). Fourth, it establishes isomorphism relation by relation: order-preserving similarity, corresponding dimensions, and one-to-one topology and singularities, while making explicit its relation-by-relation fallibility. Fifth, it uses the correspondence of singularities (saturation $\to$ zero converging into gray $\leftrightarrow$ magnitude $\to$ zero converging into the zero vector) as strong evidence of unarranged coincidence, while stating its limits. Sixth, it directly addresses the three objections of conventionality, neural coding, and forced interpretation. This chapter strictly observes the two separations between structure/given-ness and structural axis/perturbation axis.

The failure conditions of this chapter are as follows (summarized in Appendix E).

- F6.1 The order of similarity is not preserved. If there exist three intra-qualitative states whose order of experiential similarity is inconsistent with the distance order of the corresponding single-frequency gradient configurations (failure of order preservation), then the carrying of the order of similarity fails.

- F6.2 Dimensionality does not correspond. If the dimensionality and organization of experiential qualitative space cannot correspond one-to-one to the independent degrees of freedom of single-frequency gradient-configuration space (whether in number or in relation), then the carrying of dimensionality fails.

- F6.3 Topology or singularities do not correspond. If a topological feature of experiential qualitative space (ring, boundary, singularity) has no counterpart in single-frequency gradient-configuration space, or if the correspondence is misplaced, then the isomorphism at the topological level breaks. This is the most serious failure of this chapter and cannot be remedied by parameter adjustment.
- F6.4 The structure is in fact a coordinate convention. If the qualitative-space structure carried in this chapter is shown to be a coordinate convention rather than a coordinate-independent structure intrinsic to experience, then the object to be carried collapses, and Objection 1 succeeds.

Table 6.1.. The Boundary between What This Chapter Establishes and What It Cannot Derive

What this chapter establishes	What this chapter cannot derive
The continuous order within qualities is carried relation by relation by single-frequency gradient configurations (order preservation, dimensionality, topology/singularities).	It cannot carry the type differences between qualities (this belongs to Chapter 7, the frequency-family dimension).
A model of "how carrying is fulfilled": first characterize the structures on both sides, then establish relation-by-relation correspondence and provide failure conditions.	It cannot address in this chapter whether or how this quality is experienced (this belongs to given-ness in Part III).

What this chapter establishes	What this chapter cannot derive
The correspondence of singularities as strong evidence of unarranged coincidence, and its limits.	It cannot determine in this chapter which concrete state is seen at this moment (this belongs to the perturbation axis in Chapter 11).
Responses to the three objections of conventionality, neural coding, and forced interpretation (neural coding is the implementation level and is compatible with the ρ_f level).	It cannot provide in this chapter the observable proxies and measurement of qualitative-space geometry (this belongs to Part IV).

7. The Richness of the Frequency Dimension: Type Differences among Qualities

7.1. The Carrying Task of This Chapter: From "Within a Quality" to "Between Qualities"

Chapter 6 fulfilled the carrying of the continuous order within a quality—the gradation within red from vermilion to crimson, carried by single-frequency gradient configurations. This chapter advances one level further and addresses a structure of a different and more fundamental kind: the type differences among qualities.

This difference is both ordinary and profound. Red and pain are not more and less of the same thing—pain is not "very strong red", nor is red "very weak pain"; they are two mutually irreducible kinds, belonging to different "categories" of experience. Likewise, a humming sound and a bitter taste are not two points on the same dimension, but two different kinds: auditory quality and gustatory quality. Variation within a quality (Chapter 6) is quantitative, continuous, and capable of gradual change within the same space; differences among qualities (this chapter) are categorical, discontinuous, and cross different spaces. A complete theory of consciousness must be able to carry both the former and the latter—and the latter is precisely where most theories fail.

This book maintains that the type differences among qualities are carried by the energy quantum frequency dimension of ρ_f: different kinds of qualities correspond to different frequency families; red and pain are two kinds because they occupy two separated regions on the frequency axis. This chapter will argue for this point and show that it is precisely because the ontology of EQT contains a frequency dimension from the outset that this book can carry type differences, whereas this is exactly the resource fundamentally lacking in all identity theories that appeal only to a single scalar.

The distinction from Chapter 6 must first be made clear so that their scopes do not overlap. Chapter 6 concerns "gradation within red", using gradient configurations within a single frequency family; this chapter concerns "the difference between red and pain", using the frequency distinction among families. Together, the two chapters carry the complete structure of experiential qualities: both continuous gradation within qualities (gradient configurations within a frequency family) and categorical discontinuity among qualities (distinctions among frequency families).

7.2. Why "Type Difference" Is the Most Difficult Threshold for Identity Theories

Before arguing how the frequency dimension carries type differences, one must first explain why this threshold is the most difficult—because only by seeing its difficulty can one assess the weight of the frequency dimension as a resource. Chapter 5 already asserted that single-scalar identity theories must fail here; this section develops that assertion.

Imagine a theory that identifies consciousness with a certain scalar—some quantity of information, some field intensity, or some single "consciousness value". The structure of a scalar is extremely poor: it has only one dimension, namely "greater or smaller". Such a quantity can at most distinguish the strength of experience—consciousness is stronger or weaker, wakefulness is higher or lower. But the type difference among experiences is not a difference of strength: the difference between red and pain is not that one numerical value is larger and the other smaller. Even if the "intensity of red" and the "intensity of pain" are adjusted to be equal, red remains red and pain remains pain; their type difference does not disappear because their intensities are equal. A scalar with only the one dimension of "magnitude" cannot produce "two different kinds" no matter what value it takes—it can only produce "different magnitudes of the same kind". This is the deadlock of single-scalar identity theories: from the outset they lack a dimension capable of accommodating "kind".

From this one can extract the real requirement that type difference imposes on any identity theory: to carry "red and pain are two kinds", the physical side must provide an additional dimension orthogonal to intensity—a dimension not of "more and less", but of "this kind and that kind". Along this dimension, red and pain are not two points at different distances (which would still be quantitative difference within the same space), but two kinds located in different intervals and separated from each other. Without such a dimension, type difference has nowhere to be placed, and an identity theory can only retreat to a "correlation theory of consciousness intensity".

This is precisely the fundamental advantage of EQT over single-scalar theories. The basic ontology of EQT is not a scalar field, but a field ρ_f partitioned by frequency families (Chapter 2)—frequency is, from the beginning, an intrinsic dimension of this ontology independent of density magnitude. The magnitude of density gives "strength" (quantity), while frequency gives "family" (kind). EQT therefore contains from the outset the dimension orthogonal to intensity that single-scalar theories lack: the frequency dimension. This book relies on this intrinsic frequency dimension of the ontology to carry type differences. It is not a parameter introduced ad hoc to explain consciousness, but a structure already present in the physical ontology of EQT. The next section explains how it carries type differences relation by relation.

7.3. What Carries It: What Kind of Structure Is the Frequency Dimension?

For the frequency dimension to carry the type differences among qualities, it must itself possess the corresponding structure. This section characterizes the structural features of the frequency dimension as the carrier, so that they can be set against the structure of type differences.

First, the frequency dimension is orthogonal to the intensity dimension. In the ρ_f ontology, the frequency of an energy quantum and the amplitude of its density are two independent quantities: the same frequency can have different amplitudes, and the same amplitude can belong to different frequencies. This orthogonality is crucial—it ensures that "kind" (determined by frequency) and "strength" (determined by amplitude) are two irreducible dimensions, so that type difference will not be reduced to intensity difference. This corresponds precisely to the experiential fact that "kind and intensity are orthogonal": the intensity of red and the intensity of pain can be equal, while red and pain still remain two kinds—because kind is not determined by intensity.

Second, the frequency dimension has a discontinuous structure of "families". The frequency axis itself is continuous, but ρ_f is not spread uniformly along that axis; it is organized by families—energy quanta aggregate into several relatively separated frequency regions (families), with relative intervals between one family and another. This "family" structure is crucial: it makes the frequency dimension not merely a continuous line, but a line organized into several segments. The type differences among qualities—red, pain, sound, and taste as mutually discontinuous categories—correspond precisely to this discontinuous family structure: each qualitative kind corresponds to a frequency family, and the discontinuity among kinds corresponds to the intervals among families.

Third, intra-family continuity and inter-family discontinuity coexist. Within a single family, frequency (and the accompanying gradient configurations, Chapter 6) can vary continuously, carrying gradation within a quality; between families, there are intervals, carrying discontinuities among qualities. This dual structure of "continuity within families and discontinuity between families" corresponds exactly to the dual structure of experience: "continuous gradation within qualities and discontinuous classification among qualities". This is the complete picture that Chapter 6 and this chapter together seek to carry: Chapter 6 carries intra-family continuity, this chapter carries inter-family discontinuity, and both are provided uniformly by the same "frequency dimension organized by families".

When these three items are placed beside what is to be carried, the qualification of the frequency dimension as carrier becomes clear: it has a dimension orthogonal to intensity (carrying the fact that "kind is not determined by strength"), a discontinuous structure of families (carrying the fact that "kinds are discontinuous from one another"), and the coexistence of intra-family continuity and inter-family discontinuity (carrying the dual structure of "gradation within qualities and classification among qualities"). The next section makes these correspondences concrete relation by relation.

7.4. Relation-by-Relation Isomorphism: How Type Difference Is Carried

Following the criterion of Chapter 5, this section relates the type differences among qualities to the structure of the frequency dimension relation by relation.

Correspondence of orthogonality. On the experiential side: the kind of a quality is orthogonal to its intensity—the type difference between red and pain does not change as their respective intensities increase or decrease. On the gradient side: frequency is orthogonal to density amplitude—family membership does not change as amplitude increases or decreases. The isomorphism here requires that the experiential relation "adjusting intensity does not change kind" correspond to the gradient relation "adjusting amplitude does not change frequency-family membership". This correspondence is testable: if some experience were found whose qualitative kind continuously changed into another kind as its intensity continuously increased or decreased (for example, if "red" gradually became "pain" after being intensified to a certain degree, passing through a series of "half-red, half-pain" transitional kinds), then kind and intensity would not be orthogonal, and this correspondence would fail. But experiential facts do not behave this way: even the strongest red does not gradually become pain. This corresponds to the fact that frequency families do not continuously migrate with amplitude.

Correspondence of discontinuous structure. On the experiential side: qualitative kinds are discontinuous—there is no continuous spectrum of "intermediate kinds" filling the space between red, pain, sound, and taste (there is no continuous transitional quality between "red" and "pain", half red and half pain). On the gradient side: frequency families are discontinuous—there are intervals between families, not continuously filled. The isomorphism here requires that the discontinuity pattern among experiential kinds correspond to the interval pattern among frequency families: several mutually discontinuous major qualitative classes correspond to several mutually separated frequency families; the absence of any "intermediate transition" between two kinds corresponds to the interval between their two families not being crossed.

Correspondence of the dual structure. On the experiential side: continuity within qualities (Chapter 6) and discontinuity among qualities (this chapter). On the gradient side: continuity within families and discontinuity between families. The isomorphism here requires the dual structures on the two sides to correspond precisely. This item stitches Chapter 6 and this chapter together: the same "frequency dimension organized by families" provides continuity within families (the continuous variation of single-frequency gradient configurations within a family, Chapter 6) and discontinuity between families (the intervals among families in this chapter), corresponding exactly to the overall structure of experiential qualities as "continuous within, discontinuous between". This stitching is itself a strong support: the two chapters do not separately carry two structures in isolation; rather, one and the same frequency-family structure uniformly carries the entire structure of experiential qualities. Such unity supports the truth of the isomorphism more strongly than two unrelated carrying relations.

If these three correspondences hold one by one, then according to the criterion of Chapter 5, the frequency dimension is sufficient to carry the type differences among qualities. The strictness of relation-by-relation mapping must again be emphasized: this chapter does not say "experience has kinds, and frequency has families, so the two are similar". Rather, it requires that "the orthogonality of kind and intensity", "the discontinuity pattern among kinds", and "the dual structure of continuity within and discontinuity between" each find a corresponding relation in the frequency dimension. If any one of them fails, the isomorphism breaks at that point.

7.5. The Hardest Objection: Why Is "This" Frequency Band Red and "That" Frequency Band Pain?

The sharpest and deepest objection faced by this chapter does not concern whether the correspondences above hold, but a more fundamental question. Even if one grants that red corresponds to one frequency band, pain to another, and the two are discontinuous as described, one can still ask: why does "this" frequency band correspond to red rather than pain? Why does a certain frequency band "feel like" red rather than something else? If this cannot be answered, then is the carrying of qualities by the frequency dimension not merely the attachment of arbitrary frequency labels to qualities—renaming "red" as "frequency band A" and "pain" as "frequency band B", while explaining nothing about why red is red? If this objection is not addressed directly, the carrying claim of this chapter risks collapsing into "arbitrary labeling".

The response of this book must proceed on two levels, and the second level must be stated with particular honesty, because it touches the boundary guarded by the C3 prohibition throughout the book.

First level: what this chapter carries is the structure of type differences, not the given qualitative character of a particular kind. The question must be decomposed. "Red and pain are two discontinuous kinds; kind does not change with intensity; there is no continuous transition between kinds"—this is the structure of type difference, and this chapter has argued that it is carried relation by relation by the family structure of the frequency dimension. But "why does this frequency band feel like red?" asks about the given qualitative character of a particular kind itself, namely "the what-it-is-like of red as red". These two belong to different levels: the former is structure (relations among kinds), while the latter already touches given-ness (what a kind is experienced as). This chapter (and all of Part II) carries only the former—the structure of type differences. It does not claim to explain the latter—the specific what-it-is-like of a given quality. Therefore, if the question "why is this frequency band red?" asks "why is the what-it-is-like of red this way?", it does not belong to the carrying scope of this chapter, but to given-ness, and ultimately touches the final cell of Chapter 15. To take this chapter's carrying claim (type structure carried by the frequency dimension) as if it had "explained the what-it-is-like of red" would overstep the boundary—and this chapter does not make that overstep.

Second level: at the structural level, the correspondence between frequency band and kind is not arbitrary, because it must preserve the entire structure. The first level above must not be misused to suggest that "the correspondence is arbitrary". At the structural level, which frequency band corresponds to which kind cannot be freely assigned. The reason is that isomorphism requires preserving all relations (Chapter 5). Red is not an isolated kind; it occupies a definite position in the entire structure of qualities—it belongs with certain kinds (such as other colors) to a larger category (vision), and is separated from other kinds (such as pain and taste); its qualitative space has specific dimensions and topology (Chapter 6). For the frequency dimension to carry red, it must use a family that occupies the corresponding position in the frequency structure: this family must share a larger frequency region with the "families carrying other colors" (corresponding to the category of vision), must be separated from the "families carrying pain and taste", and its intra-family structure must be isomorphic to the qualitative space of red. Thus the correspondence between frequency band and kind is constrained by the whole structure: it cannot be assigned arbitrarily, just as on a distance-preserving map the position of a city cannot be moved arbitrarily while still preserving its distances to all other cities. "Red corresponds to frequency band A" is not arbitrary labeling because frequency band A must occupy a position in the entire structure of the frequency dimension that corresponds precisely to the position red occupies in the qualitative structure. If another frequency band were substituted, all its relations with other kinds would be broken, and the isomorphism would rupture.

Taken together, the two levels yield the answer of this chapter to "why is this frequency band red": at the structural level, the correspondence is not arbitrary; it is uniquely constrained by the isomorphism that preserves all relations. At the level of given qualitative character (why the what-it-is-like of red is this way), this chapter does not answer; that is the final cell and does not belong to this chapter. This answer preserves both "carrying is not labeling" (structure is not arbitrary) and "no counterfeit explanation of what-it-is-like" (the C3 prohibition). A theory that conflates these two levels will either, because of the second level, mistakenly think that the carrying relation of this chapter is empty labeling, or, because of the first level, mistakenly think that this chapter explains what-it-is-like. This chapter separates the two levels precisely to avoid both misunderstandings.

7.6. Two Concrete Tests: Synesthesia and Cross-Modal Correspondence

The claim that the frequency dimension carries type differences among qualities can be tested through two concrete phenomena. They are both potential objections and possible strong confirmations.

Synesthesia. In synesthetes, a stimulus of one kind triggers experience of another kind—for example, hearing a certain sound (auditory kind) while seeing a certain color (visual kind), that is, "colored hearing". Does this constitute a counterexample to "kinds are discontinuous from one another"? This book holds that it does not; on the contrary, it is a confirmation. According to this chapter, different kinds correspond to different frequency families; under normal conditions, families are relatively separated, and kinds therefore do not cross. Synesthesia corresponds to abnormal cross-frequency coupling (the mechanism of cross-frequency resonance in Chapter 9) between certain families that should have been separated: activation of the auditory family, through abnormal cross-frequency coupling, also activates the visual family, so one hears sound and sees color. The key point is that in synesthetic colored hearing, the color seen is still color (a visual kind), and the sound heard is still sound (an auditory kind); the two do not fuse into an intermediate "half-sound, half-color" kind. This precisely confirms the discontinuity of kinds: even when cross-kind co-activation occurs, what is activated remains two complete and mutually discontinuous kinds, rather than a continuous transitional mixture. Synesthesia is therefore not a counterexample to "kind discontinuity", but a confirmation that "kinds are carried by families; families can be co-activated through abnormal cross-frequency coupling; yet they still each maintain their discontinuity".

Cross-modal correspondence. Ordinary people also exhibit systematic cross-kind associations—for example, associating high pitch with "bright", "small", and "sharp", and low pitch with "dark", "large", and "blunt" (correspondences between sound and visual/spatial attributes). Does this show that there is in fact a continuous bridge between kinds, and that they are not discontinuous? This book holds that cross-modal correspondence does not correspond to continuous transition between kinds, but to parallels in certain structural dimensions between different families. The "brightness" of high pitch and the "brightness" of a bright color belong to the internal structures (the qualitative spaces of Chapter 6) of two different families (auditory and visual), where a structurally corresponding dimension exists (such as each family's "intensity/sharpness" dimension), so the two can be systematically associated. This association is a parallel between structural dimensions, not a continuous transition of kind: high pitch remains an auditory kind, and bright color remains a visual kind; they do not merge into one kind in between, but simply have corresponding dimensions in their respective spaces. Cross-modal correspondence is therefore also not a counterexample to kind discontinuity, but an expression of the fact that "different families each have internal structures, and corresponding relations can exist between structural dimensions across families". This cross-family structural correspondence itself also falls under the mechanism of cross-frequency resonance in Chapter 9.

The treatment of the two phenomena follows the same logic: discontinuity among kinds (separation of families) and relations among kinds (cross-frequency coupling and structural-dimensional parallelism) are not contradictory. The former concerns the separation and completeness of each family; the latter concerns coupling and structural parallelism among families. The two can coexist, just as different instruments are each complete (discontinuous) and yet can couple in an ensemble and share certain structures (relations). This chapter carries discontinuity (kinds carried by families), while Chapter 9 carries relation (cross-frequency coupling). Each has its proper role, and they do not conflict.

7.7. Conclusion of This Chapter

This chapter has fulfilled the second carrying task of Part II: the type differences among qualities are carried relation by relation by the family structure of the energy quantum frequency dimension. The key point is that type difference is the most difficult threshold for identity theories because it requires a "kind" dimension orthogonal to intensity, and single-scalar theories lack this dimension from the outset and must therefore fail. The ontology of EQT contains the frequency dimension intrinsically (orthogonal to density amplitude and organized as a discontinuous family structure), and precisely provides this resource. This chapter has established three correspondences relation by relation: orthogonality, discontinuous structure, and the dual structure of continuity within and discontinuity between; it has also used as a support the stitching by which "one and the same frequency-family dimension uniformly carries intra-qualitative continuity (Chapter 6) and inter-qualitative discontinuity (this chapter)". The chapter directly addressed the sharpest objection—"why is this frequency band red?"—and preserved the boundary through a two-level response: at the structural level, the correspondence is not arbitrary (it is uniquely constrained by the isomorphism that preserves all relations); at the level of given qualitative character (the what-it-is-like of red), no answer is given (it belongs to the final cell, the C3 prohibition). Finally, synesthesia and cross-modal correspondence served as tests, showing that discontinuity among kinds and relations among kinds can coexist without contradiction.

Together with Chapter 6, this chapter carries the complete structure of experiential qualities: within qualities, continuous gradation is carried by single-frequency gradient configurations; between qualities, categorical discontinuity is carried by frequency families. Chapter 8 will turn to another class of structure—the temporal evolution of experience (the stream of consciousness)—and will there formally situate the axis of "perturbation" that drives experiential content (already cited in advance in Chapter 11).

7.8. Chapter Summary

This chapter is the second isomorphic carrying task in Part II. First, it delimits its scope: it carries the type differences among qualities (red and pain as two kinds), distinguished from the continuous order within qualities in Chapter 6. Second, it argues that type difference is the hardest threshold for identity theories: it requires a "kind" dimension orthogonal to intensity, whereas single-scalar theories have only the one dimension of "magnitude" and therefore must fail by retreating into a correlation theory of intensity. Third, it shows that the frequency dimension of EQT (orthogonal to amplitude and organized as a discontinuous family structure) is precisely this resource, and that it belongs intrinsically to the ontology rather than being introduced ad hoc. Fourth, it establishes the isomorphism relation by relation: orthogonality (kind does not change with intensity $\leftrightarrow$ family does not migrate with amplitude), discontinuous structure (there is no intermediate transition among kinds $\leftrightarrow$ there are intervals among families), and the dual structure of continuity within and discontinuity between (stitching Chapter 6 and this chapter together through the same frequency dimension organized by families). Fifth, it directly responds to the deepest objection, "why is this frequency band red?", on two levels: at the structural level, the correspondence is not arbitrary (it is uniquely constrained by preserving all relations through isomorphism); at the level of given qualitative character, no answer is given (the final cell, C3). Sixth, it uses synesthesia (co-activation through abnormal cross-frequency coupling, while kinds remain discontinuous) and cross-modal correspondence (parallelism of structural dimensions among families, not continuous transition among kinds) as tests, showing the coexistence of discontinuity among kinds and relations among kinds.

The failure conditions of this chapter are as follows (summarized in Appendix E).

- F7.1 Kinds transform continuously with intensity. If there exists an experience whose qualitative kind continuously transforms into another kind as its intensity continuously increases or decreases (producing a continuous spectrum of half-red, half-pain transitions), then kind and intensity are not orthogonal, and the claim that the frequency dimension carries type differences fails.

- F7.2 The discontinuity patterns do not correspond. If the discontinuity pattern among experiential qualitative kinds (how many major classes there are, and between which classes there is no transition) cannot correspond to the interval pattern among frequency families, then the carrying of discontinuous structure fails.

- F7.3 The stitching of the dual structure fails. If the dual structure of qualities as "continuous within, discontinuous between" cannot be uniformly carried by "one and the same frequency dimension organized by families" (continuous within families, discontinuous between families), and instead requires two unrelated mechanisms, then the stitching between Chapter 6 and this chapter fails, and the unity of the carrying relation is lost.
- F7.4 The correspondence is arbitrary labeling. If the correspondence between frequency bands and kinds is shown not to be constrained by the isomorphism that preserves all relations, and can be assigned arbitrarily without breaking any structural relation, then the correspondence degenerates into arbitrary labeling and the carrying relation of this chapter collapses (note: this is a structural-level failure; "why the what-it-is-like of red is this way" is not included here and belongs to the final cell).

Table 7.1.. The Boundary between What This Chapter Establishes and What It Cannot Derive

What this chapter establishes	What this chapter cannot derive
The type differences among qualities are carried relation by relation by the frequency-family structure (orthogonality, discontinuity, dual structure).	It cannot carry the continuous order within qualities (this belongs to Chapter 6, single-frequency gradient configurations).
The frequency dimension is the "kind" dimension intrinsic to the EQT ontology, which single-scalar theories lack; therefore EQT can carry kinds, whereas they cannot.	It cannot explain the given qualitative character of a particular kind ("why the what-it-is-like of red is this way", which belongs to the final cell, C3).

What this chapter establishes	What this chapter cannot derive
The two-level response to "why is this frequency band red": at the structural level the correspondence is not arbitrary; at the qualitative level no answer is given.	It cannot address in this chapter whether or how a quality is experienced (this belongs to given-ness in Part III).
Synesthesia and cross-modal correspondence as tests of the coexistence of discontinuity among kinds and relations among kinds.	It cannot provide in this chapter the observable proxies and measurement of frequency families (this belongs to Part IV).

8. The Richness of Evolution under the Constraint of the Master Equation: The Stream of Consciousness and the Perturbation Axis

8.1. The Dual Status of This Chapter

This chapter has a special status in Part II: it both undertakes the third carrying task among the five sources and serves as the formal source of an important axis in the whole book.

As a carrying chapter, this chapter argues that the temporal-dynamic structure of experience as a "stream"—its flow, coherence, non-arbitrary transitions, and the movement of the attentional focus—is carried by the evolution of ρ_f, while the evolution of ρ_f is constrained by the unified EQT Master Equation (cited in Chapter 2). The preceding two chapters carried structures of experience at a given moment: gradation within qualities (Chapter 6) and types among qualities (Chapter 7). These are structures at the level of a "snapshot". This chapter carries the structure of experience across moments: how one snapshot transitions into the next, and how the entire stream of consciousness flows.

As the source of an axis, this chapter formally situates the "perturbation axis", which has already been cited in advance in Chapter 11 and will also be relied upon in Chapter 14. Chapter 11 indicates that experience has two orthogonal axes: structural sources (how rich the gradient configuration is, determining what experience can possibly be like) and perturbation sources (what perturbation drives the gradient, determining what experience is actually like at this moment). "Perturbation-driven gradient evolution" is precisely the concern of the master equation—the master equation not only constrains how gradients evolve on their own, but also describes how gradients change under perturbative driving. The formal entry of the perturbation axis therefore most naturally falls in this chapter.

These two statuses are not two separate matters, but a pair: the master equation gives the "law" of evolution (how the stream of experience can flow, the structural axis), while perturbation gives the "drive" of evolution (where the stream of experience actually flows at this moment, the perturbation axis). Law and drive are like the shape of a river channel and the water entering it: the channel (the master equation) determines how water can flow, while the source of water (perturbation) determines which current is actually flowing at this moment. This chapter will first treat how the "law" carries the structure of the stream of consciousness (Sections 8.2 to 8.4), then formally situate the "drive" as the perturbation axis (Sections 8.5 and 8.6), and finally draw the boundary between this chapter and Chapter 10 (Section 8.7).

A boundary with Chapter 10 must first be drawn, so that the two do not overlap. This chapter concerns the law of evolution of the stream of consciousness—the coherence of flow, the constraints on transition, and directionality; Chapter 10 concerns the temporal thickness and memory hierarchy of experience—retention and protention in the present, and the short and long timescales of memory. To use an analogy: this chapter concerns "how water flows", while Chapter 10 concerns "how deeply and how long the flow leaves traces in the riverbed". Both involve time, but one concerns the law of flow, and the other the timescale of retention.

8.2. What Is to Be Carried: What Kind of Temporal Structure Is the Stream of Consciousness?

To carry the stream of consciousness, one must first characterize its temporal structure on the experiential side. William James used the phrase "the stream of consciousness" to capture several temporal features of experience. This book takes those features that can be structurally specified as the object to be carried in this chapter.

First, continuity (flow rather than jump). Experience is continuous in time, not a splice of discrete frames. Even when content changes sharply (as when one suddenly remembers something), the transition itself occurs within a continuous experiential field; there is a transition that, though rapid, is not an instantaneous break. There is no "absolute, transitionless instantaneous rupture" in experience—even the most abrupt thought arises from the existing field of consciousness rather than replacing it out of nowhere. This continuity is the most basic temporal structure of the stream of consciousness.

Second, non-arbitrariness of transition (constrained succession). The transition from one experiential state to the next is not arbitrary—the experience of the next moment is constrained by the experience of the present moment, and there is a kind of "succession" between them. The continuation of thoughts has its context (one thought often leads to another in an intelligible relation), and the flow of perception has its inertia (when the gaze moves, the scene changes continuously rather than jumping randomly). The stream of consciousness is not an irregular random sequence, but a constrained evolution: the present state restricts the possible range of the next state.

Third, directionality (flowing toward the future). The stream of consciousness has an intrinsic temporal direction—it flows from the past toward the future and is not reversible. This directionality is not only the asymmetry that "we remember the past rather than the future" (that part belongs to Chapter 10), but the orientation of the flow itself:

the stream of experience has a tendency of “flowing toward”, an evolutionary direction that cannot be played backward.

Fourth, the movement of attention. Within the continuous stream, there is a focus—attention—that is constantly shifting: it moves from one content to another, sometimes captured by external objects, sometimes actively redirected. The movement of attention is a characterizable dynamic feature in the structure of the stream of consciousness: where the focus is, how it shifts, and the distinction between focus and background.

These four items—continuity, non-arbitrariness of transition, directionality, and the movement of attention—constitute the temporal structure of the stream of consciousness on the experiential side. They are what this chapter must carry by the evolution constrained by the master equation. The separation established in Chapter 5 must be reiterated: this chapter carries the structure of the stream of consciousness (how it flows), not its given-ness (that this flow is experienced, which belongs to Part III), and it does not specify in advance which concrete content it flows toward (that is determined by perturbation; see Section 8.5).

8.3. What Carries It: What Kind of Structure Is Evolution under the Constraint of the Master Equation?

We now turn to the gradient side. The evolution of ρ_f is constrained by the unified EQT Master Equation (Chapter 2 cites its form; this chapter does not re-prove it). As the carrier, the master equation has structural features that correspond to the four items of the stream of consciousness.

First, continuity of evolution. The master equation is an equation that describes the continuous temporal evolution of ρ_f—it gives continuous evolutionary trajectories, not discrete state jumps. Gradient configurations evolve continuously along the master equation; even if there are rapid transformations (sharp changes under the dominance of nonlinear terms), they are steep segments of a continuous trajectory rather than genuine instantaneous breaks. This corresponds to the continuity of the stream of consciousness: the continuous flow of experience corresponds to the continuous evolution of gradient configurations along the master equation.

Second, deterministic constraint of evolution (the present constrains the next moment). The master equation is a law of evolution: given the present gradient configuration, the equation constrains how it can change in the next moment. The state of the next moment is not arbitrary, but jointly restricted by the present state and the equation. This corresponds to the non-arbitrariness of transition in the stream of consciousness: the present experience restricts the possible range of the next experience, just as the present gradient configuration restricts the possible range of the next configuration through the master equation. The contextual succession of thoughts and the inertia of perceptual flow correspond to gradient evolution constrained by the equation rather than random evolution.

Third, directionality of evolution. The master equation contains dissipative terms (Chapter 2), and dissipation gives evolution an irreversible direction—the system evolves in the direction of increasing dissipation, and this direction cannot spontaneously reverse. This corresponds to the directionality of the stream of consciousness: the flow of experience toward the future and its irreversibility correspond to the irreversible direction of gradient evolution under dissipative constraints. The "arrow" of the stream of consciousness corresponds to the "arrow" of evolution; both share their origin in the direction set by dissipation.

Fourth, focus dynamics within evolution. The nonlinear terms of the master equation allow locally enhanced, movable structures to form within gradient configurations—a local region is amplified and maintained through nonlinear self-sustaining, and can move within the field. This corresponds to the movement of attention: the location of the attentional focus corresponds to the locally enhanced configuration in the gradient field that is amplified and maintained by nonlinearity; the shift of the focus corresponds to the movement of that enhanced configuration in the field; the distinction between focus and background corresponds to the distinction between the enhanced configuration and the rest of the field.

Placed side by side, the qualification of master-equation-constrained evolution as a carrier becomes clear: it has continuous evolution (carrying flow), deterministic constraint (carrying non-arbitrary transition), dissipative directionality (carrying directionality), and nonlinear movable local enhancement (carrying the movement of attention). The next section makes these correspondences concrete relation by relation.

8.4. Relation-by-Relation Isomorphism: How the Stream of Consciousness Is Carried

Following the criterion of Chapter 5, this section relates the temporal structure of the stream of consciousness to the structure of master-equation evolution relation by relation.

Correspondence of continuity. On the experiential side: the stream of consciousness is continuous and has no instantaneous break; even abrupt changes are continuous transitions. On the gradient side: the master equation gives continuous evolutionary trajectories, and sharp changes are steep segments of continuous trajectories rather than instantaneous breaks. The isomorphism requires that the experiential property of "no absolute instantaneous rupture" correspond to the gradient-evolution property of "continuous trajectory". Its testability lies here: if experience contains a genuine instantaneous rupture without any transition (with no continuous bridge between the experience of the previous moment and that of the next), while master-equation evolution is everywhere continuous, then this correspondence fails.

Correspondence of transition constraints. On the experiential side: the present experience restricts the possible range of the next experience (succession has context, and flow has inertia). On the gradient side: the master equation makes the present configuration restrict the possible range of the next configuration. The isomorphism requires that the constraint pattern of experiential transition correspond to the constraint pattern of gradient evolution—what successions in experience are natural and what successions are abrupt should correspond to which transitions in gradient evolution are low-resistance paths permitted by the equation and which require overcoming greater obstacles. A testable implication is that naturally fluent successions of thought should correspond to gradient evolution along low-resistance paths permitted by the equation, whereas abrupt and effortful transitions should correspond to those that require overcoming greater evolutionary obstacles.

Correspondence of directionality. On the experiential side: the stream of consciousness flows one-way toward the future and cannot be played backward. On the gradient side: dissipative terms make evolution irreversible and give it a definite direction. The isomorphism requires that the arrow of experiential time correspond to the arrow of evolution, and that the two share their origin: the physical counterpart of the directionality of the stream of consciousness is not an additional time direction, but precisely the direction set by dissipation in the evolution of ρ_f. This correspondence reduces "why experience has a temporal direction" to "why evolution has a dissipative direction", and provides a substantive unification.

Correspondence of attentional movement. On the experiential side: the attentional focus constantly shifts within the stream; it has a location, it moves, and there is a focus–background distinction. On the gradient side: the nonlinear amplified local enhancement has its position, can move, and is distinguished from the background. The isomorphism requires the three features of attention (location, shift, focus–background) to correspond one by one to the three features of the enhanced configuration (position, movement, distinction from the background). A testable implication is that the speed of attentional shift and the number of foci that can be simultaneously maintained should correspond to the dynamical constraints on the movement and coexistence of enhanced configurations (for example, the resources required for nonlinear maintenance of a high-intensity local region restrict the number of foci that can be simultaneously maintained).

If these four correspondences hold one by one, then according to the criterion of Chapter 5, evolution constrained by the master equation is sufficient to carry the temporal structure of the stream of consciousness. The strictness of relation-by-relation mapping must again be emphasized: this chapter does not say "experience flows and gradients evolve, so they are similar". Rather, it requires continuity, transition constraints, directionality, and attentional movement each to find a corresponding relation in master-equation evolution and preserve that relation. If any one of them fails, the isomorphism breaks at that point.

8.5. The Formal Entry of the Perturbation Axis: The Difference between Law and Drive

At this point, this chapter has completed one side of the "law"—the structure of the stream of consciousness is carried by the evolutionary law of the master equation. But there is one thing that the evolutionary law itself does not determine, and it is crucial for understanding the concrete content of experience: toward which content does the stream of consciousness actually flow at this moment?

The master equation gives the law of evolution—given the present configuration, how the next moment can change. But "how it can change" is a range, not a determinate result; where within that range it actually flows depends on what perturbation is injected at this moment. When one sees a flash of light, the stream of consciousness is pushed by this external perturbation toward the content of "seeing light"; when a thought arises, the stream of consciousness is pushed by this internal perturbation toward the content of "thinking of this matter". The law defines the river channel, and perturbation injects the water flow—the channel determines how water can flow, while the water flow determines which current actually flows at this moment. This is the perturbation axis, which this chapter formally situates.

The distinction between "law" and "drive" must be made clear, because confusing the two leads to two errors. First, if one thinks that the law of the master equation itself determines all concrete contents of experience, one falls into a rigid determinism—as if the stream of experience were a track already written in advance, leaving no room for external or spontaneous input. This ignores the driving role of perturbation. Second, if one thinks that the concrete content of experience is wholly determined by perturbation and has nothing to do with law, one ignores the structural constraints of the stream of experience—as if any perturbation could push experience toward any state, without continuity or succession. This ignores the constraining role of law. The correct picture is the cooperation of the two: law constrains the possible forms of evolution (the structural axis), and perturbation drives evolution along those possible forms toward actual content (the perturbation axis). Every moment of the stream of experience is the joint result of law and drive—it flows along a continuous, constrained, and directed path permitted by the master equation (law), while where it flows is determined by the perturbation injected at that moment (drive).

This distinction also clarifies the exact scope of the carrying claim of this chapter. What Sections 8.2 to 8.4 carry is the structure of the stream of consciousness (continuous, constrained, directed, and involving movement of focus)—this is the carrying of law and belongs to the structural axis. The actual content of experience at this moment belongs to the perturbation axis and is not part of this chapter's structural carrying. This chapter does not claim that "the master equation predetermines what you are thinking at this moment"; it claims only that "what you are thinking at this moment arrives along a structure permitted by the master equation and is driven there by the perturbation of the moment".

8.6. Two Sources of Perturbation: Exogenous and Endogenous

Since perturbation is the driver of experiential content, where perturbation comes from determines the source of experiential content. Chapter 11 has already used this to situate the sense of self, the sense of agency, and affective tone. As the source of the perturbation axis, this chapter must formally state the source structure of perturbation.

Perturbations are divided into two classes according to their source. Exogenous perturbations are perturbations that come from outside the system—the environment, through sensory channels, transforms external changes into perturbations injected into the evolution of ρ_f. Seeing, hearing, and touching are all cases in which exogenous perturbations drive the stream of consciousness toward corresponding contents. Endogenous perturbations are perturbations spontaneously generated by the system itself—not caused by external objects, but initiated by internal states of the system. Remembering, imagining, spontaneous thoughts, and the arising of emotion are all cases in which endogenous perturbations drive the stream of consciousness.

The key commonality between the two classes of perturbations is this: they drive the same evolutionary law and flow through the same river channel. Whether the perturbation comes from outside or inside the system, what it drives is the same ρ_f evolution constrained by the same master equation. The difference between exogenous and endogenous perturbations lies only in the source of perturbation, not in the evolutionary mechanism they drive. This commonality is significant: it shows that experience triggered by external stimuli (the world seen) and experience triggered spontaneously from within (imagined scenes, arising emotions) are mechanistically the same kind of process—both are perturbation-driven evolution under the master equation, which then becomes given through internal–

external co-evolution (Chapter 14); they differ only in where the perturbation comes from. This is the basis on which Chapter 11 situates the sense of self, the sense of agency, and affective tone as "types of endogenous perturbation": they are not a sixth kind of structure, but different modes of driving within the class of endogenous perturbations—emotion as a diffuse-modulatory endogenous perturbation, agency as a pre-initiating endogenous perturbation, and self as the system's continuous self-perturbation (see Chapter 11).

It must be noted here that the distinction between the inside and outside of perturbation sources (this chapter) is different from the "inside" and "outside" of internal–external co-evolution in Chapter 14, so as to avoid confusion (Chapters 11 and 14 will also state this). The inside/outside of perturbation sources refers to whether experience is triggered from outside or inside the system; the inside/outside of the two poles of co-evolution (Chapter 14) refers to the two poles in the co-evolutionary relation on which given-ness depends. The two are orthogonal: whether the perturbation comes from inside or outside, given-ness is completed through the same two poles of co-evolution. This chapter concerns only the level of perturbation source, and does not concern the inside/outside of given-ness (which belongs to Part III).

Thus the perturbation axis is fully situated in this chapter: perturbation drives master-equation evolution and determines the content of experience at this moment; perturbations are divided into exogenous and endogenous classes, both sharing the same evolutionary law; and several modes of endogenous perturbation carry the sense of self, the sense of agency, and affective tone (Chapter 11). The perturbation axis and the structural axis (the five sources) are orthogonal and complementary—the structural axis determines what experience can possibly be like, while the perturbation axis determines what experience is actually like at this moment.

8.7. The Boundary with Chapter 10, and the Conclusion of This Chapter

This chapter must finally draw the boundary with Chapter 10, because both concern time and are therefore most easily conflated. This chapter carries the evolutionary law of the stream of consciousness—continuity of flow, constraints on transition, directionality, and the movement of attention; it deals with "how experience flows". Chapter 10 carries the temporal thickness and memory hierarchy of experience—retention and protention in the present, the short and long timescales of memory, and forgetting; it deals with "how deeply and how long experience is retained in time". Flow and retention are two different matters: flow is the progression of evolution through time (the master equation), while retention is the stable duration of transients at different frequencies (the transient timescale spectrum, Chapter 10). This chapter concerns the former, and Chapter 10 concerns the latter. Together, they carry the complete structure of experiential temporality—both the law of flow and the timescale of retention.

This chapter has fulfilled the third carrying task of Part II and has situated the perturbation axis. The temporal structure of the stream of consciousness (continuous, constrained, directed, and involving movement of focus) is carried relation by relation by evolution constrained by the master equation; and what content experience flows toward at this moment is driven by perturbation. Perturbations are divided into exogenous and endogenous classes, both sharing the same evolutionary law, while the modes of endogenous perturbation carry the sense of self, the sense of agency, and affective tone. The cooperation of law and drive constitutes every moment of the stream of experience. This chapter strictly observes several separations: what is carried is the structure of the stream of consciousness (not given-ness, which belongs to Part III; not temporal retention, which belongs to Chapter 10), and it clearly distinguishes law (the structural axis) from drive (the perturbation axis), and the inside/outside of perturbation sources from the inside/outside of co-evolution.

8.8. Chapter Summary

This chapter has the dual status of a carrying chapter and the source of the perturbation axis. First, as a carrying chapter, it argues that the temporal structure of the stream of consciousness is carried by evolution constrained by the master equation: continuity (continuous without instantaneous rupture $\leftrightarrow$ continuous evolutionary trajectory), non-arbitrariness of transition (the present constrains the next moment $\leftrightarrow$ the equation constrains the next configuration), directionality (irreversible flow toward the future $\leftrightarrow$ dissipative orientation), and attentional movement (location/shift/focus–background of the focus $\leftrightarrow$ position/movement/distinction from background of the nonlinear enhanced configuration), with all four correspondences established relation by relation. Second, as the source of the perturbation axis, it formally situates the distinction between "law and drive": the master equation gives the law of evolution (the structural axis, what experience can possibly be like), while perturbation gives the drive of evolution (the perturbation axis, what experience is actually like at this moment), like a river channel and water flow. Third, it states the two sources of perturbation: exogenous (the environment through the senses) and endogenous (spontaneous within the system). They share the same evolutionary law and differ only in source; modes of endogenous perturbation carry the sense of self, the sense of agency, and affective tone (Chapter 11). Fourth, it distinguishes the inside/outside of perturbation sources from the inside/outside of the two poles of co-evolution (Chapter 14); the two are orthogonal. Fifth, it draws the boundary with Chapter 10: this chapter concerns the law of flow, while Chapter 10 concerns the timescale of retention.

The failure conditions of this chapter are as follows (summarized in Appendix E).

- F8.1 Continuity does not correspond. If experience contains genuine transitionless instantaneous ruptures while master-equation evolution is everywhere continuous (or conversely), then the carrying of continuity fails.
- F8.2 Transition constraints do not correspond. If the constraint

pattern of experiential transition (which successions are natural and which are abrupt) cannot correspond to the constraint pattern of master-equation evolution (which paths are low-resistance and which require overcoming obstacles), then the carrying of transition constraints fails.

- F8.3 Directionality does not share a source. If the arrow of experiential time is shown to have no systematic relation to the dissipative direction of ρ_f evolution (for example, if the two directions can be separated or opposed), then the carrying of directionality fails, and the claim that "the arrow of time shares its origin with dissipation" collapses.
- F8.4 Attentional movement does not correspond. If the dynamics of attention (location, speed of shift, number of foci that can be maintained simultaneously) cannot correspond to the dynamics of nonlinear enhanced configurations (position, movement, coexistence constraints), then the carrying of attentional movement fails.
- F8.5 Law and drive cannot be separated. If the "law of evolution" (structural axis) and the "drive of perturbation" (perturbation axis) are shown to be inseparable (for example, if experiential content is completely predetermined by law, or completely determined by perturbation without any constraint from law), then the law/drive framework of this chapter fails, and the placement of the perturbation axis must be reassessed.

Table 8.1.. The Boundary between What This Chapter Establishes and What It Cannot Derive

What this chapter establishes	What this chapter cannot derive
The temporal structure of the stream of consciousness is carried relation by relation by master-equation evolution (continuity, constraint, direction, focus).	It cannot carry the temporal thickness of experience and the timescales of memory (these belong to Chapter 10, the transient timescale spectrum).

What this chapter establishes	What this chapter cannot derive
The distinction between law and drive: the master equation gives the law (structural axis), and perturbation gives the drive (perturbation axis).	It cannot predetermine by law the concrete content of experience at this moment (content is driven by perturbation and belongs to the perturbation axis).
The formal placement of the perturbation axis: exogenous/endogenous classes, sharing the same evolutionary law; endogenous modes carry self/agency/emotion.	It cannot give a detailed reduction of self/agency/emotion in this chapter (that belongs to Chapter 11; this chapter is its source).
The distinction between the inside/outside of perturbation sources and the inside/outside of the two poles of co-evolution (orthogonal).	It cannot treat given-ness or the inside/outside of co-evolution (these belong to Part III, Chapter 14).

9. The Richness of Resonance Modes: Unity, Binding, and the Dissolution of the Combination Problem

9.1. The Place of This Chapter in the Isomorphism Program and Its Dual Task

The structural sufficiency criterion established in Chapter 5 requires that, for every class of structural features of experience, a specific source must be identified on the gradient side and must carry that feature relation by relation. The preceding three chapters have already fulfilled three such classes: the continuous order within qualities is carried by single-frequency gradient configurations (Chapter 6), the type distinction among qualities is carried by the frequency-family dimension (Chapter 7), and the temporal evolution of the stream of consciousness is carried by evolution constrained by the master equation (Chapter 8). These three classes all concern structures at the level of experiential "content": what experience is like, what quality it belongs to, and how it flows.

This chapter turns to a deeper and more difficult structural feature: the organizational level of experience. At any given moment, why are the diverse contents of experience organized into a unified experiential field belonging to one subject, and why, within that field, are certain features bound as aspects of the same object while others belong to different objects? This is the unity and binding of experience. This book maintains that these two organizational structures are carried by the resonance modes of ρ_f gradients: the unity of experience corresponds to the coherence established by same-frequency resonance, while the binding of experience corresponds to the coupling established by cross-frequency resonance. This is the first task of this chapter, namely, to fulfill the isomorphic carrying of the fourth source according to the structural sufficiency criterion.

This chapter also has a second task, and this is why it is one of the heaviest-bearing chapters in Part II. Chapter 1 decomposed the hard problem and anticipated that the combination problem—"how multiple interiors become one 'I'"—would be dissolved rather than solved. The reason the combination problem is difficult lies in an apparently unavoidable picture: the brain is composed of tens of billions of cells; if each cell has its own minimal interior (which this book does maintain; see Chapter 6 and the discussion of boundary in Chapter 3), then the "I" seems able to be only some kind of sum of these tens of billions of micro-interiors. Yet William James already pointed out that many independent feelings do not automatically add up to one unified feeling. This chapter will show that once the unity of experience is correctly understood as resonant coherence, the "I" is not a sum of microscopic subjects at all, but a single coherent resonance. The combination problem therefore loses its object and is dissolved without requiring a solution. These two tasks are not separate matters: it is the same claim—that unity is coherent resonance—which both fulfills the isomorphic carrying relation and dissolves the combination problem.

9.2. The Unity of Experience: A Phenomenological Characterization

Before arguing how the gradient side carries unity, one must first characterize clearly what "unity" means on the experiential side; otherwise, we will not know what gradients are supposed to carry.

The unity of experience does not mean that experiential content is singular or impoverished. On the contrary, it means that although, at any given moment, experience is extremely rich and differentiated in content—I may at this moment simultaneously see the tree outside the window, hear the sound of a distant car, feel the pressure of the chair back, and think of an unfinished matter—all these diverse contents are nevertheless given as contents of one and the same experiential field belonging to one and the same experiential subject, rather than as several isolated experiences occurring separately. Vision does not occur in one place, hearing in another, and then I learn from outside that the two have occurred. Rather, seeing and hearing coexist from the beginning within the same "my experience at this moment". This feature, that "multiple contents co-belong to one field", is the unity of experience.

The philosophical tradition has characterized this with considerable precision, and this book adopts its central elements. First, there is the singleness of the subject: multiple experiences co-belong to one and the same subject; there is no "half of me seeing and half of me hearing". Second, there is co-consciousness: any two simultaneous experiential contents are not only each given, but given as being experienced "together with" one another—between seeing and hearing, there is a relation of "being jointly aware", not merely two isolated awarenesses that happen to occur at the same time. Third, there is the singleness of perspective: the unified experiential field has a single perspectival origin, and its diverse contents are organized from the same "here, now, for me".

It must be emphasized that the unity of experience is a structural feature, not given-ness itself. "Multiple contents co-belong to one field" is an organizational relation within the experiential field, and it can be structurally characterized (which contents co-belong, and in what relation they co-belong); by contrast, "this field is experienced" is given-ness and belongs to Part III. This chapter carries only the former—unity as a structure. This is consistent with the methodological separation in Chapter 5 between "structure and given-ness": the question here is "what gradient structure corresponds to the organizational relation that multiple contents co-belong to one field", not "why this field is awake".

9.3. Same-Frequency Resonance Carries Unity: Coherence as the Physical Counterpart of "Co-Belonging to One Field"

This book maintains that the gradient-side counterpart of the unity of experience—multiple contents co-belonging to one field—is the coherence established by same-frequency resonance.

The precise physical meaning of "coherence" must first be clarified, so that it does not become a vague metaphor. In wave and oscillatory systems, for several oscillations to be coherent means that they maintain definite and stable phase relations among themselves: they do not oscillate independently with randomly drifting phases, but are locked into a durable phase coordination. A coherent set of oscillations forms a single whole that cannot be divided at will—one cannot extract an oscillation from a coherent set without disrupting the overall phase relation, just as one cannot remove a voice from a harmonious chorus without changing the overall harmony. Whether coherence obtains is a fact with definite physical criteria (measurable by the degree of phase locking, coherence length, coherence time, and so on), not a vague descriptive degree.

When this physical fact is related to the unity of experience, the isomorphism holds relation by relation. Following the criterion of Chapter 5, this book examines it item by item.

Correspondence to the singleness of the subject: a coherent resonant set is a single whole—its phases are locked as one, and therefore it "is one" rather than "many". The singleness of the experiential subject (the absence of half a self) corresponds to the coherence of the gradient resonance that carries experience: precisely because multiple oscillations are coherent as one, the experiences they carry co-belong to one subject. The higher the degree of coherence, the stronger the singleness of the subject; once coherence breaks down (phase locking is lost), the unified subject also breaks down into multiple parts or disperses. This correspondence will be given observable tests in Chapter 24 through split-brain, anesthesia, dissociation, and related states.

Correspondence to co-consciousness: in a coherent set, any two oscillations are related to each other through shared phase locking. They are not two independent oscillations that merely happen to occur simultaneously, but two components within the same coherent structure, jointly participating in the overall phase coordination. The co-consciousness relation in experience, whereby "seeing and hearing are jointly aware", corresponds to the physical relation in which "two oscillations are within one coherent structure and share its phase coordination". Co-consciousness is not the external juxtaposition of two isolated awarenesses, just as coherence is not the accidental simultaneity of two independent oscillations.

Correspondence to the singleness of perspective: a coherent resonance has its single overall phase structure, and the diverse component oscillations are organized relative to this overall phase. The single perspectival origin of the experiential field (here, now, for me) corresponds to the single reference provided by this overall phase structure.

If these three correspondences hold one by one, then according to the structural sufficiency criterion, the "coherence of same-frequency resonance" is sufficient to carry the structural feature of "the unity of experience". One point must be emphasized to avoid misunderstanding: "same-frequency" here does not mean literally one single frequency, but refers to the frequency relations among oscillations that can phase-lock with one another and thereby enter the same coherent structure. They may belong to the same family or to neighboring relations that allow stable phase locking. The emphasis is not on numerical identity of frequency, but on whether phase can be locked as one: those that can lock as one are coherent as a unified field; those that cannot lock belong to different fields or remain mutually dispersed.

9.4. The Binding of Experience: A Phenomenological Characterization

Unity answers the question of "why multiple contents co-belong to one field". Binding answers a more fine-grained question: within that unified field, why are certain features organized as attributes of the same object, while others belong to different objects?

Consider an ordinary visual experience: I see a red ball rolling to the left. In this experience, the three features "red", "round", and "moving leftward" are given as three attributes of the same object—that ball. I do not separately experience a free-floating red, a free-floating roundness, and an ownerless motion, and then infer from outside that they belong to the same thing. From the beginning, they are bound as aspects of one object. At the same time, if there is also a stationary blue square in the visual field, then "blue", "square", and "stationary" are bound as another object and are not confused with the features of the ball. This organizational relation—which features are bound as one thing, and which belong to different things—is binding.

Binding constitutes an independent structural feature (and cannot be reduced to unity) because unity guarantees that red, roundness, motion, blue, squareness, and stillness all co-belong to one experiential field, but it does not determine how they are grouped. The six features in one and the same field could in principle be grouped in many ways; actual experience, however, gives one definite grouping (red-round-moving together, blue-square-stationary together), and this grouping itself is experienced and has phenomenological reality. The well-known phenomenon of illusory conjunction—under specific experimental conditions, subjects mistakenly bind the color of one object with the shape of another object (for example, binding the redness of a red square with the roundness of a blue disk as a "red circle")—precisely shows that binding is an independent process that can go wrong separately: grouping can be mistaken, and what goes wrong is binding, not unity.

Therefore, the structural sufficiency criterion requires the gradient side to provide not only "multiple components co-existing in one coherent whole" (that is unity), but also a structure of "how, within this whole, components are grouped into sub-wholes". This book maintains that this finer structure is carried by cross-frequency resonance.

9.5. Cross-Frequency Resonance Carries Binding: Coupling as the Physical Counterpart of "Belonging to the Same Object"

The three features red, round, and moving belong, according to Chapter 7, to different qualitative dimensions and therefore correspond on the gradient side to configurations in different frequency families (color, shape, and motion each have their own frequency structures). Their being bound as the same object therefore cannot be carried by same-frequency coherence, because they are already different in frequency. What carries their binding must be a mechanism capable of crossing different frequency families and linking configurations located at different frequencies into one sub-whole. This is precisely cross-frequency resonance.

According to the Lorentzian resonance kernel of the foundational EQT volume (cited in Chapter 2), cross-frequency resonance refers to the stable coupling established between oscillations located in different frequency families when certain frequency relations are satisfied (for example, simple integer ratios between frequencies, or low-frequency phase modulating high-frequency amplitude). This coupling differs from same-frequency coherence: same-frequency coherence locks multiple oscillations into a fully phase-coordinated whole, whereas cross-frequency resonance preserves the frequency differences among the families while establishing definite and stable phase–amplitude relations among them, thereby linking configurations located at different frequencies into one coordinated sub-whole. Their frequencies remain different (so the features remain different: red remains red, roundness remains roundness), but they are tied together by the same cross-frequency coupling relation (so they belong to the same object).

This physical structure also corresponds relation by relation to experiential binding:

Correspondence to "belonging to the same object": configurations linked by the same cross-frequency coupling relation form a coordinated sub-whole; the features bound as one object in experience (red, round, moving) correspond to multiple hetero-frequency configurations linked by the same cross-frequency coupling. The multiple aspects of one object co-belong to that object; correspondingly, multiple hetero-frequency configurations co-belong to one cross-frequency coupling.

Correspondence to "belonging to different objects": configurations not linked by the same cross-frequency coupling, or linked by another independent cross-frequency coupling, form different sub-wholes; features belonging to different objects in experience (the redness of the ball vs. the blueness of the square) correspond to configurations belonging to different cross-frequency couplings.

Correspondence to the fallibility of binding: the establishment of cross-frequency coupling depends on frequency matching and timing; under poor conditions (for example, when features are presented too quickly or attention is insufficient), the coupling may mistakenly link configurations that should not have been linked. This corresponds exactly to the phenomenological fact of illusory conjunction. The fallibility of binding corresponds to the possibility of mismatching in cross-frequency coupling. Unity (same-frequency coherence) does not go wrong in such cases—what goes wrong is grouping (cross-frequency), not co-belonging to one field (same-frequency). This relation-by-relation correspondence, together with its correspondence in error mode, is a strong example of the strictness of isomorphism: not only normal cases correspond, but the modes of error correspond as well.

Through the two layers of unity (same-frequency coherence) and binding (cross-frequency resonance), the organizational structure of experience receives relation-by-relation carrying on the gradient side: coherence locks multiple components into one unified field, while cross-frequency coupling groups components within the field into objects. According to the structural sufficiency criterion of Chapter 5, the carrying of the fourth source—resonance modes—for the organizational structure of experience is thereby fulfilled.

9.6. The Combination Problem: Why It Seems Unavoidable

After completing the isomorphic carrying, this chapter turns to its second task: the dissolution of the combination problem. The strongest form of the combination problem must first be stated clearly, so that one can see where its dissolution lies; a "dissolution" that avoids its force would be dishonest.

The difficulty of the combination problem can be displayed through a stepwise argument. First step: this book maintains that living systems have interiors because they have boundaries, and that the cell, as the most basic bounded unit of life, has its own minimal interior (Chapters 3 and 6). Second step: the brain is composed of tens of billions of cells; since each cell has its membrane and its interior, each seems to be a minimal experiential subject. Third step: the "I"—the unified first person—is clearly related to the whole brain and is the joint product of these tens of billions of cells. Thus the fourth step seems unavoidable: the "I" must be some kind of sum or synthesis of these tens of billions of cellular subjects.

It is precisely this fourth step that gives rise to James's famous objection. James imagines taking one hundred people and giving each of them a word, arranged in order; no matter how closely they stand together, and no matter how simultaneously they utter their respective words, there will never thereby emerge a one-hundred-and-first consciousness that "understands the whole sentence". One hundred independent consciousnesses do not synthesize into a consciousness that grasps the whole simply by being juxtaposed or adjacent. Applied to the brain: how could tens of billions of cellular subjects, however tightly connected and however synchronized in their firing, synthesize into a unified "I", rather than always remaining tens of billions of tiny subjects, each governing itself and mutually inaccessible? Between independent subjects there seems to lie a gap that cannot be crossed merely by "tight connection". This is the combination problem, and panpsychism, along with every approach that seeks to synthesize consciousness from the bottom up, runs aground here.

One must see clearly that the entire force of this objection comes from the unchecked assumption in the fourth step: that the "I" is a sum of cellular subjects. The combination problem is a problem precisely because it assumes that the unified subject must be "added up" from microscopic subjects; and adding independent subjects together is indeed impossible. This book dissolves the combination problem not by searching for a mysterious "addition" that allows multiple subjects to be synthesized (that path is bound to fail), but by denying the fourth-step assumption itself: this book will argue that the "I" is not a sum of cellular subjects at all.

9.7. Dissolution: The "I" Is a Single Coherent Resonance, Not a Sum of Micro-Subjects

The key to dissolution lies in carrying through the claim of Section 9.3 that "unity is same-frequency coherence".

This chapter has argued that the unity of experience corresponds to the coherence of gradient resonance—multiple oscillations phase-lock into one indivisible whole. Now ask: what exactly is this unified "I" on the ρ_f side? According to the identity-level claim of this book, it is what this coherent resonance itself is from within. And a coherent resonance, by its very nature, is not a sum of parts—it is a single and indivisible whole mode formed after multiple oscillations enter phase locking. A coherent state is less "the sum of many oscillations" than "one single state jointly entered by many oscillations"; in this state, the question "which part of the phase belongs to which component oscillation" loses its meaning, just as it loses its meaning to ask, in a beam of coherent light, "which photon carries coherence". Coherence is a property of the whole, not the sum of properties of components.

The picture of the combination problem is thereby fundamentally changed. This book does not deny that cells each have their own interior. But the "I" is not the result of adding up those cellular interiors. The "I" is the single coherent resonance mode formed when multiple gradient oscillations in the multi-cellular space of the brain enter same-frequency coherence and are then bound through cross-frequency resonance. Cells, relative to this resonance, are not "little selves that make me up", but the medium and carrier through which this resonance is maintained—just as a standing wave is related to the string that carries it. Every segment of the string participates in the vibration, but the standing wave itself is a single mode of the whole string, not the "sum" of the vibrations of its segments, and still less a "large standing wave" synthesized from the "small standing waves" of the segments. To ask "from which small standing waves is this standing wave on the string added together" is a misplaced question; likewise, to ask "from which little cellular selves is the unified I added together" is also a misplaced question.

James's objection therefore loses its object. What James proves is that independent subjects cannot add up to a new subject. That is correct—but this book never claims that the "I" is produced by addition. In this book's picture, there simply is no operation waiting to be completed in which "tens of billions of cellular subjects are added together"; what exists is the physical fact that "multiple gradient oscillations become coherent as one single mode within a multi-cellular medium". The former (addition of independent subjects) is indeed impossible; the latter (many oscillations becoming coherent as a single mode) is a familiar occurrence in coherent phenomena. From lasers to superconductivity, examples in nature where "many components enter a single coherent state" are abundant, and in them there is no problem of "subject addition", because the coherent state is from the beginning holistic rather than additive.

This dissolution is consistent with, rather than isolated from, the isomorphic carrying relations of Chapters 7 and 8. The "I", as a coherent resonance, still contains within itself the qualitative gradations given by single-frequency configurations, the qualitative types given by the frequency dimension, the flow of consciousness given by the master equation, and the object binding given by cross-frequency resonance. All of these unfold within the same coherent resonance. Unity does not "put" these contents into a container; rather, these contents are from the beginning internal structures of the same coherent mode. The "I" is therefore not a collection of contents, but the single coherent resonance that carries and organizes these contents. Its "oneness" is the "oneness" of coherence, not the "sum" of addition.

9.8. An Objection That Must Be Addressed Directly: Why Coherence Is Not Merely "Tight Connection"

A strong objection to the above dissolution is the following: by calling the "I" a coherent resonance, is one merely replacing the "tight connection" rejected by James with a physical term? Did James not precisely deny that however tight the connection may be, a subject cannot be added up from it? Is coherence not simply an especially tight kind of connection? If so, the dissolution has not succeeded; it has only changed the wording.

This objection must be answered directly because it touches the success or failure of the dissolution. The answer is this: coherence differs from "tight connection" at a crucial ontological level. "Tight connection" describes relations among parts—the parts remain parts; they are merely more closely related. However tight those relations may be, ontologically there is still "many", and James's objection applies. Coherence, by contrast, does not describe the degree of tightness among parts; it describes the entry of components into a new whole state, in which the "parts" no longer exist as independently separable items. A coherent state is a single mode, and its identity is no longer constituted by the identities of the components. This is the difference between "the emergence of a new whole" and "a tighter relation among parts": a superconducting state is not "electrons connected more tightly", but electron pairs entering a single macroscopic quantum state, in which asking "which electron" has already lost its meaning.

Therefore, the dissolution proposed by this book is not that "if connection is tight enough, a subject is synthesized"—that would indeed fall under James's rejection. Rather, it is that "when components enter a coherent state, there are no longer 'multiple subjects' to be synthesized, but only one single mode". James's objection targets the "addition from many into one", whereas coherence provides a "whole state that is one from the beginning". The two belong to different ontological levels. The combination problem presupposes "first many, then seeking unity", whereas the picture of this book is that, at the level of coherence, unity is not a later synthesis, but what the coherent state already is. The dissolution of the combination problem ultimately depends on this ontological distinction: once unity is placed at the level of "the wholeness of a coherent state", rather than at the level of "the synthesis of parts", the "multiple subjects" that would need to be combined do not exist from the outset, and the combination problem cannot arise.

An honest qualification must be added in order to respect the separation discipline established in Chapter 5. What this chapter dissolves is the structural aspect of the combination problem—"how multiple interiors become structurally one". It shows that unity is the wholeness of a coherent state and does not require addition. But this chapter has not, and cannot, answer "why this unified coherent state is experienced". That is given-ness and belongs to Part III. The structural aspect of the combination problem is dissolved here; its given-ness aspect will be treated together with the account in Part III that "given-ness is internal–external co-evolution of a bounded interior". This chapter strictly observes this boundary: it dissolves "how it becomes one", and does not touch "why this one is experienced". To conflate the two would violate the separation established in Chapter 5 and the prohibition running through the whole book.

9.9. Conclusion of This Chapter

This chapter has completed a dual task. In terms of isomorphic carrying, it has argued that the unity of experience is carried by the coherence of same-frequency resonance (corresponding relation by relation to the singleness of the subject, co-consciousness, and the singleness of perspective), and that the binding of experience is carried by the coupling of cross-frequency resonance (corresponding relation by relation to belonging to the same object, belonging to different objects, and the fallibility of binding corresponding to the possibility of mismatching in cross-frequency coupling). According to the structural sufficiency criterion of Chapter 5, the carrying of the fourth source—resonance modes—for the organizational structure of experience is thereby fulfilled. In terms of the combination problem, this chapter first stated the strongest form of the problem through James's objection, identified that its entire force comes from the unexamined assumption that "I am a sum of cellular subjects", and then rejected that assumption by arguing that "unity is coherence, and a coherent state is a single mode rather than a sum of parts". The combination problem thereby loses its object and is dissolved. The chapter also directly answered the objection that "coherence is merely tight connection", pointing out that coherence and tight connection belong to different ontological levels: the former is the entry of components into a single whole state, while the latter remains a relation among multiple parts.

The thesis established in this chapter—that the "I" is a single coherent resonance mode formed by multiple gradient oscillations in the multi-cellular medium of the brain—will be repeatedly used in the following parts: it is the ontological basis of Chapter 16 on "subjective boundary and the I" (the boundary is the boundary of the coherent resonance), the basis of Chapter 24's explanation of split-brain, anesthesia, and dissociation (these states correspond to the splitting, collapse, and reorganization of coherence), and the premise on which Part III anchors given-ness in the "internal–external co-evolution of this coherent resonance".

9.10. Chapter Summary

This chapter is one of the heaviest-bearing chapters in Part II, completing the dual task of isomorphic carrying and the dissolution of the combination problem. First, the unity of experience (multiple contents co-belonging to one field: the singleness of the subject, co-consciousness, and the singleness of perspective) is carried by the coherence of same-frequency resonance, and the relation-by-relation correspondences hold. Second, the binding of experience (how features within the field are grouped into objects) is carried by the coupling of cross-frequency resonance, and the fallibility of binding corresponds relation by relation to mismatching in cross-frequency coupling. Third, the force of the combination problem arises from the assumption that "I am a sum of cellular micro-subjects"; this book rejects that assumption and argues that the "I" is a single coherent resonance mode. A coherent state is by its nature not a sum of parts, so James's objection loses its object and the combination problem is dissolved. Fourth, the chapter directly answers the objection that "coherence is tight connection", distinguishing the ontological levels of coherence (components entering a single whole state) and tight connection (relations among multiple parts becoming tighter). Fifth, it strictly observes the separation between structure and given-ness: this chapter dissolves the structural aspect of the combination problem, while its given-ness aspect is left to Part III.

The failure conditions of this chapter are as follows (summarized in Appendix E).

- F9.1 The isomorphic carrying of unity fails. If a structural feature of experiential unity (the singleness of the subject, co-consciousness, or the singleness of perspective) has no corresponding relation on the side of same-frequency coherence, or if the existence of coherence and the existence of experiential unity have no systematic relation (for example, coherence remains intact while unity collapses, or coherence collapses while unity remains unchanged), then the isomorphic carrying of unity fails.

- F9.2 The isomorphic carrying of binding fails. If feature binding has no systematic correspondence with the coupling of cross-frequency resonance, especially if the error modes of binding (illusory conjunctions) cannot correspond to mismatching in cross-frequency coupling, then the isomorphic carrying of binding fails.

- F9.3 The dissolution of the combination problem is invalid. If the unified first person is shown to have to be understood as the real sum of multiple microscopic subjects rather than as a single coherent mode (that is, if the ontological distinction that "a coherent state is a whole rather than a sum" does not hold), then the combination problem cannot be dissolved and must be solved directly; this book would return to James's difficulty.

- F9.4 The distinction between coherence and connection does not hold. If "coherence" in the argument of this chapter does not in fact go beyond "tight connection" (that is, if it fails to identify the ontological-level difference between a coherent state and the relations among parts), then the response in Section 9.8 fails, and the dissolution degenerates into a change of wording.

Table 9.1.. The Boundary between What This Chapter Establishes and What It Cannot Derive

What this chapter establishes	What this chapter cannot derive
The unity of experience is carried relation by relation by the coherence of same-frequency resonance.	It cannot answer in this chapter "why the unified coherent state is experienced" (this belongs to givenness in Part III).
The binding of experience is carried by the coupling of cross-frequency resonance, including correspondence in error modes.	It cannot provide observable proxies and predictions for unity and binding in this chapter (these belong to Chapters 23 and 24 in Part IV).
The dissolution of the combination problem: the "I" is a single coherent resonance mode, not a sum of micro-subjects.	It cannot define in this chapter the precise location of the subjective boundary (this belongs to Chapter 16).
The ontological-level distinction between coherence and tight connection.	It cannot prove in this chapter how a coherent state is physically maintained in a multi-cellular medium (this belongs to Part IV and the appendices).

10. The Timescale Spectrum of Gradient Transients: Temporal Thickness and Memory Hierarchy

10.1. The Carrying Task of This Chapter, and the Closure of Part II

This chapter undertakes the fifth and final carrying task among the five sources, and Part II reaches its closure here. The preceding four tasks have respectively carried the intra-qualitative order of experience (Chapter 6), qualitative types (Chapter 7), the evolutionary law of the stream of consciousness (Chapter 8), and unity and binding (Chapter 9). What this chapter carries is the remaining side of experiential temporality that has not yet been treated—the temporal thickness of experience and the hierarchy of memory.

A boundary must first be drawn with Chapter 8, because both chapters concern time and are therefore most easily conflated. Chapter 8 carries the evolutionary law of the stream of consciousness—how experience flows: continuously, under constraint, directionally, and with shifting attentional focus. This chapter, by contrast, carries the retention of experience in time—why the "present" of experience has thickness (rather than being an extensionless point), and why past experience is retained across different timescales (from the fleeting to the long-lasting). Flow and retention are two different matters: flow is the progression of evolution through time (the master equation, Chapter 8), whereas retention is the duration for which experiential content is stably maintained in time (this chapter). To use an analogy: Chapter 8 concerns how river water flows, while this chapter concerns how deeply, and for how long, the water leaves traces in different riverbeds.

The core claim of this chapter, and also one of EQT's distinctive insights on the problem of time, is this: the gradient configurations of ρ_f exist in the form of "transients"—stable but not eternal; and transients in different frequency families have systematically different stable durations: high-frequency transients are short, while low-frequency transients are long. This "frequency-dependent timescale spectrum" carries the temporal thickness of experience and the hierarchy of memory—from fleeting present retention to long-term memory retention, corresponding to a continuous spectrum from high-frequency short-duration transients to low-frequency long-duration transients. Especially important is that this "frequency-dependent" tendency is not merely qualitative, but is supported by an explicit scaling law drawn from EQT freezing dynamics (Section 10.3). This allows the carrying relation in this chapter to move from a qualitative direction to a computable and testable quantitative relation. This chapter will first characterize the temporal thickness and memory structure to be carried on the experiential side (Section 10.2), then characterize the transient timescale spectrum on the gradient side and its scaling law (Section 10.3), establish the correspondences relation by relation (Sections 10.4 and 10.5), and directly address objections (Section 10.6).

10.2. What Is to Be Carried: What Kind of Structure Are Temporal Thickness and Memory Hierarchy?

In addition to flow (Chapter 8), the temporality of experience has two structural features that this chapter must carry: the thickness of the present and the hierarchy of memory.

First, the temporal thickness of the present. The experiential "now" is not an extensionless instantaneous point. This point has been characterized with particular precision in the phenomenological tradition, especially by Husserl: present experience always carries retention of what has just passed—the note that has just sounded has already passed, yet remains within present experience as "just sounded", enabling me to hear a "melody" rather than isolated notes; at the same time, the present also carries protention toward what is about to come—a kind of anticipatory extension toward the next moment. Thus the "now" is a temporally thick domain: it is not a point, but an "extended present" with an internal temporal structure composed of retention–present–protention. Precisely because of this thickness, experience is a coherent flow rather than a series of discrete flashes: retention connects what has just passed to the present, while protention extends the present toward the future.

This thickness has its internal gradient: the retention of what has just passed is most vivid; what lies slightly farther back becomes gradually dimmer and more blurred, until it finally slips out of the present and becomes something that must be actively recalled. Retention is not all-or-nothing, but a gradient that continuously decays with "how long ago" something has passed—the immediate aftersound is clear, the slightly earlier one fades, and the more distant one disappears. This structure of continuous decay is the core feature of the thickness of the present.

Second, the hierarchy of memory. Once beyond the thickness of the present, the retention of experience enters memory, and memory is layered according to its retention duration. The psychological tradition distinguishes several timescales: fleeting sensory registration (hundreds of milliseconds), short-term working memory (seconds to tens of seconds, capacity-limited and requiring maintenance), and long-term memory (potentially lifelong, with nearly unlimited capacity). These timescales are not sharply separated drawers, but are more like several segments on a continuous spectrum: from the most fleeting to the most persistent, retention duration unfolds continuously. The structure of memory hierarchy is this "spectrum of retention duration": experiential contents are distributed along a continuous spectrum from the momentary to the long-lasting, according to how long they are retained.

Taken together, the two are in fact the same structure extended across different scales: the thickness of the present (the continuous decay of retention) deals with retention "below the second", while the hierarchy of memory deals with retention "above the second and up to a lifetime". From the thickness of protention–present–retention, to working memory, and then to long-term memory, there is a continuous retention-timescale spectrum—the duration for which experiential content is stably maintained unfolds continuously from the shortest (protention and immediate retention) to the longest (lifelong memory). This continuous timescale spectrum is what this chapter must carry by gradients. The separation established in Chapter 5 must be reiterated: this chapter carries the structure of temporal thickness and memory (the spectrum of retention duration), not their given-ness (that this retention is experienced, which belongs to Part III), and it does not specify what concrete content is retained (that is determined by perturbation, Chapter 8).

10.3. What Carries It: The Timescale Spectrum of Gradient Transients and Its Scaling Law

We now turn to the gradient side. This section characterizes the timescale spectrum of ρ_f gradient transients as the carrier, and gives its explicit frequency-dependent scaling law.

According to the EQT ontology, the gradient configurations of ρ_f are not things that reside eternally, but exist in the form of transients. Chapter 2 has already cited this point: a transient is a mode of existence that is stable but not eternal; it forms, is maintained for a period of time, and eventually decays. Matter (such as membrane) is a highly stable transient, whereas experiential configurations are more easily changing transients; yet every structure of ρ_f is a transient of some duration, differing only in the length of its stable duration.

The key structural fact is this: the stable duration of a transient systematically depends on the frequency family in which it is situated—high-frequency transients are short, and low-frequency transients are long. This dependence is not merely a qualitative direction, but has an explicit scaling law from EQT freezing dynamics. According to the freezing criterion in *Frozen Cosmology: A Natural Philosophy of Cosmic Evolution in the Perspective of Energy Quantum Theory*, the freezing time of a ρ_f structure (the characteristic duration over which a structure exits active evolution and is stably retained) follows the scaling relation

$$t_{\text{freeze}}(f) \sim \tau_0(f)/f,$$

where $\tau_0(f)$ is a slowly varying effective background factor—it absorbs corrections from the background, coupling, diffusion, decay, and source terms. It is not an arbitrary free parameter, but an effective quantity that must be slowly varying, calibratable, and testable. Its core tendency is $1/f$: high-frequency structures freeze earlier (with shorter stable duration), while low-frequency structures freeze later (with longer stable duration).

Why does this scaling law apply to memory transients in the brain? This book maintains that memory transients in the brain are directly frozen structures of ρ_f under the brain's background, and therefore obey the same scaling law—except that the "background" is no longer cosmic expansion, but the dissipative and competitive background within the brain (neural decay, noise, metabolic constraints, and so on), absorbed by $\tau_0(f)$. Denoting their retention duration as $\tau_{\text{retain}}(f)$, we have

$$\tau_{\text{retain}}(f) \sim \tau_0(f)/f$$

(freezing under the brain's background).

This is not an analogical transfer of a cosmological formula, but the same scaling law of the same ρ_f freezing mechanism under different backgrounds: under the cosmic background, it gives the cosmological freezing time of structures; under the brain's background, it gives the retention duration of memory transients. Memory retention and the freezing of cosmic structures are therefore manifestations of the same ρ_f freezing mechanism under two different backgrounds. This is a concrete fulfillment, on the problem of memory, of the book's broader ambition of mind–matter unity. In the later part of Part III (Chapter 17), this book will formally use this scaling law to characterize "memory as gradient transient"; this chapter introduces it first at the level of structural carrying.

From this scaling law, transients unfold along the frequency axis into a timescale spectrum: from extremely short-duration transients at the high-frequency end (τ_{retain} short, carrying fleeting immediate retention), to extremely long-duration transients at the low-frequency end (τ_{retain} long, carrying long-term memory), stable duration increases continuously as frequency decreases. This timescale spectrum has several structural features that this chapter must use. First, it is a continuous spectrum rather than a set of discrete levels: from the shortest to the longest, transient duration changes continuously with frequency, without sharp breaks. Second, it has an internal decay gradient: any transient, within its stable duration, does not remain constant, but goes through a process of formation–maintenance–decay, and its "vividness" continuously decays with time until it dissipates. Third, it spans an extremely wide range of scales: from below milliseconds at the high-frequency end to very long durations at the low-frequency end, the timescale spectrum covers the entire range from the fleeting to the persistent.

A qualification concerning $\tau_0(f)$ must be stated in accordance with the discipline of the whole book: in the context of consciousness, this book preserves the status of $\tau_0(f)$ as a slowly varying correction factor (consistent with *Frozen Cosmology: A Natural Philosophy of Cosmic Evolution in the Perspective of Energy Quantum Theory*); it is not treated as an exact constant, nor is it here specifically tied to any particular neural frequency band or metabolic rate. It is an effective background quantity that must be slowly varying, calibratable, and testable. What this book claims is the form of the scaling relation ($\tau_{\text{retain}} \sim \tau_0/f$, with $1/f$ dominant and τ_0 as a slowly varying correction) and its direction (high frequency short, low frequency long), not a zero-free-parameter exact law. This restraint preserves the honesty of the whole book in not claiming zero free parameters (Chapter 2).

Placing Sections 10.2 and 10.3 side by side, the qualification of the carrier becomes clear: on the experiential side, there is a continuous retention-timescale spectrum from protention to lifelong memory, together with a continuous decay gradient of retention; on the gradient side, there is a continuous transient timescale spectrum from high-frequency short duration to low-frequency long duration, obeying $\tau_{\text{retain}} \sim \tau_0/f$, together with a continuous decay gradient of transients. The next section makes the correspondence concrete relation by relation.

10.4. Relation-by-Relation Isomorphism: How Temporal Thickness and Memory Are Carried

Following the criterion of Chapter 5, this section establishes the correspondences relation by relation.

Correspondence of the retention-timescale spectrum. On the experiential side: experiential contents are distributed, by retention duration, along a continuous timescale spectrum from protention/immediate retention (shortest) to long-term memory (longest). On the gradient side: transients are distributed, by stable duration ($\tau_{\text{retain}} \sim \tau_0/f$), along a continuous timescale spectrum from high-frequency short-duration transients (shortest) to low-frequency long-duration transients (longest). The isomorphism here requires an order-preserving correspondence between the experiential retention-timescale spectrum and the gradient transient-timescale spectrum: experiential contents retained for longer durations correspond to transients with longer stable durations (lower frequencies), and the entire order of durations is preserved one-to-one. With the scaling law, this correspondence goes beyond order preservation and gives a quantitative direction: retention duration is approximately inversely proportional to the carrying frequency (dominated by $1/f$, with τ_0 as a slowly varying correction). This correspondence is testable: if some class of experience is found whose retention duration has no order-preserving relation to the frequency of the transient carrying it (for example, an extremely fleeting experience being carried by an extremely low-frequency long-duration transient), or if it systematically violates the inverse tendency of $\tau_{\text{retain}} \sim \tau_0/f$, then the correspondence breaks.

Correspondence of the decay gradient. On the experiential side: retention is not all-or-nothing, but decays continuously with how long something has passed (what just happened is clear, what happened slightly earlier fades); the same holds for memory, where more distant contents become more easily blurred and forgotten. On the gradient side: a transient undergoes continuous decay within its duration (vividness decreases with time until it dissipates). The isomorphism here requires that the experiential decay gradient correspond to the decay gradient of transients—the continuous experiential decay from "clear to blurred to disappearing" corresponds to the continuous transient process from "vivid to decaying to dissipated". In particular, forgetting receives a natural counterpart here: forgetting is not the "deletion" of some content, but the decay of the transient that carries it until it dissipates. Forgetting corresponds to transient decay, and the gradual fading of memory corresponds to the decrease in transient vividness.

Correspondence of continuity between present thickness and memory hierarchy. This is the most distinctive and powerful item in this chapter. On the experiential side: the thickness of the present (sub-second retention in protention–retention) and the hierarchy of memory (seconds and above, up to lifelong retention) are not two separate mechanisms, but extensions of one and the same retention-timescale spectrum across different scales. On the gradient side: high-frequency short-duration transients (carrying the sub-second thickness of the present) and low-frequency long-duration transients (carrying long-term memory) are likewise not two separate mechanisms, but extensions at different frequencies of the same transient timescale spectrum (the same scaling law $\tau_{\text{retain}} \sim \tau_0/f$). The isomorphism here requires that the experiential structure in which "present thickness and memory hierarchy are one continuous timescale spectrum" correspond to the gradient structure in which "short-duration and long-duration transients are one continuous frequency–timescale spectrum". The weight of this correspondence lies in its unity: this chapter does not use two mechanisms to carry present thickness and memory separately, but uses **one and the same transient timescale spectrum (one and the same scaling law)** to uniformly carry the entire retention structure from protention to lifelong memory—just as Chapter 7 used the same frequency dimension organized by families to carry both intra-qualitative continuity and inter-qualitative discontinuity. This unity, in which "one structure uniformly carries an entire side of experience", supports the truth of the isomorphism more strongly than the patching together of two mechanisms.

If these three correspondences hold one by one, then according to the criterion of Chapter 5, the timescale spectrum of gradient transients is sufficient to carry the temporal thickness of experience and the hierarchy of memory.

10.5. A Verificatory Detail: The Frequency–Timescale Correspondence of Memory

Just as Chapter 6 used the correspondence of singularities, and Chapter 7 used the stitching of the dual structure, as strong supports, this chapter also has a correspondence that can serve as a strong support—the systematic relation between different timescales of memory and different frequencies. This section explains its weight and its limits.

The core isomorphic claim of this chapter is that there is an order-preserving correspondence between memory timescale and transient frequency, and that, because of the scaling law $\tau_{\text{retain}} \sim \tau_0/f$, this correspondence further has a quantitative inverse tendency: short-term memory is carried by relatively higher-frequency transients, and long-term memory by relatively lower-frequency transients. This claim is a strong and testable support because it makes a concrete prediction that can be independently examined: distinguishable timescale levels in experience (immediate retention, working memory, long-term memory, and so on) should correspond to independently measurable, systematically different levels of neural oscillation frequency—shorter retention corresponding to higher-frequency activity, longer retention corresponding to lower-frequency activity, with an approximate inverse relation between the two (within the slowly varying correction of τ_0).

It is worth noting that the direction of this prediction is consistent with several independently established findings in neuroscience: rapid, fleeting neural activity is often observed in higher-frequency bands (for example, fast gamma-band oscillations are often associated with immediate and easily vanishing processing), while slower, more sustained activity is often observed in lower-frequency bands (for example, low-frequency oscillations are often associated with maintenance and integration). This book does not examine the specific correspondences in detail here (that belongs to the observable-proxy work of Part IV; the complete falsifiable formulation is given in Chapter 25 under the unique prediction of "frequency–timescale correspondence"). It only points out that the frequency–timescale correspondence of this chapter is not an arbitrary stipulation, but a prediction with a scaling law as its backbone, consistent in direction with existing evidence, and open to further precise testing. An isomorphism that can make this kind of concrete, testable prediction carries far more weight than a vague resemblance claiming only that "experience has timescales, and transients also have durations".

Its limits must be stated honestly. First, the order-preserving correspondence of frequency and timescale (and its inverse tendency) is a strong support, but strict carrying requires the entire timescale spectrum (from protention to lifelong memory) to correspond segment by segment, rather than only several levels matching. Moreover, the slow variation of $\tau_0(f)$ means that this inverse relation is not an unconditionally exact f^{-1} law; the test must therefore concern the whole structure of "$1/f$ dominant tendency + slowly varying τ_0 correction", rather than demanding an exact pointwise inverse relation. The full test must rest on all three correspondences in Section 10.4. Second, memory contains much more than "retention duration": it also involves encoding, retrieval, reconstruction, and content selection. This chapter carries only the timescale structure of memory (the spectrum of retention duration), not the entire mechanism of memory. Encoding and retrieval belong to perturbation-driven evolution (Chapter 8), and the selection of content belongs to the perturbation axis. This chapter does not overstep into carrying these; it carries only the timescale structure of "how long something is retained". This boundary must be maintained, so as not to exaggerate "carrying the timescale structure of memory" into "explaining all of memory".

10.6. Objections That Must Be Addressed Directly

The carrying claim of this chapter will face several strong objections that must be addressed directly.

Objection 1: long-term memory is structural storage (such as synaptic weights), not a "persistent transient", and therefore cannot be carried by transients. This is the sharpest objection. Long-term memory can be retained for decades, whereas "transient" seems to mean something decaying and non-persistent. To call lifelong memory a "low-frequency long-duration transient" therefore seems self-contradictory—it is clearly stable structural storage, not some "transient". The response must appeal to the ontology of Chapter 2: in EQT, all ρ_f structures are transients, differing only in stable duration. Matter itself (such as membrane or synaptic structure) is a highly stable, extremely low-frequency transient (Chapters 3 and 4). This is precisely supported by the scaling law $\tau_{\text{retain}} \sim \tau_0/f$: the lower the frequency, the longer the stable duration; material transients such as synaptic structures are extremely low in frequency, and their τ_{retain} is therefore extremely long, allowing them to be maintained for decades. Thus "long-term memory is structural storage" and "long-term memory is a low-frequency long-duration transient" are not contradictory, but two ways of saying the same thing: so-called "structural storage" (such as physical changes in synaptic weights) is, in the EQT sense, an extremely low-frequency and extremely long-duration transient. It is stable enough to be maintained for decades, but is still a transient of ρ_f, and will still decay on longer timescales (memory eventually dissipates as neural structures degrade, which is consistent with the scaling law: even at very low frequency, τ_{retain} remains finite). This chapter does not deny that long-term memory has a material basis (such as synapses); it maintains that this material basis itself is a low-frequency long-duration transient—which is continuous with the ontology of the whole book that "matter is stable transient". Objection 1 therefore does not hold: it presupposes an opposition between "transient" and "stable structure", whereas in EQT they are simply two ends of the timescale spectrum (the scaling law).

Objection 2: retention/protention is a psychological construction and need not correspond to any physical process. One might say that Husserlian retention–protention is a phenomenological description and need not correspond to any identifiable physical process. Response: this chapter carries the structure characterized by retention–protention (the thickness of the present and the continuous decay of retention), and this structure is coordinate-independent and consistently confirmed in experience (we do in fact hear melody rather than isolated notes, and we do in fact have an "extended present"). This chapter maintains that this structure is carried by the formation–decay process of high-frequency short-duration transients. This does not require retention–protention to be an independently named "psychological organ"; it requires only that its structure (thickness and continuous decay) correspond relation by relation to the structure of high-frequency transients (short duration and continuous decay). The name of the psychological construction may be disputed, but the temporal-thickness structure it characterizes is a reality of experience. This chapter carries the latter.

Objection 3: the frequency–timescale correspondence is only a rough correlation and is insufficient to support isomorphism. Response: this chapter does not merely claim a rough correlation, but an order-preserving, segment-by-segment correspondence, with a quantitative inverse tendency supplied by the scaling law $\tau_{\text{retain}} \sim \tau_0/f$ as its backbone. It also gives explicit failure conditions (Sections 10.4 and 10.7): if retention duration and transient frequency fail to preserve order, if they systematically violate the $1/f$-dominant tendency, if decay gradients do not correspond, or if the thickness of the present and the hierarchy of memory cannot be uniformly carried by one timescale spectrum, then the isomorphism breaks. A correspondence that provides segment-by-segment order preservation, a quantitative tendency, and clear failure conditions is not a "rough correlation"—rough correlation is not falsifiable, whereas the correspondence of this chapter can be touched by counterexamples at every point.

10.7. Conclusion of This Chapter and the Completion of Part II

This chapter has fulfilled the fifth and final carrying task of Part II: the temporal thickness of experience and the hierarchy of memory are carried relation by relation by the timescale spectrum of gradient transients (high-frequency short-duration, low-frequency long-duration, obeying $\tau_{\text{retain}} \sim \tau_0/f$). The key point is that the thickness of the present (the continuous decay of retention–protention) and the hierarchy of memory (from working memory to long-term memory) are one continuous retention-timescale spectrum, and are uniformly carried by one and the same transient timescale spectrum (one and the same freezing scaling law). Forgetting corresponds to transient decay, and long-term memory corresponds to extremely low-frequency long-duration transients (consistent with the ontology that "matter is stable transient"). This chapter uses the order-preserving frequency–timescale correspondence (and its $1/f$ inverse tendency) as a strong support (consistent in direction with neuroscientific findings linking high frequency with rapidity and low frequency with maintenance; its complete falsifiable formulation appears in Chapter 25), and directly addresses the three objections that "structural storage is not transient", "retention–protention is a psychological construction", and "this is only rough correlation". This chapter strictly observes its separations: what is carried is the structure of temporal retention (not given-ness, which belongs to Part III; not the law of flow, which belongs to Chapter 8; not the encoding, retrieval, or content of memory, which belong to the perturbation axis and Chapter 8).

With this chapter completed, all five carrying tasks of Part II have been fulfilled: intra-qualitative order (single-frequency gradient configurations), qualitative types (frequency-family dimension), the stream of consciousness (master-equation evolution), unity and binding (resonance modes), and temporal thickness and memory (transient timescale spectrum). The entire structure of experience has now been carried relation by relation by the fivefold richness of ρ_f gradients. Chapter 11 has clarified that the sense of self, the sense of agency, and affective tone belong to the perturbation axis rather than to a sixth structural source, and the completeness claim is therefore stable yet open with respect to the structural axis. The structural sufficiency required by the identity level has thus been fulfilled. But it must be recalled once more: Part II has from the beginning carried only the structure of experience. The given-ness of experience is not contained within the isomorphism of this part; it is the subject of Part III. Part II has provided the complete score of experience (including how it is played by internal and external perturbations and how long its notes are retained), but why this played music is heard awaits the answer of Part III.

10.8. Chapter Summary

This chapter is the fifth and final isomorphic carrying task of Part II, and it closes Part II. First, it draws the boundary: this chapter carries the retention of experience (temporal thickness and memory hierarchy), as distinct from the flow of experience (the evolutionary law) in Chapter 8. Second, it characterizes the structure to be carried on the experiential side: the thickness of the present (retention–present–protention, and the continuous decay of retention), and the hierarchy of memory (from working memory to long-term memory), which together form one continuous retention-timescale spectrum from protention to lifelong memory. Third, it characterizes the carrying structure on the gradient side: the frequency-dependent timescale spectrum of gradient transients, following the scaling law $t_{\text{freeze}}(f) \sim \tau_0(f)/f$ drawn from *Frozen Cosmology: A Natural Philosophy of Cosmic Evolution in the Perspective of Energy Quantum Theory*; memory transients in the brain are directly ρ_f freezing under the brain's background and obey $\tau_{\text{retain}}(f) \sim \tau_0(f)/f$ (τ_0 as a slowly varying correction factor, with $1/f$ dominant)—memory retention and cosmic freezing are manifestations of the same ρ_f mechanism under two backgrounds. Fourth, it establishes the isomorphism relation by relation: order-preserving correspondence of retention-timescale spectra (with a quantitative $1/f$ tendency), correspondence of decay gradients (forgetting corresponds to transient decay), and the continuous correspondence by which the thickness of the present and the hierarchy of memory are uniformly carried by the same timescale spectrum. Fifth, it uses the order-preserving frequency–timescale correspondence (and the $1/f$ inverse tendency) as a strong support (consistent with the direction of neuroscientific findings associating high frequency with rapidity and low frequency with maintenance; the complete prediction appears in Chapter 25), while stating its limits (the whole spectrum must correspond segment by segment, and slow variation in τ_0 means it is not an unconditionally exact inverse relation; the chapter carries only the timescale structure, not the entire mechanism of memory). Sixth, it responds to three objections: long-term memory as extremely low-frequency long-duration transient (consistent with the ontology that "matter is stable transient", so "structural storage" and "transient" are not contradictory), retention–protention as a structure rather than a psychological organ, and segment-by-segment order-preserving correspondence (with a scaling-law backbone) rather than rough correlation.

The failure conditions of this chapter are as follows (summarized in Appendix E).

- F10.1 The retention-timescale spectrum does not preserve order. If the retention duration of experiential content and the frequency of the transient carrying it do not stand in an order-preserving relation (for example, an extremely fleeting experience being carried by an extremely low-frequency long-duration transient), then the carrying of the retention-timescale spectrum fails.

- F10.2 The decay gradients do not correspond. If the continuous decay of experience (retention fading and memory gradually being forgotten) cannot correspond to the continuous decay of transients, especially if forgetting cannot correspond to transient decay and dissipation, then the carrying of the decay gradient fails.

- F10.3 The continuous correspondence fails. If the thickness of the present and the hierarchy of memory cannot be uniformly carried by "one and the same transient timescale spectrum", but instead require two unrelated mechanisms, then the unified continuous carrying relation fails.

- F10.4 Long-term memory cannot be incorporated as transient. If long-term memory is shown in principle to be impossible to understand as low-frequency long-duration transient (that is, if the ontology that "all structures are transients" fails on the problem of memory), then Objection 1 succeeds and the carrying relation of this chapter collapses.

- F10.5 The scaling law systematically fails. If the retention duration of memory transients in the brain systematically violates the $1/f$-dominant tendency of $\tau_{\text{retain}} \sim \tau_0/f$ with respect to carrying frequency (and this violation cannot be accommodated by the slowly varying correction of τ_0), then the freezing scaling law cited in this chapter fails for consciousness and memory, and the quantitative backbone of the frequency–timescale correspondence collapses (for its measurable test, see F25.2 in Chapter 25).

Table 10.1.. The Boundary between What This Chapter Establishes and What It Cannot Derive

What this chapter establishes	What this chapter cannot derive
Temporal thickness and memory hierarchy are carried relation by relation by the gradient transient timescale spectrum (high-frequency short-duration, low-frequency long-duration, obeying $\tau_{\text{retain}} \sim \tau_0/f$).	It cannot carry the evolutionary law of the stream of consciousness (this belongs to Chapter 8).
The thickness of the present and the hierarchy of memory are uniformly carried by the same transient timescale spectrum (the same scaling law); forgetting corresponds to transient decay.	It cannot carry the encoding, retrieval, or content selection of memory (these belong to the perturbation axis and Chapter 8).
Long-term memory as extremely low-frequency long-duration transient (consistent with the ontology that "matter is stable transient", supported by the scaling law).	It cannot address whether or how temporal retention is experienced (this belongs to given-ness in Part III).
The order-preserving frequency–timescale correspondence (and $1/f$ inverse tendency) as a strong support (consistent with the direction of neuroscientific findings), and its limits.	It cannot provide in this chapter the observable proxies and complete falsifiable predictions of the timescale spectrum (these belong to Chapter 25).
The scaling law $\tau_{\text{retain}} \sim \tau_0/f$ is drawn from *Frozen Cosmology: A Natural Philosophy of Cosmic Evolution in the Perspective of Energy Quantum Theory* (τ_0 slowly varying); memory retention and cosmic freezing share the same source.	It cannot provide the exact form of τ_0 or a zero-free-parameter law (τ_0 is a slowly varying calibratable factor).

11. The Overall Test of the Isomorphism Program and the Boundary of Completeness

11.1. Taking Stock of Part II, and a Misplaced Question

Since Chapter 5 established the structural sufficiency criterion, Part II has fulfilled, chapter by chapter, the isomorphic carrying of five classes of experiential structure by five gradient sources: single-frequency gradient configurations carry the continuous order within qualities (Chapter 6), the frequency-family dimension carries the type distinction among qualities (Chapter 7), evolution constrained by the master equation carries the temporal order of the stream of consciousness (Chapter 8), same-frequency and cross-frequency resonance modes carry the unity and binding of experience (Chapter 9), and the timescale spectrum of gradient transients carries temporal thickness and memory hierarchy (Chapter 10). Together, these five items cover the main aspects of experiential structure—what quality experience has, how it gradually varies within a quality, how qualities are classified across types, how it flows through time, how it is organized into unified yet differentiated objects, and what temporal thickness and memory it possesses. With respect to these aspects, the structural sufficiency required by the identity level has been fulfilled.

Chapter 5 stated the claim that "the five classes of experiential structure are exhausted by the five sources" as a falsifiable completeness claim, and identified in advance three candidates most likely to become counterexamples—the sense of self, the sense of agency, and affective tone. A responsible program cannot declare completion simply after fulfilling the five items; it must examine directly whether these three constitute a sixth class of structure outside the five sources, requiring supplementation.

The core argument of this chapter is that these three do not constitute a "sixth source" problem at all—because they do not belong on the axis of "structural sources" in the first place. To place them alongside the five sources and ask whether they are a sixth structural source is itself a misplaced question. To see this clearly, one must first clarify a distinction that Part II has not yet sufficiently made explicit: the structural sources of experience and the perturbation sources of experience are two different axes.

11.2. Two Axes: Structural Sources and Perturbation Sources

The five gradient sources answer one and the same question: from what kind of richness of gradient configurations is the structure of experience carried—why it has continuous gradation, types, flow, unity and binding, and temporal thickness. This is the structural axis: it characterizes how rich the gradient configuration itself is, such that it can be isomorphic to the structure of experience.

But experience also has another aspect, one that the five sources do not directly answer and that is precisely the key to understanding the three candidates: any concrete experience is always triggered by some perturbation. A gradient configuration does not present determinate content out of nothing; it presents this determinate experience at this moment because it is driven and altered by a particular perturbation. This is the perturbation axis: it characterizes not "how rich the gradient structure is", but "what perturbs the gradient, such that it presents this concrete content at this moment".

These two axes are orthogonal and cannot be reduced to each other. The structural axis determines what experience can possibly be like (the richness of gradient-configuration space), while the perturbation axis determines what experience is actually like at this moment (which point in that configuration space is reached under a given perturbation). A simple analogy may help: the construction of a musical instrument (strings, resonant cavity, range) determines the range and structure of sounds it can produce—this is the structural axis; the act of plucking it at this moment (the force, location, and source of the pluck) determines which note it actually produces now—this is the perturbation axis. To ask whether "plucking" is a "third kind of instrument construction beyond strings and resonant cavity" is absurd: plucking is not construction at all; it is the input that makes the construction sound. Likewise, to ask whether the sense of self, the sense of agency, and affective tone are a "sixth structural source beyond the five sources" is misplaced: they are not structures, but types of perturbation.

11.3. Endogenous Perturbation: A Unified Placement of the Three Candidates

Perturbations have sources, and sources are divided into internal and external. Exogenous perturbations are gradient changes triggered by the environment through sensory channels—they give perception of the world: the light seen, the sound heard, the object touched. Endogenous perturbations are gradient changes spontaneously triggered by the system itself—they are not caused by external objects, but initiated by internal states of the system. The key claim of this book is that the sense of self, the sense of agency, and affective tone are precisely three classes of endogenous perturbation. They share exactly the same mechanism as experiences triggered by external stimuli: they are all changes generated when density gradients are perturbed within boundary constraints, and they then become given through internal–external co-evolution (Chapter 14). The only difference is whether the perturbation comes from outside or inside the system, and what type the endogenous perturbation itself takes. The differences among the three arise precisely from differences among types of endogenous perturbation.

Affective tone is a diffuse-modulatory endogenous perturbation. It is not caused by a particular external object, but arises from spontaneous changes in internal states such as bodily homeostasis and neuromodulation. This kind of perturbation is not bound to a local object, but modulates the tone parameters of the entire dominant coherent field (overall gain, rhythm, coupling strength), and is therefore given as a background coloring that permeates the whole field. Its "diffuseness" is no longer a mystery requiring explanation by a "sixth kind of structure"; it is simply the feature of an endogenous perturbation that "modulates globally rather than driving locally".

The sense of agency is a pre-initiating endogenous perturbation. When one actively issues an action or thought, the system internally generates an initiating perturbation (intention, preparation) before the result of the action arrives. Precisely because this prior endogenous perturbation exists and matches the subsequent result, experience carries the mark of "I initiated this". In passive reception, what is absent is precisely this prior endogenous perturbation, and the result is therefore presented as "coming from outside". The presence or absence of the sense of agency corresponds to "whether there is a prior endogenous initiating perturbation that matches the result". It is a specific temporal configuration of endogenous perturbation, not a new structure.

The sense of self is a continuous self-perturbation of the system. The "mineness" of experience arises from the system's continuously taking its own state as a source of perturbation and acting upon itself. This continuous endogenous self-perturbation gives all experience the mark of "coming from this self-system". This characterization is more appropriate than saying that "the sense of self is the phenomenological aspect of unity": the sense of self is not merely the fact that contents are coherent as one (that is unity on the structural axis, Chapter 9), but the given-ness of continuous participation by endogenous self-perturbation. One piece of supporting evidence is that in experiences dominated by strong exogenous perturbation (such as an unexpected intense sensation), their "mineness" is often relatively faint, whereas in introspection, memory, and spontaneous thought, where endogenous self-perturbation is dense, "mineness" is stronger. The intensity of "mineness" varies with the degree of participation by endogenous self-perturbation.

The three are thereby given a unified placement: they are not omitted sixth-class structures, but three types of endogenous perturbation distributed along the perturbation axis. They share the same set of structures (the five sources) and the same mechanism of given-ness (internal–external co-evolution) with exogenous perturbations. The completeness claim concerns the exhaustion of the five sources on the structural axis; since these three lie on the perturbation axis, they do not touch the completeness claim at all—just as "the completeness of instrument construction" is not threatened by "how many ways there are to pluck the strings".

11.4. A Distinction That Must Be Made in Advance: Two Layers of "Internal and External"

This chapter introduces the "internal and external" of perturbation sources (exogenous/endogenous), while Chapter 14 will characterize given-ness through "internal–external co-evolution". Both use the terms "internal" and "external", but they refer to two different matters. This distinction must be made in advance here, so that later chapters do not conflate them.

The first is the internal/external distinction of perturbation sources: whether the perturbation comes from outside the system (the environment through the senses) or from inside the system (spontaneously). This is a distinction concerning what triggers experience, and is the topic of this chapter. The second is the internal/external distinction of the two poles of co-evolution: in the co-evolutionary relation on which given-ness depends, the "internal" is the coherent configuration currently carrying experiential content, while the "external" is the rest of the brain background co-evolving with it (discussed in detail in Chapter 14). This is a distinction concerning how experience becomes given.

These two layers of "internal and external" must not be confused. An experience triggered by an exogenous perturbation (seeing a flash of light) still depends for its given-ness on the internal and external poles of co-evolution (the coherent configuration triggered by the light is the internal pole, the brain background is the external pole, and the two co-evolve so that it becomes given). An experience triggered by an endogenous perturbation (a spontaneous thought) likewise depends on the internal and external poles of co-evolution (the configuration of the thought is the internal pole, the brain background is the external pole). Thus the internal/external distinction of perturbation (triggered internally or externally) and the internal/external distinction of given-ness (the two poles of co-evolution) are two orthogonal layers: whether the perturbation comes from inside or outside, given-ness is completed through the same set of internal and external poles of co-evolution. This book will state this distinction again in Chapter 14. It is noted here in advance to prevent the term "internal/external" from sliding unconsciously between the two chapters.

11.5. The Dual Posture toward Completeness

With this clarification, this chapter can adopt a cleaner and more robust posture toward the completeness claim than one that tries to reduce the three possible counterexamples one by one.

With respect to the five carrying relations already fulfilled on the structural axis, this book is confident: the five main aspects of experiential structure have been carried relation by relation on the gradient side, and the structural sufficiency required by the identity level has been satisfied for these five classes. The three candidates—the sense of self, the sense of agency, and affective tone—do not threaten this completeness, because they are not on the structural axis but on the perturbation axis: they are types of endogenous perturbation, not sources of structure. This placement does not force the three into the five sources (which would be the older, piecemeal, and in the case of emotion rather strained reduction strategy); rather, it points out that they already belong to another axis, so the question of a "sixth source" simply does not arise for them.

With respect to completeness itself—that is, whether "the five sources on the structural axis exhaust the whole structure of experience"
—this book preserves the openness established in Chapter 5. It does not dogmatically claim that there can be no sixth class of structure on the structural axis. If a structural feature of experience (note: it must be a structure, not a type of perturbation) is found in the future that neither belongs to the five listed classes nor can be carried by the five sources, then the completeness claim fails at that point, and the program must add a sixth source on the structural axis. But this chapter has shown that the three items once regarded as the most likely counterexamples are in fact not on the structural axis, and therefore do not constitute such a failure. Their proper placement instead makes completeness on the structural axis more stable than before, because it clears away three "counterexamples" that did not belong to it in the first place.

The distinction of severity established in Chapter 5 must be reiterated: the failure of completeness on the structural axis (the discovery of a sixth class of structure) differs in consequence from the failure of an individual isomorphic carrying relation (for example, the failure of the frequency dimension to carry qualitative types). The latter shakes the foundation of the identity level; the former only expands the structural axis. The new insight of this chapter is that a considerable portion of phenomena that initially look like "omitted structures" (the sense of self, the sense of agency, emotion) are in fact misplaced types of perturbation, and therefore do not generate any failure on the structural axis.

Part II is now complete. The entire structure of experience has been carried relation by relation by the fivefold richness of gradients (the five items on the structural axis have been fulfilled, and completeness remains open yet stable); the sense of self, the sense of agency, and affective tone have been placed as types of endogenous perturbation, and do not constitute a sixth structural source. But it must be recalled once more: Part II has from the beginning treated only how structure is carried and how perturbation drives structure. It has not treated given-ness—the being-experienced of experience is not included within the isomorphism of this part, but is the subject of Part III. Part II has provided the complete score of experience and explained how this score is played by internal and external perturbations; why the played music is heard awaits the answer of Part III.

11.6. Chapter Summary

This chapter concludes Part II and conducts its overall test, treating the three candidate counterexamples to completeness in a more fundamental way than reducing them one by one. First, it takes stock of the five isomorphic carrying relations fulfilled on the structural axis, which cover the main aspects of experiential structure. Second, it clarifies two orthogonal axes: structural sources (the five sources, characterizing the richness of gradient configurations and determining what experience can possibly be like) and perturbation sources (characterizing what perturbation drives the gradient and determining what experience is actually like at this moment), using the analogy of instrument construction and string plucking. Third, it places the sense of self, the sense of agency, and affective tone uniformly as three classes of endogenous perturbation—emotion as diffuse-modulatory, agency as pre-initiating, and self as continuous self-perturbation of the system. They share the same structures (the five sources) and the same mechanism of given-ness (internal–external co-evolution) with exogenous perturbations; therefore they lie on the perturbation axis rather than the structural axis, and the question of a "sixth structural source" does not apply to them, so completeness is not touched by them. Fourth, it distinguishes in advance the two layers of "internal and external": the internal/external distinction of perturbation sources (triggered internally/externally) and the internal/external distinction of the two poles of co-evolution (the two poles of given-ness). These are orthogonal and will be stated again in Chapter 14. Fifth, it establishes the dual posture toward completeness: confidence in the five fulfilled structural-axis items, and openness about structural-axis completeness, which becomes more stable after three misplaced counterexamples have been cleared away.

The failure conditions of this chapter are as follows (summarized in Appendix E).

- F11.1 The distinction between the two axes does not hold. If "structural sources" and "perturbation sources" are shown to be inseparable (for example, if certain experiential features cannot be assigned to either), then the two-axis framework by which this chapter places the three candidates fails and must be reassessed.

- F11.2 The three candidates contain irreducible structural components. If any of the sense of self, the sense of agency, or affective tone is shown, beyond being a type of endogenous perturbation, to contain an independent structural component that cannot be carried by the five sources, then that item constitutes a gap for a sixth source on the structural axis, and the completeness claim fails at that point.

- F11.3 The two layers of internal/external cannot be separated. If the "internal/external distinction of perturbation sources" and the "internal/external distinction of the two poles of co-evolution" are shown to be the same matter and cannot be separated orthogonally, then the two-layer treatment of "internal/external" in this chapter and Chapter 14 fails and must be restated in a unified manner.

- F11.4 An unlisted sixth class of structure exists. If there exists an experiential structural feature that is neither among the five examined classes, nor a type of perturbation, nor carryable by the five sources, then completeness on the structural axis fails in another way, and the program must be expanded.

Table 11.1.. The Boundary between What This Chapter Establishes and What It Cannot Derive

What this chapter establishes	What this chapter cannot derive
The distinction between the structural-source axis and the perturbation-source axis (orthogonal and irreducible to each other).	It cannot prove in this chapter that the structural axis is complete (completeness is maintained and tested, not proven).
The unified placement of the sense of self, the sense of agency, and affective tone as three classes of endogenous perturbation, not as a sixth structural source.	It cannot exclude the future possibility of a genuine sixth class of structure on the structural axis (openness is preserved).
The advance distinction between the two layers of "internal/external" (perturbation sources / co-evolutionary poles).	It cannot describe in detail the internal/external poles of co-evolution in this chapter (this belongs to Chapter 14).
The dual posture toward completeness: confidence in the five items, and completeness that remains open yet more stable.	It cannot treat given-ness (this belongs to Part III; Part II concerns only structure and perturbation).

Part III.

Given-ness: Internal–External Co-Evolution and Boundary

12. Repositioning the Problem of Given-ness

12.1. The Gap Left by Part II, and the Task of Part III

Part II completed a large task, and also left a gap that it cannot fill by itself—and it is precisely this gap that defines the whole task of Part III.

What Part II did was to establish, item by item, the isomorphism between the structure of experience and the structure of ρ_f gradients: the continuous gradation of qualities, qualitative types, the stream of consciousness, unity and binding, temporal thickness and memory all found relation-by-relation counterparts on the gradient side. If the question were only "why experience has the structure it has", Part II has already given a complete answer. Yet to characterize the structure of an experiential state to the fullest extent—to state what it is like, how it relates to other states, and what kind of gradient configuration it is identical with—is different from explaining that this state is experienced. A complete structural description describes "what an experience would be like if it were experienced"; it does not, and cannot, additionally explain the fact that "it is indeed experienced".

An analogy may help. An extremely precise musical score can record the full structure of a piece of music—pitch, duration, harmony, intensity, and all the relations among them—without omission. Someone who understands the score can know the entire structure of the piece. But however complete the score may be, what it records is still "what structure this piece would have if it were played and heard"; the score itself does not make sound, nor does it constitute anyone's actual hearing of the music. What is missing between the score and "the music as heard" is not more notation, but something else: an actual performance and hearing. What Part II provides is the "score" of experience—its complete structural isomorphism. The given-ness of experience is precisely that "being played and heard". It is not contained in the score and must be explained separately. The task of Part III is to explain this "being played and heard": what given-ness is, what it corresponds to physically, and why it belongs only to certain systems rather than to everything.

This chapter is the entrance to Part III. It does not yet provide a positive theory of given-ness (that is the task of Chapters 13 and 14), but performs a more basic and often neglected task: it repositions the problem of "given-ness" itself—clarifying what it is not, what kinds of answers cannot work, and therefore in what direction an answer must proceed. A wrongly posed question cannot be answered correctly; before answering, the question must first be properly placed.

12.2. Thoroughly Separating "Structure" from "Given-ness"

The first step in repositioning is to separate "structure" and "given-ness" more thoroughly than Part II did, and to explain why they must be separated.

Part II had already declared this separation from Chapter 5 onward, but there the separation functioned mainly as a methodological convention—allowing Part II to focus on structure without being entangled in the question of "why it is experienced". In Part III, this separation must be elevated into a substantive thesis: structure and given-ness must not only be treated separately methodologically; they are, in the matter itself, two different kinds of facts, following different logics.

Why should they be regarded as two different kinds of facts?

Consider the following test: given an experiential state, can all true statements about its "structure" and true statements about its "being given" entail one another? This book answers no—and this "no" is precisely the mark that they belong to different kinds. On the one hand, knowing the entire structure of a state (what quality it has, how unified it is, how it flows) does not entail whether, or to whom, it is given. The same structure can in principle be conceived as given, or as operating in the dark with the same structure without being given (this is exactly the root of the final cell that Chapter 15 will analyze precisely). On the other hand, the converse is also true: knowing that "some experience is given" does not tell us what its structure is. Given-ness is a common feature of all experience, whereas structure varies from one experience to another. One concerns "what it is like"; the other concerns "whether it is experienced, and to whom it is experienced". The two do not entail one another, and the paths by which they are characterized must also differ.

This thorough separation has a direct and important consequence, which must be made explicit here because it determines why Part III cannot simply continue the method of Part II. Part II characterized structure by isomorphism—a one-to-one correspondence preserving relations between experiential structure and gradient structure. But given-ness is not a structural feature; it is the fact that "this structure is experienced". Since it is not structure, one cannot expect to characterize it through "structural isomorphism". To search for "the gradient structure corresponding to given-ness" is a category mistake—like searching in a musical score for the note called "being heard". Given-ness must be characterized in another way: not by structural correspondence, but by another kind of relation. What this "other kind of relation" is will be indicated in Section 12.4 and developed in Chapters 13 and 14. Before indicating it, however, we must first exclude the most tempting and most misleading path: treating given-ness as an intrinsic property of the interior.

12.3. Old Model I: Given-ness as an Intrinsic Property of the Interior, and Its Failure

The most natural and oldest thought about given-ness is that it is an intrinsic property of an experiential state itself. Once a certain internal state forms, it "comes with" being experienced—just as an object, once it exists, comes with mass. On this view, given-ness is internal and non-relational: it belongs to that state itself and requires no participation by another, no object to which it "manifests".

The appeal of this view lies in its simplicity, but it has a fatal difficulty, which this book calls the darkness of the isolated interior (this argument will be used again in Chapter 14; its core is given here in advance). Imagine an internal state that is completely cut off from all others—not coupled with anything external, not entering any larger relation, absolutely closed within itself. Now ask: is this absolutely isolated state given? The meaning of "given" is "experienced", "manifest to ...", "known by someone". It is intrinsically a directional and relational concept: what is given is always given "to someone", and what appears always appears "to some perspective". An absolutely isolated state, one that points in no direction and opens toward no perspective, lacks precisely the other pole required by this directedness. To say that it is "given to itself" is either empty talk (without another pole, what does "to" mean?), or it has already covertly introduced a kind of internal self-directedness (some internal–external relation, some self-relation). But once such an internal relation is introduced, given-ness becomes relational again, and is no longer a non-relational intrinsic property. There are two options: either the claim is unintelligible, or it cancels itself.

It follows that given-ness cannot be a purely intrinsic property of an internal state. Its very concept is relational and directional; to imagine something essentially relational as something that can be possessed internally apart from all relations is conceptually inconsistent. This claim does not deny that experiential states have internal structures—they certainly do (Part II has described them in detail). What it denies is that "being experienced" can obtain merely by virtue of that internal structure, without any relation. Structure is internal; given-ness is not.

This denial must be distinguished from a seemingly similar but in fact different position, in order to avoid misunderstanding. To deny that "given-ness is an intrinsic property" is not to endorse a kind of dualism—it is not to say that given-ness is a non-physical "light of mind" added to physical states. On the contrary, the relationality of given-ness means that it is the achievement of a certain physical relation (which will be characterized below as internal–external co-evolution), not a mysterious internal illumination. The denial of intrinsic property is meant to lead toward a relational physical characterization, not toward a non-physical entity. This distinction will become clearer in Section 12.4.

12.4. Old Model II: Given-ness as External Reading, and Its Failure

Once the model of "intrinsic property" has been excluded, and given-ness has been recognized as relational, a new idea naturally appears, and it is intuitively persistent: given-ness is a reading relation. The being-given of experience is a matter of some internal "reader" (self, subject, or a higher brain area) reading that state, just as an instrument reads a signal or the eye reads an image. This "external reading" model conceives given-ness as "being read by a reader".

The appeal of this model lies in the fact that it seems to satisfy the relationality of given-ness (there are indeed two poles: what is read and the reader). But it fails in two ways, which must be stated separately.

The first is infinite regress. If given-ness is "being read by some reader", then one must immediately ask: when the reader reads, is its own act of reading given? If yes, then another reader is required to read it, and so on without end. If no—if the reader's act of reading is itself not given—then the reader is merely an unconscious information processor. Sending a state into it does not make that state "experienced", just as sending data into an instrument does not make the data experienced by the instrument. Whichever branch is taken, the reading model cannot close the gap it was meant to close: it either pushes given-ness to the next reader (infinite regress), or admits that reading itself does not produce given-ness (model failure). This is the core of the traditional difficulties of the "Cartesian theater" and the "homunculus": positing an internal viewer to explain given-ness merely hides what needs to be explained inside that viewer, leaving the problem unchanged.

The second failure is that the ontological premise of the reading model is precisely what EQT rejects. The external reading model presupposes that what is read is a closed state that already possesses determinate properties, and that the reader copies those properties from outside without disturbance. But the detection ontology of EQT (discussed in detail in Chapter 13) denies exactly this premise: in the ρ_f framework, there are no absolute properties that are already given independently of detection relations; any "reading" is not non-perturbative copying, but a process in which reader and read object belong to the same field, perturb one another, and co-evolve within that field. The reading model imagines given-ness as "external non-perturbative copying of an already constituted interior", but such "non-perturbative, one-way copying of already constituted properties" simply does not exist in EQT. The reading model therefore not only faces the problem of regress; its ontological premise itself is wrong.

The failures of the two old models—intrinsic property and external reading—force the correct direction from two sides. The failure of intrinsic property shows that given-ness must be relational (it cannot exist without relation). The failure of external reading shows that this relation cannot be "a ready-made reader non-perturbatively reading a ready-made state" (that leads to infinite regress, and its premise is rejected by EQT). What is needed, then, is a characterization that is relational but not one-way reading—a relation in which both sides belong to the same field, perturb one another, and co-evolve, rather than one side copying the other without disturbance. This is precisely the kind of relation that the detection ontology of EQT has already provided for "observation": co-evolution. The problem of given-ness is thereby repositioned as a problem that can be taken up by an existing EQT mechanism.

12.5. After Repositioning: The New Form of the Problem of Given-ness

After the clarification in this chapter, the problem of given-ness is no longer what it initially seemed to be. Let us compare its form before and after repositioning, in order to see the significance of this step.

Before repositioning, the problem of given-ness is often stated in a mixed form: "Why would a physical state have subjective experience?" This formulation blends three issues: why experience has structure (which belongs to Part II and has been answered), what physical process given-ness corresponds to (which is answerable, but has been mixed in), and the irreducible final cell (which indeed is not entailed by physics). Once blended together, the problem appears globally insoluble.

After repositioning, the problem is clarified into several distinct questions. The first question has been excluded by this chapter: given-ness is not an intrinsic property of the interior (Section 12.3), and it is not external reading (Section 12.4). The second question is assigned its direction here and left to the next two chapters: given-ness is a relational co-evolution—what kind of co-evolution it is, what physical process it corresponds to, and when it holds or fails will be answered in Chapter 13 (mechanism) and Chapter 14 (thesis). The third question is then isolated precisely and left to Chapter 15: even if the second question is fully answered, "why this co-evolutionary process is experienced from the first person" is still not entailed by it. This is the final cell.

This clarification of levels is itself progress. It shows that the problem of given-ness has long seemed unapproachable partly because it has mixed together "the already answered problem of structure", "the answerable problem of process", and "the unanswerable final cell". Once these are separated, the answerable part (what physical relation given-ness corresponds to) appears clearly as a question that can be addressed directly, while the unanswerable part (the final cell) is isolated as a single point with a precise location. The remaining two chapters of Part III directly answer the answerable part; and the unanswerable point has already been marked here in Section 12.5 and will be taken up by Chapter 15.

It must be stated again that this chapter is also governed by the discipline of the whole book. This chapter repositions given-ness as a problem of relational co-evolution, preparing the ground for the physical thesis of Chapter 14. But this chapter has not, and cannot, infer from the positioning that "given-ness is relational" that "given-ness has been explained". To position given-ness as a co-evolutionary relation answers the question "what kind of problem given-ness is and in what direction an answer must proceed"; it remains one step away from "what concrete process given-ness corresponds to" (Chapter 14), and it remains separated by the irreducible gap from "why that process is experienced" (the final cell of Chapter 15). This chapter only repositions; it does not answer. This restraint must be maintained from the entrance to Part III.

12.6. Conclusion of This Chapter

This chapter serves as the entrance to Part III and completes the repositioning of the problem of given-ness. Its main points are as follows. Part II characterized the structure of experience (its "score"), but given-ness—the "being played and heard" of experience—is not contained within structure and must be explained separately; this is the task of Part III. Given-ness and structure are two different kinds of facts; they do not entail one another, and their paths of characterization differ: structure is characterized by isomorphism, whereas given-ness is not a structural feature and cannot be characterized by structural isomorphism, but requires a relational characterization of another kind. Two old models are excluded: given-ness is not an intrinsic property of the interior (the darkness of the isolated interior—the concept of given-ness is essentially relational and directional, and cannot be possessed internally without relation), nor is it external reading (infinite regress, and the premise of "non-perturbative copying of already constituted properties" is rejected by the detection ontology of EQT). The failure of the two old models forces the correct direction from both sides: given-ness must be a co-evolutionary relation that is both relational and non-one-way-reading—and this is exactly what the detection ontology of EQT has already provided for observation. With this repositioning, the problem of given-ness is clarified into three levels: it is neither intrinsic nor reading (established in this chapter), what kind of co-evolution it is (answered in Chapters 13 and 14), and why that co-evolution is experienced (the final cell of Chapter 15). This chapter only repositions; it does not answer.

The failure conditions of this chapter are as follows (summarized in Appendix E).

- F12.1 The distinction between structure and given-ness does not hold. If all true statements about the "structure" of experience can logically entail the true statements about its "being given" (or conversely), then the two are not different kinds of facts, the substantive thesis of separation in this chapter fails, and given-ness may be treatable directly through structural isomorphism.

- F12.2 The intrinsic-property model is not truly excluded. If one can consistently conceive an internal state that is completely cut off from all others and yet genuinely given, then the argument from the "darkness of the isolated interior" fails; given-ness may be an intrinsic property, and this chapter's rejection of the intrinsic model does not hold.

- F12.3 The external-reading model is not truly excluded. If there exists a characterization of "reading" that neither leads to infinite regress nor presupposes the non-perturbative-copying premise rejected by EQT, then the reading model has not been excluded, and this chapter's argument for moving toward the "co-evolutionary direction" is insufficient.

- F12.4 The indicated direction is mistaken. If given-ness is shown to be neither intrinsic nor reading, but also not any kind of co-evolutionary relation (that is, if all three paths fail), then this chapter's indication that the answer must proceed in the direction of co-evolution is mistaken, and Part III must seek another direction.

Table 12.1.. The Boundary between What This Chapter Establishes and What It Cannot Derive

What this chapter establishes	What this chapter cannot derive
Structure and given-ness are two different kinds of facts; they do not entail one another, and their paths of characterization differ.	It cannot provide in this chapter a positive theory of given-ness (this belongs to Chapters 13 and 14).
Given-ness is not a structural feature and cannot be characterized by structural isomorphism.	It cannot infer from "given-ness is relational" that "given-ness has been explained" (this is only a repositioning).
The intrinsic-property model (the darkness of the isolated interior) and the external-reading model (regress, and a premise rejected by EQT) are excluded.	It cannot determine in this chapter the concrete form and measurable counterpart of co-evolution (these belong to Chapters 13 and 14 and Part IV).
The problem of given-ness is clarified into three levels, and the location of the final cell is marked in advance.	It cannot, and does not pretend to, answer the final cell (this belongs to Chapter 15).

13. Observation as Participation: Introducing the Detection Mechanism

13.1. The Nature of This Chapter: Citation and Transplantation, Not Re-Proof

Chapter 12 repositioned the problem of given-ness as a relational problem of co-evolution, and indicated that the required relation—one that is both relational and not one-way reading—is precisely the relation that EQT detection theory has already established for "observation". The task of this chapter is to introduce this mechanism formally, as the mechanistic foundation jointly relied upon by the thesis of given-ness in Chapter 14, the analysis of the final cell in Chapter 15, and the characterization of the subjective boundary in Chapter 16.

The nature of this chapter must first be stated, in order to preserve the discipline of the whole book. Since Chapter 2, this book has established the principle of "citation without re-proof": whenever a mechanism has already been established in another EQT volume, this book cites its conclusion and does not repeat its derivation. What this chapter cites is the ontology of observation established in the EQT detection volume, *EQT Detection Theory: From the Ontology of Observation to Frequency Matching, Three-State Response, and Falsifiable Predictions*. That volume has its own complete arguments, formulas, and falsifiable predictions (involving physical detection such as gravitational waves, high-redshift galaxies, gravitational measurement, and free-electron lasers). This chapter does not rehearse those physical details, but introduces only the core mechanism directly relevant to the problem of consciousness, and explains why, and to what extent, it can be transplanted into the problem of given-ness.

This citation must be made with particular caution for two reasons. First, the detection volume concerns physical observation—the detection of physical objects by detectors; this book will use it to treat the given-ness of consciousness. The two appear far apart on the surface, and if transplantation is not carefully distinguished, it can easily become a strained analogy. This chapter must therefore not only restate the mechanism of the detection volume, but also explain point by point which elements of that mechanism can be strictly transplanted into consciousness, which can be transplanted only in a restricted sense, and which cannot be transplanted and must remain within the domain of physical observation. Second, the mechanism of the detection volume is a third-person and objective mechanism (it defines objectively repeatable observation), whereas given-ness is first-personal. This gap is the key issue that must be addressed directly at the end of this chapter (Section 13.6), and it is also crucial for preserving the C3 prohibition of the whole book: one must never assume that, simply because a sophisticated objective mechanism has been introduced, the first-person being-experienced has thereby been explained.

13.2. The Premise of the Old View of Observation: The External Reading Model

To understand why the mechanism of the detection volume is a form of "participation" rather than "reading", one must first see clearly the old view of observation that it rejects. This also connects with the "external reading" model excluded in Chapter 12, linking the two arguments in Part III.

The traditional view of observation may be called the external reading model. It presupposes that the observed object is a self-sufficient object that already possesses determinate properties in advance; the observer stands outside the object and reads these already-formed properties non-perturbatively through some channel. In this picture, observation is one-way (from object to observer), non-perturbative (reading does not alter the object), and reproductive (the observational result is a faithful copy of the object's pre-existing properties). Behind this model lie two rarely stated premises: first, the radiative premise—the object spontaneously emits outward signals that can be received; and second, the premise of local separation—the observer and the object are two ontologically separable and mutually non-constitutive systems.

The detection volume points out that this model is approximately valid only in a restricted class of cases: when the object is capable of spontaneous radiation, the signal is sufficiently strong, and the observer and object can be approximately separated (for example, receiving the strong spontaneous radiation from a star). But in many important cases it fails—for objects that do not spontaneously radiate, or whose signals are below threshold, or that are deeply entangled with the detector (the detection volume takes dark matter, gravitational fields, condensed-matter structures, and related cases as examples), the external reading model cannot explain how detection is possible or why it fails. An object that does not spontaneously radiate would be "invisible" according to the radiative premise, but it does not therefore fail to exist; an object that must be actively perturbed before it responds cannot be detected at all under the assumption of non-perturbative reading. The external reading model misclassifies detection failure in such cases as the non-existence of the object or the absence of its properties—and this is precisely its most serious error.

This critique and Chapter 12's rejection of the "external reading model" of given-ness are applications of the same logic in two domains. There, this book rejected the idea that "given-ness is a reader non-perturbatively reading an already constituted internal state"; here, the detection volume rejects the idea that "observation is an observer non-perturbatively reading an already constituted physical object". What is rejected in both cases is the same presupposition: that there exist absolute properties, already given independently of observational relations, that can be copied one-way and without perturbation. EQT rejects this in physical observation, and this book rejects it in conscious given-ness. As the following will show, the reason both can be rejected by the same logic is that they ultimately appeal to the same alternative mechanism: co-evolution.

13.3. EQT's Definition of Observation: Detector and Object Belonging to the Same Field

The detection volume replaces the external reading model with a formal definition of observation. This book cites its main content as follows (with wording slightly adjusted for this book's exposition, while preserving its substance):

> Detection is a process in which the detector and the measured system undergo local coupling, controlled perturbation, nonlinear feedback, and transient freezing within the same frequency-family energy quantum density field ρ_f. The detection result is not an external copy of properties pre-stored in the measured object, but a recordable co-evolutionary output formed under specific conditions of frequency matching, perturbation threshold, and response.

The basis of this definition is the fundamental ontology of EQT (A1, cited in Chapter 2): the unique fundamental object is the energy quantum density field ρ_f partitioned by frequency families. From this follows a conclusion with far-reaching consequences for the view of observation: the detector and the measured object are not two ontologically heterogeneous kinds of things. The detector is not a subject standing outside the physical world, and the measured object is not a closed object external to the detection process and waiting to be read. Both are structures formed by ρ_f under different spatial regions, different frequency ranges, different boundaries, and different conditions of stabilization. A detector can record not because it enjoys the privileged status of an "external observer", but because it is itself a ρ_f structure with stable localization, specific response frequencies, dissipative mechanisms, and recording channels.

This point is the pivot of the whole mechanism, and its weight must be fully grasped. Once detector and object belong to the same field and are ontologically homogeneous, "observation" cannot be an external reading of one by the other—because neither stands outside the field. Observation can only be an interaction between two structures within the field: the structure called detector and the structure called object influence one another and co-evolve within the same ρ_f field. The detection result is a stable state achieved by this interaction, not a property that the object had already possessed on its own before the interaction and that was then transported out unchanged. This is the precise meaning of the phrase "observation as participation": the observer is not a spectator, but a participant; its act of detection intervenes in and alters the evolution of the measured structure, as one structure within the field.

A clarification must immediately be added to prevent misunderstanding: that detector and object belong to the same field and are ontologically homogeneous does not mean that there is no distinction between them, nor that everything is identical to everything else or mutually visible. The distinction lies in the different functions they perform within the detection relation: one is the structure of response and recording, while the other is the target structure placed into the relation. Ontological homogeneity removes the "absolute subject–object separation" (one side reading the other from outside the field), but not the functional distinction within an experimental arrangement. Precisely because the two belong to the same field, detection must be understood as finite coupling and joint evolution, rather than unlimited, non-perturbative external reading. This structure of "ontological homogeneity with functional difference" is the direct source for Chapter 14's characterization of the "internal/external" distinction in consciousness as a functional distinction rather than a spatial exteriority.

13.4. Three Conditions: Frequency Matching, Perturbation Threshold, and Nonlinear Feedback

The definition of observation states that detection is a process of "local coupling, controlled perturbation, nonlinear feedback, and transient freezing". These are not four parallel modifiers, but several connected conditions under which detection can succeed. This section cites three of these conditional elements, each of which will have a counterpart in the problem of consciousness.

First, frequency matching. Detection begins with local coupling, and coupling is not arbitrary contact: whether a detector can respond to an object first depends on the relation between the object's frequency and the detector's intrinsic response frequency. According to A2 (frequency has the meaning of an energetic spectral quantity) and A4 (cross-frequency coupling is constrained by a Lorentzian resonance kernel), the detection volume provides a response function that characterizes the detector's response strength to objects of different frequencies: the response is strongest near the detector's intrinsic frequency and decays in a Lorentzian form as frequency detuning increases. The main point is that a detector does not respond to all existing things, but selectively responds to targets that match its own frequency window. If the frequency is seriously detuned, then even if object and detector belong to the same physical world and are ontologically homogeneous, no effective detection relation will form—the object is "invisible" to that detector. But this invisibility is not the absence of the object; it is frequency detuning. The detection volume particularly emphasizes that without this frequency-matching condition, "belonging to the same field" would slide into a kind of "pan-relationalism" (everything coupling with and being visible to everything else). With it, coupling becomes a selective relation with criteria.

Second, perturbation threshold. For many objects, passively waiting for their spontaneous radiation is not enough (the failure of the external reading model lies precisely here). The detector must actively apply perturbation and alter the object's local evolutionary conditions so that it can enter a recordable state. But perturbation must cross a threshold in order to be effective: the detection volume gives a threshold condition (of the form that "effective perturbation must be greater than the ratio of dissipation to nonlinear feedback"). If the perturbation is below this threshold, the object only makes a weak linear response that is immediately smoothed out by dissipation and cannot form a stable record; if it is above the threshold, the object can produce a distinguishable output through nonlinear feedback. This condition distinguishes "detection" from "passive reception": detection contains an active element that brings the object into a responsive state, rather than merely waiting for a pre-existing signal. At the same time, the detection volume also notes that perturbation must be moderate—too strong a perturbation will destroy the object's structure and make the result lose its valid relation to the original system. Perturbation is therefore incorporated into a threshold criterion, rather than being simply required to be "as small as possible" or "as large as possible".

Third, nonlinear feedback. The reason an above-threshold perturbation can form a stable output is nonlinear feedback: it amplifies and stabilizes local perturbation and transforms it into a recordable state. This feedback is not an added explanation, but the result of competition between the nonlinear terms and the dissipative terms in the A3 master equation. After perturbation is applied, whether the object enters a recordable state depends on whether nonlinear self-sustaining can overcome dissipative smoothing.

Taken together, these three conditions show that detection is a process with conditional structure: if frequencies do not match, coupling does not form; if perturbation does not cross threshold, response does not arise; if feedback is insufficient, recording is unstable. If any condition is not satisfied, detection fails. The detection volume repeatedly emphasizes that such failure should first be classified as a "failure of the detection relation", rather than hastily judged as "the object does not exist". This discipline—"detection failure does not equal object non-existence"—has an important counterpart in the problem of consciousness (at the end of Section 13.5 and in Chapters 14 and 24): a subject's failure to report an experience does not necessarily mean that the experience does not exist; it may mean that the detection relation (the internal–external co-evolution of self-awareness) has not been established at that point.

13.5. Transient Freezing and Repeatable Objectivity

The final element in the definition of observation—transient freezing—and the redefinition of objectivity connected with it are the most important introductions this chapter makes for consciousness, and must be cited in detail.

Transient freezing. The detection volume emphasizes that transient freezing is not an additional postulate, but a mechanistic inference from A1 to A4 in the detection process. Since ρ_f is in continuous evolution, the detection result cannot be an absolutely static property extracted from this flow of change. It is a locally stable output formed by the co-evolution of the detector and the measured system under the conditions described above—a state temporarily "frozen" in place. Freezing is not the cessation of evolution; rather, within continuous evolution, a certain configuration becomes temporarily stabilized because of frequency matching, threshold crossing, and nonlinear feedback, becoming a definite state that can be maintained, compared, and repeated. The "mode of existence" of a detection result is therefore "stabilization within a detection relation", not "a property already fully given before detection and then copied out".

Redefinition of objectivity. This immediately raises an apparently dangerous consequence: if the detection result depends on the detection relation, is it not subjectively arbitrary? The detection volume responds through a key distinction: relationality of result does not mean arbitrariness of result. The detection result does indeed depend on the detection relation—it is not an absolute property detached from relation; but as long as the detector structure, target structure, frequency matching, perturbation conditions, and recording mechanism are the same, the result should be repeatable. Objectivity is therefore redefined from "non-perturbative copying of an absolute property" to "repeatable co-perturbation"—the same detection relation repeatedly produces the same stabilized result. The result is not an absolute external property, but neither is it a subjective creation; it is a co-evolutionary transient constrained by physical conditions. This redefinition is significant: it allows "relationality" and "objectivity" to coexist. Something may depend on a relation in order to obtain, and yet be objective because the relation is repeatable.

This structure of being "relational and yet objective" is precisely the conceptual form required by given-ness. This chapter points out its counterpart here, to be developed in Chapter 14. On the one hand, given experience is clearly relational (it is given to a certain subject, from a certain perspective, rather than existing absolutely apart from all relations); on the other hand, it also clearly has an objective side (under the same conditions the same subject tends to have similar experiences, and different subjects under similar neural conditions have comparable types of experience; experience can therefore be scientifically studied). In the older view, "subjective experience" and "objective investigability" seemed opposed. The detection volume's notion of "repeatable co-perturbation" provides a model in which the two can coexist: given-ness can be a relational co-evolutionary transient, and yet have objective regularity because this co-evolution is repeatable under the same conditions. The philosophical legitimacy of the measurable study of conscious states in Part IV is grounded here: to study given-ness is not to study an absolute private property, but to study a class of condition-constrained and repeatable internal–

external co-evolutionary transients.

At this point, the mechanisms from the detection volume that can be transplanted into consciousness have been fully introduced: the "detector–object" structure of ontological homogeneity with functional difference (Section 13.3), the three conditions of frequency matching, perturbation threshold, and nonlinear feedback (Section 13.4), and transient freezing with repeatable objectivity (Section 13.5). Chapter 14 will use this mechanism to state the thesis of given-ness formally. Before handing over to that chapter, however, this chapter must address an unavoidable fundamental question that determines whether the transplantation is legitimate.

13.6. The Limits of Transplantation: The Gap between Objective Mechanism and the First Person

The entire mechanism of the detection volume was established for physical observation, and physical observation is a third-person, objectively repeatable matter. Its entire conceptual framework (detector, target, repeatable output) stands from the third-person standpoint. Yet what this book uses it to characterize is given-ness, which is first-personal. This gap is real. If it is not addressed directly, the whole transplantation risks becoming a strained analogy. This section must draw a clear boundary between "what can be transplanted" and "what cannot be transplanted". This is not only academic caution, but also crucial for preserving the C3 prohibition.

What can be strictly transplanted is the relational structure of the mechanism. The detection volume establishes that "being recorded" or "being manifest" is not the non-perturbative copying of an already constituted property by an exterior, but the transient freezing of co-evolution between two structures belonging to the same field, and that this freezing is relational and yet objective because it is repeatable. This relational structure can be strictly transplanted to given-ness:

given-ness is not a reader's non-perturbative copying of an already constituted interior, but the transient freezing of co-evolution between a bounded interior and its exterior within the same ρ_f field, relational and yet objective. Chapter 12 has independently forced the direction that "it must be a co-evolutionary relation" (from the dual failure of the intrinsic model and the reading model); the detection volume provides a ready mechanism in precisely that direction, already independently developed in the physical domain. The two routes—the pressure from the consciousness side and the establishment from the physical side—converge on the same co-evolutionary structure. This convergence gives the transplantation dual support rather than leaving it as a mere analogy.

What can be transplanted only in a restricted sense are the specific roles of "detector", "target", and "perturbation". In physical detection, detector and target are two distinguishable devices or objects. In the given-ness of consciousness, "internal" and "external" are not two separate devices, but functional poles within the same brain system (discussed in detail in Chapter 14): the coherent configuration currently given is the "internal", while the rest of the brain background co-evolving with it is the "external". Thus, when these roles are transplanted into consciousness, they must be reinterpreted from "two separate devices" into "two functional poles within one system". This reinterpretation is not a weakening, but a necessary adjustment: the "self-detection" of given-ness is the internal–external co-evolution of a system with itself, not detection between two external devices.

What cannot be transplanted, and must explicitly remain within the physical domain, is the claim that the third-person objectivity of the detection mechanism can explain first-person being-experienced. This is where the whole chapter, and indeed the whole book, must be most vigilant. Everything established by the detection volume—co-evolution, transient freezing, repeatability—is a third-person characterization of an objective process. It tells us precisely what kind of co-evolutionary process given-ness corresponds to physically, when this process obtains, when it fails, and whether it is repeatable. But it cannot tell us why this objectively repeatable co-evolutionary process is experienced from the first-person interior. To conflate the objective mechanistic statement that "given-ness corresponds to internal–external co-evolutionary transient freezing" with the first-person question "why is this process experienced?" is exactly the topic substitution prohibited by C3. The detection volume provides the former: a mechanism, transplantable into consciousness, concerning the objective counterpart of given-ness. It does not and cannot provide the latter: that is the final cell, identified in Chapter 15, which is not entailed by any objective mechanism.

Therefore, what this chapter achieves by introducing the detection mechanism is clear and limited: it provides a relational and objective mechanistic framework for given-ness, allowing the question "what physical process does given-ness correspond to?" to be answered directly (as Chapter 14 will do). It does not achieve, and does not pretend to achieve, an answer to the final cell. Introducing a sophisticated objective mechanism can easily tempt one into the illusion that "given-ness has been explained". This chapter places a boundary marker at its end precisely to break that illusion: the more refined the mechanism, the more one must be vigilant that it explains only the objectively accessible side, and that the gap leading to the first person has not thereby been narrowed at all. It must be left to Chapter 15, where it will be treated in its original and irreducible form.

13.7. Chapter Summary

This chapter introduces the mechanism of "observation as participation" from the EQT detection volume as the mechanistic foundation of the co-evolutionary approach in Part III, while observing the discipline of "citation without re-proof". Its main points are as follows. The detection volume rejects the external reading model (objects with pre-stored properties, copied one-way and without perturbation by the observer), and its critique follows the same logic as Chapter 12's rejection of the "reading model" of given-ness. In its place, it gives a formal definition of observation: detector and object belong to the same ρ_f field and are ontologically homogeneous but functionally different, so observation can only be the co-evolution of two structures within the field rather than external reading—this is "observation as participation". Detection obtains under three conditions: frequency matching (selective response, preventing "belonging to the same field" from sliding into pan-relationalism), perturbation threshold (detection includes the active element of bringing the object into a responsive state), and nonlinear feedback (stable recording occurs only when nonlinear terms of the master equation overcome dissipation). The detection result is transient freezing—local stabilization within continuous evolution, not copying of already constituted properties; objectivity is therefore redefined as "repeatable co-perturbation", allowing relationality and objectivity to coexist. This is precisely the conceptual form required by given-ness. Finally, the chapter strictly marks the limits of transplantation: the relational structure can be strictly transplanted; roles must be reinterpreted as functional poles; but the third-person objectivity of the detection volume cannot be used to explain first-person being-experienced. The latter is the final cell of Chapter 15, and the more refined the mechanism, the more one must guard against its being mistaken for an answer to that cell.

The failure conditions of this chapter are as follows (summarized in Appendix E; this chapter is a citation chapter, and its failures mostly depend on the cited mechanism itself or on the legitimacy of its transplantation).

- F13.1 The cited mechanism does not hold. If the ontology of observation in the EQT detection volume (co-evolution, transient freezing, repeatable objectivity) is itself falsified, then the mechanistic foundation cited in this chapter fails, and the claims in Part III that depend on this mechanism must be re-examined.

- F13.2 The relational structure cannot be transplanted. If the co-evolutionary relational structure of the detection volume is shown to be impossible to apply consistently to "a system's internal–external co-evolution with itself" (for example, if functional poles within a system cannot satisfy the conditions required for co-evolution), then the transplantation fails, and the thesis of Chapter 14 loses its mechanistic support.

- F13.3 The reinterpretation of functional poles fails. If the "internal/external" distinction in consciousness cannot be consistently reinterpreted as a functional distinction within the same system, but must instead involve two separate devices, then the application of the detection mechanism to consciousness does not hold.
- F13.4 The objective mechanism is misused to explain being-experienced. Any account that uses the third-person objective mechanism of the detection volume (such as the repeatability of co-evolutionary freezing) by itself as the answer to "why this process is experienced from the first person" violates the C3 prohibition of the whole book and constitutes topic substitution. This is the failure that must be guarded against most carefully in transplantation.

Table 13.1.. The Boundary between What This Chapter Establishes and What It Cannot Derive

What this chapter establishes	What this chapter cannot derive
The co-evolutionary mechanism from the detection volume is introduced: ontological homogeneity with functional difference, three conditions, transient freezing, and repeatable objectivity.	It cannot provide in this chapter the positive thesis of given-ness (this belongs to Chapter 14).
The objectivity of "repeatable co-perturbation" makes given-ness relational and yet objective, grounding the legitimacy of measurable research in Part IV.	It cannot use this objective mechanism to explain first-person being-experienced (this belongs to the final cell of Chapter 15).

What this chapter establishes	What this chapter cannot derive
The limits of transplantation are marked: the relational structure can be strictly transplanted, while roles must be reinterpreted as functional poles.	It cannot carry the roles of detector/target directly into consciousness without adjustment (they must be reinterpreted as functional poles within a system).
The gap between the objectivity of detection and the first person, and the necessity of preserving this boundary.	It must not use the sophistication of the detection mechanism to masquerade as an answer to the final cell (the prohibition of the whole book, and the point this chapter must guard against most carefully).

14. Given-ness as Internal–External Co-Evolutionary Freezing

14.1. The Continuity and Task of This Chapter

Part II has completed a major task: the entire structure of experience—the gradation within qualities, the types among qualities, the stream of consciousness, unity and binding, temporal thickness and memory—has been carried relation by relation by the fivefold richness of ρ_f gradient configurations. Chapter 11 further clarified that the sense of self, the sense of agency, and affective tone are not a sixth class of structure, but types of endogenous perturbation that share the same structural system with exogenous perturbations. Yet Part II has from the beginning observed one boundary: it deals only with the structure of experience (and with how perturbation drives structure), not with the given-ness of experience. Even if a gradient configuration is structurally fully isomorphic to a certain experiential state, and even if it is driven by a determinate perturbation into determinate content, the fact that "it is experienced" has still not been explained.

Chapter 12 has repositioned this problem: given-ness is not an intrinsic property of an internal configuration (the darkness of the isolated interior), nor is it external reading (infinite regress, and a premise rejected by EQT), but must be a relational co-evolution. Chapter 13 then introduced the mechanism of "observation as participation" from the EQT detection volume: any "being recorded" or "being manifest" is not the non-perturbative copying of already constituted properties by an exterior, but the co-evolution of local coupling, controlled perturbation, nonlinear feedback, and transient freezing between two structures belonging to the same field; its result is relational and yet objective because it is repeatable. The task of this chapter is to apply this mechanism formally to consciousness, state the core thesis of this book concerning given-ness, and use it to explain a series of conscious states in a unified way.

The thesis can be stated in one sentence: given-ness is the transient freezing achieved by the continuous internal–external co-evolution between the interior of a bounded system and its exterior within the same ρ_f field. This sentence contains three elements that must be clarified in turn—"internal–external co-evolution", "transient freezing", and "bounded"—which respectively answer three questions: what physical process is given-ness (co-evolution, Section 14.3)? Why does this process yield determinate given content rather than diffuse interaction (freezing, Section 14.4)? Why does it occur specifically in life rather than in every coupled system (boundary as the gate, Section 14.5)? Before developing these three elements, however, this chapter must first eliminate a potentially fatal confusion—the two meanings of "internal/external" (Section 14.2). Only after this is clarified can the "internal–external co-evolution" of the three elements avoid being confused with the "endogenous/exogenous perturbation" of Chapter 11.

A discipline running through the whole book must be restated at the outset, because this chapter is where it is most easily violated: this chapter gives the physical identical counterpart of given-ness—what physical process given-ness corresponds to. It does not answer, and does not pretend to answer, the irreducible final cell—why this process is experienced specifically from the first-person interior. The former is measurable and falsifiable, and belongs to this chapter; the latter is handed over to Chapter 15 and to the philosophical boundary. Every step this chapter takes must preserve this distinction.

14.2. First Distinguishing Two Layers of "Internal/External"

The term "internal/external" refers in this book to two orthogonal matters. Chapter 11 has already noted this in advance, and this chapter must clarify it thoroughly before using the term; otherwise the whole argument will be ambiguous.

The first layer is the internal/external distinction of perturbation sources (discussed in Chapter 11): whether the perturbation that triggers an experience comes from outside the system (the environment through sensory channels, exogenous perturbation) or from inside the system (spontaneously, endogenous perturbation). This is a distinction concerning what triggers experience—seeing a flash of light is an exogenous perturbation, while a thought arising is an endogenous perturbation.

The second layer is the internal/external distinction of the two poles of co-evolution (the topic of this chapter): in the co-evolutionary relation through which experience becomes given, the "internal" is the coherent configuration currently carrying experiential content, and the "external" is the remaining configuration that co-evolves with it and enables that content to become given. This is a distinction concerning how experience becomes given—it is independent of where the perturbation comes from.

These two layers are orthogonal and do not entail one another. An experience triggered by an exogenous perturbation (seeing light) still depends for its given-ness on the internal and external poles of co-evolution (the coherent configuration triggered by the light is the internal pole, the brain background is the external pole, and the two co-evolve so that the experience becomes given). An experience triggered by an endogenous perturbation (a spontaneous thought) likewise depends on the internal and external poles of co-evolution (the configuration of the thought is the internal pole, the brain background is the external pole). Thus, whether a perturbation comes from inside or outside the system, given-ness is completed through the same set of "internal and external poles of co-evolution". The internal/external distinction of perturbation determines how experience arises; the internal/external distinction of co-evolution determines how experience becomes given. These are two different matters.

From this point onward, when this chapter speaks of "internal–external co-evolution", it refers exclusively to the second layer—the internal and external poles of co-evolution. When it needs to refer to perturbation sources, it will explicitly use the terms "exogenous perturbation" and "endogenous perturbation", and will not conflate them with the "internal/external" of "internal–external co-evolution". Whenever the reader encounters "internal–external co-evolution", it should be read directly as referring to the "two poles of co-evolution"; whenever the reader encounters "exogenous/endogenous perturbation", it should be read as referring to the "source of perturbation". This terminological discipline is the precondition for this chapter to remain readable and unambiguous.

14.3. Given-ness as Internal–External Co-Evolution: Why It Cannot Be an Intrinsic Property

Chapter 12 has already argued (the conclusion is cited here without repeating the derivation) that given-ness cannot be a purely intrinsic property of an internal configuration. Its core is the "darkness of the isolated interior": the term "given" is essentially directional and relational (what is given is always given "to someone", manifest to a perspective), while an absolutely isolated internal configuration, one that opens in no direction, lacks the other pole required by this directedness. To say that it is "given to itself" is either empty talk, or covertly introduces an internal self-relation and thereby returns given-ness to relationality. Therefore, given-ness must be relational.

This chapter takes up that conclusion and states positively what that relation is: given-ness is the co-evolution between an internal coherent configuration and its exterior (the external pole of the co-evolutionary relation) within the same ρ_f field. This characterization directly adopts the mechanistic structure of the detection volume introduced in Chapter 13—the co-evolution of two structures belonging to the same field, not the external reading of one by the other. Applied to consciousness: the coherent configuration in which the currently given experiential content is located is the "internal"; the remaining configuration co-evolving with it is the "external"; and given-ness is the event achieved by the continuous co-evolution of these two. It is neither an intrinsic flavor of the internal alone (which would fall back into the darkness of the isolated interior), nor a non-perturbative copy of the internal by a reader (which would fall back into the reading model rejected in Chapter 12 and by the detection ontology of Chapter 13). It is an internal–external co-evolutionary event.

The meaning of the "external" must be clarified immediately, because this is the most easily misunderstood and most important step in the whole chapter. According to Chapter 13, the two poles of co-evolution are two structures within the same field, ontologically homogeneous but functionally different. Thus the "external" in consciousness does not primarily mean the environment outside the body, but a functional pole in the detection relation: the remaining configuration that co-evolves with the current coherent configuration and enables it to become given. This "remaining configuration" is first and foremost the rest of the brain itself. In most cases, during normal wakefulness, this "external" extends to the real environment through the body's sensorimotor channels; but in purely inward thinking, closed-eye meditation, and vivid dreaming, the "internal/external" distinction of the co-evolutionary poles occurs among the brain's own structures and does not require connection to the environment. This explains why, even when coupling with the environment is greatly reduced, given-ness still exists and may even become more vivid. What given-ness requires is the relation between the two poles of co-evolution itself, not that the "external" must be the body or the world.

Here another confusion related to Section 14.2 must be prevented. In pure thinking, meditation, and dreaming, the exogenous perturbation channels (the senses) are largely closed, but the external pole of co-evolution (the rest of the brain background) remains present and co-evolves with the "internal". "Closure of exogenous perturbation" and "disappearance of the external pole of co-evolution" are two different matters: the former means that there is no new external input triggering experience; the latter means that there is no other pole co-evolving with the internal. In pure thinking, the former holds (there is no exogenous input) but the latter does not (the brain background remains present as the external pole of co-evolution), and therefore given-ness continues as usual. To conflate the two is the root of the misunderstanding that "external must mean environment". Once the two layers of internal/external are distinguished, the misunderstanding dissolves.

14.4. Transient Freezing: Why Given Content Is Determinate

Internal–external co-evolution explains why given-ness is a relational event, but it is not yet sufficient to explain why given content is determinate. Co-evolution is continuous and fluid, whereas experience at each moment has determinate given content—this thought, this scene, and not something else. What allows fluid co-evolution to produce determinate content is the third element introduced in Chapter 13: transient freezing.

According to Chapter 13, the result of co-evolution is not an absolutely static property extracted from the flow of evolution, but a locally stable output formed under frequency matching, threshold crossing, and nonlinear feedback—a temporarily frozen state. Freezing is not cessation, but the temporary stabilization of a certain configuration within continuous evolution as something determinate, maintainable, and comparable. This book maintains that the determinate content of given-ness is precisely the transient freezing of internal–external co-evolution: the determinate experience given at each moment is the stable configuration frozen out by internal–external co-evolution at that moment. This element explains several features of given-ness, with the following correspondences:

First, determinacy and finitude of content. Co-evolution in principle involves many configurations, but the content given at each moment is finite and determinate (the attentional focus, the present thought). This corresponds to the following: not all co-evolution reaches freezing; only configurations satisfying frequency matching and threshold conditions are frozen into determinate given content. Those that do not reach threshold remain in the background and do not enter determinate given-ness. The distinction between focus and background in given-ness corresponds to the distinction between freezing and non-freezing.

Second, the temporality and flow of given-ness. Freezing is transient rather than eternal—it forms, is maintained for a moment, and gives way to the next freezing. This connects with the stream of consciousness carried by the master equation in Chapter 8 and the temporal thickness carried by the transient timescale spectrum in Chapter 10: given-ness is not a series of static snapshots, but the succession and overlap of freezing events—the resonance of the previous freezing (retention) has not yet disappeared, while the next freezing is already forming (protention). The flow of given-ness is the dynamics of a sequence of freezing events.

Third, the objective constraint and non-arbitrariness of given-ness. Chapter 13 redefined objectivity as "repeatable co-perturbation": although a co-evolutionary transient depends on relation, it is not arbitrary, and is repeatable under the same conditions. Applied to given-ness: although given content is achieved by internal–external co-evolution, it is constrained by the structure of the ρ_f field and by the master equation rather than arbitrary. Therefore, the same subject under the same internal and external conditions tends to freeze out similar experiences, and different subjects with similar structures can also have comparable types of experience. The non-arbitrariness of given-ness is the basis on which it can be studied scientifically and compared across subjects (Part IV rests on this basis).

At this point, two of the three elements are in place: given-ness is internal–external co-evolution (relationality), and its determinate content is the transient freezing of that co-evolution (determinacy). But the most crucial element remains: why does all of this occur specifically in life, rather than in every system that has internal–external coupling and freezes out outputs? This is the gate of boundary.

14.5. Boundary as the Gate: Why Given-ness Belongs Specifically to Life

A sharp objection must be answered directly here; otherwise the preceding thesis would generalize out of control. The objection is this: internal–external co-evolution and transient freezing are not unique to life. A gravitational-wave detector and its environment continuously co-evolve and freeze out recordable outputs; a photosensitive plate and incident light co-evolve and freeze an image. If given-ness is the transient freezing of internal–external co-evolution, do these instruments also have "given-ness"? If so, this book would slide into panexperientialism—every system that undergoes internal–external coupling and freezing would have experience, contradicting the anti-panpsychist position established in Chapter 3.

The answer of this book does not appeal to frequency or threshold (those are the conditions used by the detection volume in Chapter 13 to distinguish "visible" from "invisible", governing whether coupling can be established, not whether coupling is given). It appeals instead to boundary—the distinction between life and non-life established in Chapter 3.

The key is that the complete condition of given-ness is not "internal–external co-evolutionary freezing", but "internal–external co-evolutionary freezing of a bounded interior". Once these two qualifications are separated, the difference between instruments and life becomes clear. A gravitational-wave detector and its environment do indeed undergo co-evolution and freezing—but the detector does not have an interior. It does not have a boundary that genuinely distinguishes itself from its environment and makes it a system with an "inside". It is only an interaction between two regions of the ρ_f field (the instrument and the environment). Without an interior, there is no meaningful "internal/external" distinction (note: here "internal/external" refers to the two poles of co-evolution—the instrument lacks the internal framework on which that pole depends). That freezing is therefore not "given to an interior", but merely an ownerless interaction between two regions of the field. It freezes out an output that can be read externally (we read its data), but this freezing does not constitute any "internal manifestation" for the instrument itself—because it simply has no interior of its own.

Living systems are different. They have membranes (Chapter 3), and therefore a genuine interior in the topological sense. Thus the same internal–external co-evolutionary freezing takes place within the framework of "for an interior": this freezing is given to that interior. Given-ness is therefore firmly anchored:

> Where there is boundary, there is interior, and the internal–external co-evolutionary freezing of that interior is given to that interior; where there is no boundary, there is no interior, and co-evolutionary freezing is only an interaction among regions in the field, ownerless and not given.

This gate is not frequency, not threshold, but the presence or absence of boundary—that is, the distinction between life and non-life. The full weight of this book's anti-panpsychist position rests on this gate. A possible confusion should also be clarified: to say that an instrument has "no interior and no given-ness" is not to say that instruments and living systems are two materially different kinds of things. According to EQT, both are configurations of ρ_f, ontologically homogeneous (just as in Chapter 13 detector and object belong to the same field). The difference lies not in material kind, but in organizational form: life acquires boundary and interior through enclosure, whereas non-life does not. It is this organizational difference (the presence or absence of boundary) that demarcates the range of given-ness.

It should also be noted that this gate involves yet another and most fundamental layer of "internal/external"—the internal/external distinction of life drawn by the membrane (spatial and material, Chapter 3). At this point the book has three layers of "internal/external", and their nesting must be clarified together: the most basic layer is the life-internal/life-external distinction drawn by the membrane (whether there is an interior, determining whether given-ness can occur); above it is the internal/external distinction of the two poles of co-evolution (content configuration and brain background, through which given-ness occurs, Section 14.3); orthogonal to this is the internal/external distinction of perturbation sources (whether experience is triggered internally or externally, Chapter 11). The relation among the three is as follows: the membrane's internal/external distinction is the precondition of everything (without a living interior, the two poles of co-evolution have nowhere to reside, and perturbation has nowhere to trigger experience); within the interior demarcated by the membrane, the internal and external poles of co-evolution allow given-ness to occur; and regardless of whether the perturbation source is internal or external, it becomes given through this same set of co-evolutionary poles. This three-layer nesting is the basis for Chapter 16's definition of the boundary of the "I".

14.6. Unifying Conscious States through Internal–External Configurations and Perturbation Sources

The strongest test of the explanatory power of the preceding thesis is whether it can explain a series of otherwise scattered conscious states as different configurations of the same "internal–external co-evolutionary freezing of a bounded interior" under different conditions. These states must be read along two dimensions: whether the two poles of co-evolution unfold or collapse (determining whether given-ness exists), and whether the exogenous perturbation channel is connected (determining whether given-ness is anchored in the real world). This section examines them one by one, and these examinations also form the phenomenological basis for the measurable predictions of Part IV.

Wakefulness: the two poles of co-evolution are fully unfolded, and the exogenous perturbation channels (the senses) are connected to the real environment. The internal (current experiential configuration) and the external (the rest of the brain background, extending through the body to the world) actively co-evolve; freezing is frequent and content-rich; given-ness is strong and clear, and is anchored in the real world. This is the configuration of "co-evolution unfolded + exogenous channel connected".

Pure thinking and meditation: the two poles of co-evolution are active (internal: the current thought; external: the rest of the brain background), but the exogenous perturbation channels are largely closed, and experience is mainly driven by endogenous perturbation. Co-evolution continues, freezing continues, and given-ness therefore continues and may even become more vivid; only its content is not anchored in the present environment, but supplied by endogenous perturbation. This is the configuration of "co-evolution unfolded + exogenous channel closed, endogenous drive". It directly refutes the misunderstanding that "external must mean environment" (Section 14.3), because the "external" of co-evolution (the brain background) remains present; what is absent is only exogenous perturbation.

Dreaming (REM): similar to pure thinking, the two poles of co-evolution are active within the brain, while exogenous channels are largely closed (sensory input is blocked, motor output inhibited), and experience is driven by endogenous perturbation. Dream experience is therefore vividly given. The difference between dreaming and wakefulness does not lie in whether given-ness exists, but in whether the exogenous channel is connected: dream content is endogenously generated and therefore detached from reality constraints. This is the configuration of "co-evolution unfolded + exogenous channel closed, endogenous self-generation".

Deep anesthesia and dreamless deep sleep: the two poles of co-evolution themselves collapse—coherent configurations are difficult to maintain, cross-regional coupling breaks down, the functional distinction between "internal" and "external" cannot be established, and freezing cannot form stably. Given-ness therefore disappears as a whole. It should be noted that the disappearance of given-ness here is rooted not in the closure of exogenous channels (exogenous closure corresponds to pure thinking and dreaming, where given-ness remains), but in the collapse of the relation between the two poles of co-evolution itself. This is the configuration of "co-evolution collapsed". It is sharply distinct from mere exogenous isolation and constitutes a key testable prediction (Chapters 24 and 25): the existence or disappearance of given-ness depends on the maintenance of the two poles of co-evolution, not on the presence or absence of exogenous input.

Sensory deprivation: only the exogenous perturbation channels are cut off, while the brain's internal co-evolutionary mechanism remains intact. Initially, experience shifts to endogenous perturbation as its driver (as in pure thinking and dreaming), and given-ness is maintained by endogenous content (hallucinations, mind-wandering, inner monologue). But if deprivation is prolonged and thorough, brain-internal co-evolution without exogenous constraints may gradually become unstable, and given-ness may become disordered or partially collapse. This is the dynamic configuration of "exogenous channel cut off, endogenous drive, and gradual destabilization over time". It confirms that given-ness depends on the maintenance of the two poles of co-evolution, not on the exogenous channel itself.

Locked-in syndrome: motor output is almost completely lost, but sensory input and brain-internal co-evolution are largely intact. The two poles of co-evolution continue as usual (internal: experiential configuration; external: brain background, with exogenous channels still connected to the world), freezing continues as usual, and therefore given-ness is complete—the patient is conscious and experientially rich, but cannot express it. This is the configuration of "co-evolution intact + exogenous channel connected, but output channel severed". It shows that given-ness depends on the maintenance of the two poles of co-evolution and is independent of the presence or absence of motor output—a powerful dissociation.

Placed side by side, these six states are not six unrelated phenomena, but configurational variations of the same "internal–external co-evolutionary freezing of a bounded interior" along two dimensions (the survival or collapse of co-evolutionary poles, and the connection or disconnection of exogenous channels). Whether given-ness exists depends on whether the two poles of co-evolution are maintained (anesthesia collapses them and thus eliminates given-ness; the other states retain them and thus retain given-ness). Whether given-ness is anchored in the real world depends on whether the exogenous channel is connected (wakefulness and locked-in syndrome are connected; pure thinking, dreaming, and sensory deprivation are not). This unified explanatory power strongly supports the core thesis of this chapter. The observable proxies and quantitative predictions for each item constitute the content of Chapters 24 and 25.

14.7. The Boundary of This Chapter: What It Answers and What It Does Not Answer

This chapter has provided the physical identical counterpart of givenness and used it to explain conscious states in a unified way. But the boundary must be maintained with the greatest clarity, because the achievement of this chapter is precisely what is most easily exaggerated into something it has not achieved.

What this chapter has answered is: what physical process givenness is—internal–external co-evolutionary freezing of a bounded interior; and how the presence, absence, or strength of given-ness, and whether it is anchored in the world, vary with physical conditions—they vary systematically with the survival or collapse of the two poles of co-evolution and with the connection or disconnection of exogenous channels. Both are claims about third-person-accessible physical processes; they are measurable and falsifiable (Part IV will turn them into predictions). They constitute the substantive contribution of this book to given-ness, and also fulfill the first-level claim of "mind–matter unity" on the side of given-ness: mind (given experience) and matter (the ρ_f internal–external co-evolutionary freezing of a bounded interior) are here claimed to be two sides of one and the same process.

What this chapter has not answered, and explicitly does not pretend to answer, is this: why is this internal–external co-evolutionary freezing experienced from the first-person interior, whereas the freezing of an instrument (even if one imagined it to have some internal/external structure) is purely third-person and has no inner side? Even after we say that "given-ness is internal–external co-evolutionary freezing of a bounded interior", the final question can still be asked: "Once boundary, interior/exterior, co-evolution, and freezing—all these structural and relational facts—are given, why is the fact that all of this is experienced from within not logically entailed by them?" This question cannot be answered by this chapter, and this chapter does not pretend to answer it. It is precisely the final cell identified in Chapter 1 and compressed to the utmost by boundary and co-evolution; Chapter 15 will analyze it directly and hand it over to the philosophical boundary (Chapter 27).

The prohibition of the whole book must therefore be stated again here, because this is the chapter where it is most likely to be violated: the mechanistic statement that "given-ness is internal–external co-evolutionary freezing" must not by itself be used as the answer to the final cell. The mechanistic statement answers "which process given-ness corresponds to" (the measurable side), while the final cell asks "why this process is experienced" (the irreducible residual point). This chapter provides the former in full and leaves the latter fully and honestly open. This opening is not a weakness of the chapter, but a mark of its honesty: it pushes given-ness into physics as far as physics can take it, and clearly indicates that the next cell can no longer be taken up by physical language. To obscurely declare that this cell has also been "solved" would repeat the error of topic substitution—and that is exactly what this book rejects from beginning to end.

14.8. Chapter Summary

This chapter gives the core thesis of given-ness: given-ness is the transient freezing achieved by the continuous internal–external co-evolution between the interior of a bounded system and its exterior within the same ρ_f field. Its argument proceeds as follows. First, it distinguishes two orthogonal layers of "internal/external": the internal/external distinction of perturbation sources (whether experience is triggered internally or externally, Chapter 11) and the internal/external distinction of the two poles of co-evolution (the basis on which given-ness occurs, the topic of this chapter), and establishes terminological discipline to prevent ambiguity. Second, taking up the conclusion of Chapter 12 (given-ness is not an intrinsic property; the darkness of the isolated interior), it states positively that given-ness is the co-evolution of the internal and the external poles, and clarifies that the "external" first refers to the remaining configurations of the brain itself rather than the environment; therefore, given-ness continues as usual in pure thinking, meditation, and dreaming (exogenous channels are closed, but the external pole of co-evolution remains present). Third, the determinacy of given content is carried by transient freezing (focus corresponds to freezing, flow corresponds to sequences of freezing, and non-arbitrariness corresponds to objective constraint). Fourth, boundary is the gate: the complete condition of given-ness is "internal–external co-evolutionary freezing of a bounded interior". Instruments have no interior, so their freezing is ownerless and not given; given-ness is anchored specifically in life, preserving the anti-panpsychist position. The chapter also clarifies the nesting of three layers of internal/external (membrane / co-evolutionary poles / perturbation sources). Fifth, it uses six conscious states as a unified explanation of the same thesis along two dimensions (survival or collapse of co-evolution, connection or disconnection of exogenous channels). Sixth, it strictly preserves its boundary: this chapter answers "what process given-ness is, and how it varies with conditions" (measurable), but it does not answer, and does not pretend to answer, "why this process is experienced" (the final cell, Chapter 15).

The failure conditions of this chapter are as follows (summarized in Appendix E).

- F14.1 The relational approach does not hold. If given-ness is shown to be an intrinsic property of an internal configuration, with no systematic relation to any co-evolutionary relation (for example, if an interior that is completely disconnected from all exteriority can be constructed and yet remains given), then the relational approach fails, and the thesis of this chapter must be abandoned.

- F14.2 The two layers of internal/external cannot be separated. If the "internal/external distinction of perturbation sources" and the "internal/external distinction of the two poles of co-evolution" are shown to be the same matter and cannot be separated orthogonally, then the clarification in Section 14.2 fails, and the argument of the whole chapter becomes ambiguous.

- F14.3 The characterization of the external pole fails. If the "external" of co-evolution cannot be consistently characterized as "the remaining configurations of the brain itself (which may extend through the body to the world)"—for example, if there exists a given state with neither exogenous coupling nor any brain-internal co-evolutionary background—then the characterization of the "external" fails, and the resolution in Section 14.3 becomes invalid.

- F14.4 Freezing does not correspond to given content. If the determinacy of given content, the focus–background distinction, and the structure of flow cannot systematically correspond to the formation, non-formation, and succession of transient freezing, then the freezing element fails.

- F14.5 The boundary gate fails. If decisive evidence shows that an unbounded (non-living) system possesses first-person given-ness (rather than merely objective freezing available to external reading), then the boundary gate fails, the anchoring of given-ness collapses, and the anti-panpsychist position collapses with it.
- F14.6 The explanation of states is overturned by counterexample. If there exists a conscious state whose presence, absence, or strength of given-ness cannot be consistently explained by the two-dimensional configuration (survival/collapse of co-evolution, connection/disconnection of exogenous channels), then the unified explanatory power is weakened, and the configurational framework must be revised.
- F14.7 The final cell is counterfeited. If the mechanistic statement that "given-ness is internal–external co-evolutionary freezing" is used by itself as the answer to "why this process is experienced", then it violates the prohibition of the whole book and constitutes topic substitution.

Table 14.1.. The Boundary between What This Chapter Establishes and What It Cannot Derive

What this chapter establishes	What this chapter cannot derive
The core thesis of given-ness: internal–external co-evolutionary transient freezing of a bounded interior.	It cannot answer "why this freezing is experienced from the first person" (the final cell, Chapter 15).
The clarification and terminological discipline of two orthogonal layers of "internal/external" (perturbation sources / co-evolutionary poles).	It cannot provide a quantitative model of internal–external co-evolution in this chapter (this belongs to Appendix C and Part IV).

What this chapter establishes	What this chapter cannot derive
The "external" first refers to the remaining brain configurations; the nesting of three layers of internal/external (membrane / co-evolutionary poles / perturbation sources).	It cannot provide observable proxies and thresholds of freezing in this chapter (these belong to Chapters 23, 24, and 25).
Boundary as the gate: given-ness is anchored specifically in life, preserving the anti-panpsychist position.	It cannot define in this chapter the precise location of the boundary of the "I" (this belongs to Chapter 16).
Six conscious states as a unified explanation in terms of two-dimensional configurations.	It must not use the mechanistic statement to masquerade as an answer to the final cell (the prohibition of the whole book).

15. The Irreducible Residual Point of Given-ness

15.1. Where We Have Arrived: An Honest Inventory

By this chapter, this book has advanced quite far in its physical treatment of consciousness. It is therefore worth first taking a sober inventory, so as to see exactly where we have arrived—because only by seeing the path already taken can the final step that cannot be taken be accurately identified.

Part II fulfilled one task: the entire structure of experience is carried relation by relation by the fivefold richness of ρ_f gradients. Why experience has continuously gradated qualities, mutually heterogeneous qualitative types, a flowing stream of consciousness, a present field that is both unified and differentiated, and memory layers of different durations—all these structural "why" questions found counterparts on the gradient side. With respect to the structure of experience, the explanatory gap has been closed: if one asks "why experience has this structure", the answer is "because it is identical with this kind of gradient configuration".

Part III (especially Chapter 14) fulfilled another task: the givenness of experience—that it is experienced and manifests to someone—has been characterized as internal–external co-evolutionary freezing of a bounded interior, anchored in the boundary of life. What physical process given-ness corresponds to, and how the presence, absence, or strength of given-ness varies with conditions, have both been answered in a measurable and falsifiable way. With respect to the physical counterpart and law of variation of given-ness, the explanatory gap has also been closed: if one asks "what given-ness corresponds to, and when it appears or disappears", the answer is "it corresponds to internal–external co-evolutionary freezing, and varies with its configuration".

Putting these two achievements together, one may say that almost everything in the hard problem that can be taken up by physical language has now been taken up. The structure of experience has an identity, and the process of given-ness has a characterization. What remains is no longer a diffuse fog, but a single problem with a determinate location, forced into a corner. The task of this chapter is to bring this problem—what this book calls the "final cell"—fully into view, to state exactly what it is, why it is irreducible, why none of the previous achievements can fill it, and why acknowledging it is a strength rather than a weakness of this book.

A forward-looking clarification about this chapter's place in the whole book is needed at the outset, so that the reader does not mistake the residual point presented here for the book's final location of that residual point. What this chapter accomplishes is the first compression of the final cell: after the solvable structure and process components of the hard problem have been separated out, the residual point is compressed and located as "why the given-ness of a bounded interior is experienced", which this book calls the given-ness residual point. But Part III is not yet complete: the subsequent part of this section of the book (Chapters 16 to 20) will show that the given-ness of human experience is a given-ness that is cognized (Chapters 17 and 18) and demarcated by $\mathscr{B}$ as "mine" (Chapter 19). On this basis, Chapter 20 will make a second compression of this residual point, further compressing and locating it from the "general given-ness residual point" to the subject-attribution residual point. Thus the given-ness residual point established in this chapter is real and stands in its own right, but it is not the endpoint—it will be further reduced and more precisely located in the later part of this section. This chapter first makes the first compression solid and thorough; the possibility of the second compression depends on this chapter.

15.2. What Exactly the Final Cell Is

To state the final cell clearly already requires great precision, because it is easily misstated as a problem that has in fact already been solved. This book states it as follows:

> Even if all the structural and relational facts of a bounded interior—boundary, the internal–external distinction, internal–external co-evolution, transient freezing, together with all the gradient structures described in Part II—are fully given, the fact that "all of this is experienced from the first-person interior" is still not logically entailed by those facts.

Each part of this sentence must be held firmly. It does not say that "we still do not know what process given-ness corresponds to"—we already know that (Chapter 14). It does not say that "we still cannot measure internal–external co-evolution precisely"—that is a technical matter, an improvable temporary limitation, not a principled obstacle. It says something deeper: even if all the physical facts of that process—where the boundary is, how the internal and external co-evolve, when freezing occurs, what configuration is frozen, and what its structure is—are placed before us completely and without omission, the conclusion that "it is experienced" still does not follow from that full set of facts. Between "possessing all the structures and relations required for being experienced" and "actually being experienced", there remains a gap that is not entailed by the former.

This gap can be brought into view through a thought experiment, which is the precise version of Chalmers's "philosophical zombie" argument within the framework of this book. Imagine a system that is physically identical to a conscious living system: it has a membrane, a boundary, and an interior; within it occur gradient resonances fully isomorphic to the structure of real experience; its internal–external co-evolution and transient freezing are, in every physical detail, exactly the same as those of a conscious being. Now ask: must this system be experienced? The honest answer of this book is: from all the physical facts given, we cannot logically prove that it must be experienced—because the first-person fact of "being experienced" does not appear in any of the third-person physical vocabulary we use to describe it. We may say that its co-evolutionary freezing functionally constitutes a represented state for its interior; but between "being functionally represented" and "being phenomenally experienced" lies precisely the gap. This thought experiment need not show that zombies are genuinely possible (this book does not claim that zombies are metaphysically possible; see Section 15.4), but it precisely exposes the following point: being-experienced is not logically entailed by structural facts. This is the final cell.

Its two meanings—"final" and "cell"—must be emphasized again. It is "final" because everything before it—structure, process, and laws of variation—has been filled. It is a "cell" because it has been compressed as far as it can be: it no longer contains "why experience has structure" (already answered), "what given-ness corresponds to" (already answered), or "when given-ness occurs" (already answered). What remains is only the single question: "why is that which possesses all the structures and relations experienced?" Part of the dissatisfaction with earlier drafts of this project arose precisely because this cell had not been isolated with such precision: it was either mixed into the solvable parts and passed over ambiguously, or inflated into the whole hard problem and thereby made the problem seem hopeless. Precisely isolating it as the final cell is the primary contribution of this chapter.

A final note about the name and status of this cell is needed in order to connect with the forward-looking statement in Section 15.1. The cell presented in this chapter is the given-ness residual point obtained by the first compression: it locates the residual point at "why the given-ness of a bounded interior is experienced". But as Section 15.1 stated, this is not the endpoint. The given-ness of human experience is, more specifically, a given-ness that is cognized and demarcated by $\mathscr{B}$ as "mine". Chapter 20 will therefore use this chapter as its basis to perform a second compression of the residual point, further compressing and locating it from the "general given-ness residual point" to the "subject-attribution residual point" (why the activity of being integrated, cognized, and demarcated as "mine" is experienced). It must be noted that the second compression does not eliminate the residual point established here, but makes it smaller and more precisely located. The irreducibility diagnosed in this chapter—that being-experienced is not entailed by structure and relation—will remain after the second compression (what will then not be entailed is the more specific "being-experienced of subject-attribution activity"). This chapter first diagnoses that irreducibility thoroughly; it holds for general given-ness and will also hold for the subject-attribution residual point after the second compression.

15.3. Why None of the Previous Achievements Can Fill This Cell

A natural expectation is this: since this book has already identified experiential structure with gradient structure and characterized given-ness as internal–external co-evolution, could the sufficiently refined combination of these two items "produce" being-experienced? This section must argue that it cannot—and that this "cannot" is principled, not a matter of insufficient refinement. This argument is crucial because it is the theoretical basis of the C3 prohibition running through the whole book.

First, can the identity claim of Part II fill this cell? Part II claims that experiential structure and gradient structure are identical. But note the scope of what this identity asserts: it asserts identity of structure—the structure of experience is the structure of gradients. Identity is symmetrical and strict, but it strictly covers only what it asserts, namely structure. From "experiential structure = gradient structure", it does not follow that "gradient structure is experienced", just as from "the morning star = the evening star" it does not follow that "Venus is worshiped". Identity transfers the property that is being identified (here, structure), not a property not included in the identity (here, being-experienced). Part II separated structure from given-ness from the beginning (Chapter 5) precisely because it foresaw this point: however complete structural identity may be, it does not entail given-ness, because given-ness was never included in that structural identity.

Second, can the co-evolutionary characterization of Part III fill this cell? Chapter 14 characterizes given-ness as internal–external co-evolutionary freezing. But note the nature of this characterization: it is a third-person, objectively repeatable process description—how the internal and the external couple, when freezing occurs, and what is frozen can in principle all be described and repeated by an external observer (the detection volume defines objectivity precisely as "repeatable co-perturbation"). Being-experienced, however, is first-personal. A complete third-person process description, however refined, still describes "how this process objectively operates", not "what this process is experienced as from within". Chapter 14 honestly marks this point: it answers "what process given-ness is", not "why that process is experienced". To describe internal–external co-evolution in ever greater detail only adds more third-person detail, while the final cell requires a crossing from the third person to the first person. Increasing refinement always remains on the same side of the gap.

It should also be noted here that the later part of this section of the book (Chapters 17 to 20)—which further characterizes given-ness through cognization, comparison, and $\mathscr{B}$-demarcation—also remains on this side of the gap. How knowledge organizes sensation, how perturbation is compared with memory-like knowledge transients, and how activity is demarcated as "mine" are still characterizations of structure and process (third-person and objectively describable). They can therefore make the residual point smaller and locate it more precisely (the second compression in Chapter 20), but they likewise cannot fill the gap. After the second compression, the gap will remain, only at a smaller location. This chapter states this in advance, and Chapter 20 will preserve it explicitly.

Thus one may state the matter generally: the two major achievements of this book—structural identity and co-evolutionary characterization—as well as the later characterization by cognization and demarcation, all live on the same side of the gap (the structural side, the third-person side). However much and however finely one adds things on the same side, one does not thereby automatically cross to the other side (the side of being-experienced, the first-person side). This is not because we have done too little, but because the two sides of the gap are heterogeneous: one side is structure and process, while the other is being-experienced, and the latter is not logically entailed by the former. This is why the book establishes an iron rule: the mechanistic statement that "given-ness is internal–external co-evolutionary freezing" must not by itself be used as an answer to "why this process is experienced". This rule is not a gesture of humility, but a direct consequence of the logic just stated. Mechanistic statements live on the objective side of the gap; in principle, they cannot answer a question that belongs to the first-person side. Any theory that claims to have filled the final cell with a mechanistic statement is secretly and unnoticedly jumping from one side of the gap to the other, without any logical bridge. That jump is topic substitution, and this book rejects it from the outset.

15.4. This Book's Posture toward the Final Cell: Locating, Not Crossing or Dissolving

In the face of an irreducible final cell, a theory may adopt several postures. This book must choose one explicitly and explain why.

One posture is forced crossing: to declare that some mechanism (co-evolution, integration, broadcasting) simply is being-experienced and that the gap has been closed. This is the topic substitution rejected in Section 15.3, and this book rejects it. Another posture is dissolution into illusion: to declare that "being-experienced" is a cognitive illusion and that there is no genuine phenomenal consciousness to be explained (eliminativism or illusionism). This book does not adopt that posture either, because given-ness is the starting point and the most certain datum of the entire discussion; to declare it an illusion is to saw off the branch on which one sits. Moreover, "why this stubborn illusion of 'being-experienced' exists" itself becomes a difficult problem, and no explanatory work has been saved. A third posture is dogmatic stipulation: to set as an axiom that "all configurations of this kind are experienced", without further explanation. This may be internally consistent, but it defines the hard problem away rather than solving it. If the stipulation is too broad (all freezing is experienced), it destroys the anti-panpsychist gate that Chapter 14 carefully secured through boundary.

This book adopts a fourth posture: precise location, an acknowledgment that the cell cannot be crossed, and neither forced filling, nor dissolution, nor dogmatic stipulation. What this book does is to force the irreducible being-experienced into a precise location: it is not in structure (structure has already been identified), not in process (process has already been characterized), but precisely in the single cell of "why all the structures and processes of a bounded interior are experienced". It also argues why this cell cannot be crossed: it lies between the structural/third-person side and the being-experienced/first-person side; the two are heterogeneous, and the former does not entail the latter. This book gives no answer to this cell; it gives it a precise coordinate and a diagnosis of why no answer is available there. This coordinate will be further refined in Chapter 20 (the second compression into the subject-attribution residual point). The more precise the location, the more careful the book's honest treatment of the residual point becomes, but the answer is still not given.

This posture must be distinguished from cheap agnosticism. Cheap agnosticism says, "consciousness is mysterious; we will never understand it." That pushes the entire hard problem wholesale into darkness and says almost nothing. The posture of this book is the opposite: it understands everything in the hard problem that can be understood (structure, process, and laws of variation), and leaves as uncrossable only the one cell that truly is not entailed by physics, while being able to state clearly why this cell cannot be crossed and where the boundary lies between it and what has already been explained. The former is an admission of defeat; the latter is precise demarcation. To demarcate clearly "how far my explanation reaches, and why it stops exactly here" is itself a substantive epistemological achievement, not a failure of explanation.

It is worth adding that handing this final cell over to philosophy (its destination is Chapter 27, though before that it must undergo the second compression in Chapter 20 and be located at the more precise point of subject-attribution residuality) does not mean that it is necessarily unsolvable even in philosophy. It means only that it has moved beyond the range that physical language can take up. Chapter 27 will show that this cell corresponds to the phenomenological tradition's insight into the irreducibility of "given-ness": phenomenology has long maintained that given-ness is the starting point of all things and cannot be derived from anything more basic. Through a completely independent, bottom-up physical path, this book reaches the same irreducible point. This convergence is meaningful: it confirms, from two directions, the location of the same gap.

15.5. Why Acknowledging This Cell Is a Strength, Not a Weakness

A final concern must be addressed directly: if a book titled *Mind–Matter Unity*, and aiming to unify mind and matter, ends by acknowledging that there is one cell it cannot cross, does this not amount to admitting failure? This section argues the opposite: acknowledging this cell is the source of the book's credibility and the mark that distinguishes it from all overpromising theories.

First, it defines the true boundary of the achievement and thereby makes the achievement credible. A theory that claims to have "fully solved the mystery of consciousness" becomes less trustworthy precisely because it promises too much—the hardness of the hard problem is real, and any easy claim to have crossed it has likely substituted the topic somewhere. This book proceeds in the opposite way: it states precisely what it has solved (structure and process), what it has not solved (the final cell), and why the latter cannot be solved. A theory capable of drawing such precise boundaries makes the parts it does claim to solve more credible, because it proves that it can distinguish between "explained" and "unexplained" and will not disguise the latter as the former.

Second, it transforms the hard problem from something "unapproachable" into something of which "only one cell remains", and this is itself a major advance. Before this book, the hard problem was a diffuse fog: why experience has structure, why it is given, why it is unified, why it is experienced—all were entangled into an indivisible whole. This book has clarified that fog item by item, leaving only the single irreducible cell that "being-experienced is not entailed by structure". To compress a problem that seemed impossible to approach into a single point with a precise location and clear nature is a substantive epistemological victory, even if that point itself remains uncrossable. The later part of this section of the book will further show that this cell can be compressed once more (the second compression in Chapter 20 to the subject-attribution residual point), so the soluble range advances even beyond the inventory of this chapter. This further confirms that "compression" is a substantive result.

Third, it makes the claim of "mind–matter unity" more solid rather than weaker. The "unity" claimed by this book is fulfilled in the identity of structure and in the co-evolution of given-ness: mind and matter are identical in structure, and, in given-ness, they are two sides of the same process. These are substantive unifying claims (Chapter 26 discusses this in detail). The existence of the final cell does not deny these unities. It only marks that even if mind and matter are united so deeply, "why this unity is experienced from within" remains an independent fact not entailed by the structure of that unity. Acknowledging this makes "unity" a bounded and disciplined claim, not a slogan that devours everything and therefore truly explains nothing. A theory of unity that knows its boundary is closer to the truth than a theory of unity that claims to include everything.

The weakness of the earlier drafts of this book can now be diagnosed in general terms: the more ambitious drafts concealed this final cell with a phrase, and therefore looked as if they were cheating; the more cautious drafts, because they obeyed discipline, set aside the cell together with the solvable parts, and therefore looked as if they had conceded defeat. The posture established in this chapter is a third path: solve all that can be solved, to the point where it cannot be solved further (and the later part of this section will push the soluble range one layer further); locate what cannot be solved, to the point where it cannot be made smaller (and Chapter 20 will locate it even more precisely); and keep the boundary between the two clear. This path neither cheats nor admits defeat. It is the only defensible posture for an honest theory of unity.

15.6. Chapter Summary

This chapter directly analyzes the final cell after the hard problem has been compressed to the utmost, and, as the first compression of the residual point, lays the basis for the second compression in the later part of this section (Chapter 20). First, it takes inventory of the path already traveled: the structure of experience has been identified by Part II, the process of given-ness has been characterized by Part III, and all that can be taken up physically in the hard problem has been taken up; it also anticipates that the later part of this section will compress the residual point once more through cognization and $\mathscr{B}$-demarcation. Second, it precisely states the final cell (the given-ness residual point of the first compression): even if all the structural and relational facts of a bounded interior are fully given, the fact that "all this is experienced" is not logically entailed by them. It brings out this gap through a zombie thought experiment within the framework of this book, emphasizes the two senses of "final" (everything before it has been filled) and "cell" (it can no longer be made smaller), and explains that Chapter 20 will perform a second compression into the subject-attribution residual point (making it smaller, not eliminating it). Third, it argues that the previous achievements cannot in principle fill this cell: structural identity transfers structure, not being-experienced; co-evolutionary characterization is third-person and does not cross into the first person; the later characterizations through cognization and demarcation remain on the objective side of the gap; however much they are refined and added together, they do not cross the boundary. This is the logical basis of the C3 prohibition. Fourth, it establishes this book's posture: precise location, not forced filling, not dissolution, and not dogmatic stipulation; and it distinguishes this posture from cheap agnosticism. Fifth, it argues that acknowledging this cell strengthens rather than weakens the book's argument: it defines the real boundary of the achievement, compresses the hard problem from something unapproachable into a single remaining cell (and further compressible by one more cell), and makes "mind–matter unity" a disciplined and solid claim.

The failure conditions of this chapter are as follows (summarized in Appendix E).

- F15.1 The final cell is in fact fillable. If a valid derivation can be given, without appealing to any first-person premise and starting purely from structural and process facts, to infer that "this configuration is experienced", then the final cell is filled, and the core claim of this book concerning its irreducibility fails (this is a failure the book would welcome, because it would mean that the hard problem has truly been solved).

- F15.2 The location is imprecise. If the final cell still contains components that can in fact be explained physically (that is, if this book mistakenly includes a solvable problem within the irreducible residual point), then the location is imprecise; those components must be returned to the soluble side, and the residual point must be redefined (the second compression in Chapter 20 precisely further removes soluble components such as "cognization" and "demarcation" that had previously been mistakenly included in the residual point).

- F15.3 The demarcation is internally inconsistent. If the heterogeneity between the "structural/third-person side" and the "being-experienced/first-person side" is shown not to hold (for example, if the two can be continuously bridged by some mediating concept), then the diagnosis of the gap fails, and this chapter's argument for "why it cannot be crossed" must be redone.

Table 15.1.. The Boundary between What This Chapter Establishes and What It Cannot Derive

What this chapter establishes	What this chapter cannot derive
The precise statement of the final cell (the given-ness residual point of the first compression): being-experienced is not logically entailed by structural and relational facts.	It cannot, and does not attempt to, provide an answer to the final cell (it is explicitly handed over to philosophy, Chapter 27).
The argument that the previous achievements cannot in principle fill this cell (the logical basis of the C3 prohibition).	It cannot be used to deny the achievements of Parts II and III (structural identity and co-evolutionary characterization still hold).
This book's posture: precise location, not forced filling, not dissolution, not dogmatic stipulation; the residual point will undergo a second compression in Chapter 20 (made smaller, not eliminated).	It cannot assert that this cell is necessarily unsolvable even within philosophy (it asserts only that it exceeds physical language).
Acknowledging this cell strengthens rather than weakens the argument of the whole book.	It must not use any mechanistic statement to masquerade as an answer to this cell (the prohibition of the whole book).

16. Subjective Boundary and the "I"

16.1. A Long-Suspended Question: Where Do I End?

The preceding two chapters have respectively established that the unity of experience is the wholeness of a single coherent resonance (Chapter 9), and that the given-ness of experience is internal–external co-evolutionary freezing of a bounded interior (Chapter 14). Taken together, these two claims already point implicitly to a question that has not yet been answered directly: where, exactly, is the boundary of this unified and given "I"? Where do I end, and where does what is no longer me begin?

At first sight, this question seems to have an obvious answer: I am my body, or at least my brain; what lies inside the skull is me, and what lies outside it is not. But on closer examination, both ready-made answers immediately encounter difficulties. If I am the whole body, then a troublesome fact stands before us: I do not experience my liver metabolizing, my bone marrow producing blood, or my gut microbiota acting. These are undoubtedly events occurring within "my body", yet they are entirely outside the experience of "I". If I am the brain as an anatomical organ, that is not fully correct either: within the brain there is also a great deal of neural activity that never enters consciousness (the respiratory rhythm regulation of the brainstem, the motor fine-tuning of the cerebellum). These activities are inside the skull and within the brain, yet they are likewise not experienced by "me". Thus the boundary of the "I" is neither the skin nor the skull—it does not coincide with any ready-made anatomical structure. What, then, is it?

This chapter maintains that the boundary of the "I" is the boundary of the internal–external co-evolution of the single dominant coherent resonance that carries first-person experience—it is dynamically demarcated, not fixed by any anatomical structure. Whatever enters this dominant coherent resonance and participates in its internal–external co-evolutionary freezing belongs to the "I"; whatever does not enter it—whether or not it is inside the skull or inside the body—does not belong to the "I". This chapter will first state this characterization of the boundary directly (Sections 16.2 and 16.3), then test it through a series of boundary anomalies (split brain, dissociation, and conjoined cases) (Sections 16.4 and 16.5), and finally clarify its relation to the body and to those micro-subjects that are "non-I" (Section 16.6).

A forward-looking division of labor between this chapter and the later part of Part III must first be stated, so that the reader does not confuse this chapter with Chapter 19. What this chapter establishes is the ontology and physics of the boundary of the "I"—what the "I" physically is (the dominant coherent resonance), where it ends (the extension of coherence), and how it changes under physical conditions (splitting, collapse, reorganization). This is the question of the existence and boundary of the "I". Chapter 19, in the later part of Part III, will address another layer that is closely related but not identical: how the manifold, given, and cognized activities are demarcated by $\mathscr{B}$ as "mine"—this is the question of the attribution of the "I" (how my body, my world, my knowledge, and my judgments are unified as belonging to this I). This chapter concerns "what the I is and where its boundary lies" (the ontological side), while Chapter 19 concerns "how contents are demarcated as mine" (the attributional side). The two are not repetitive, but two sides of the same coherent boundary: viewed ontologically, the same dominant coherence is the boundary of the "I" (this chapter); viewed attributionally, it is the demarcation of $\mathscr{B}$ (Chapter 19). This chapter first stabilizes the ontology of this boundary, so that Chapter 19 can then develop the demarcation of attribution upon it.

16.2. The "I" Is the Dominant Coherent Resonance in the Brain's Multicellular Space

Chapter 9 has already argued that the "I" is not the sum of cellular micro-subjects, but a single coherent resonance mode constituted by gradient oscillations in the multicellular medium of the brain. This chapter pushes that claim one step further: within the brain, not all neural activity belongs to this same coherent resonance. The "I" is the dominant coherent resonance among them—the one that has the largest scale, the highest degree of coherence, and that either incorporates other activities into itself or excludes them from itself.

Why "dominant"? Because many oscillatory activities occur simultaneously in the brain, but only some of them can phase-lock with one another and enter the same highly coherent whole mode, thereby constituting a unified field of experience. Others either fail to match in frequency or lack sufficient coupling; they cannot be incorporated into this coherence and therefore remain outside it. The rhythms by which the brainstem regulates respiration and the activity by which the cerebellum corrects movement belong precisely to the latter category. They operate continuously inside the skull and within the nervous system, but because they cannot be incorporated into the dominant coherent resonance, they do not enter the experience of the "I". They are neighbors and background of the coherent resonance called the "I", not its components. Thus the boundary of the "I" is not drawn by the skull, but by coherence itself: whatever can lock into the dominant coherence is inside; whatever cannot lock into it is outside. This boundary runs through the interior of the brain, separating activities that enter consciousness from those that do not, even though the two classes of activity are often anatomically interwoven and have no clear spatial interface.

This characterization immediately explains the puzzle raised in Section 16.1. I do not experience my liver not because the liver is outside "my body" (it is of course inside the body), but because the activity of the liver does not participate at all in the brain's dominant coherent resonance—there is insufficient coherent coupling between it and the "I", and therefore it falls outside the boundary of the "I". The same applies to why I do not experience the respiratory rhythm of the brainstem: it is inside the skull, but not inside the dominant coherence. The boundary of the "I" is the boundary of coherent coupling, not the boundary of matter or anatomy. This is the core claim of this chapter, and the next section will explain why it is also the boundary of given-ness.

16.3. The Boundary Is the Boundary of Internal–External Co-Evolution: The Convergence of Two Layers of Internal/External

Chapter 14 distinguished two layers of "internal/external": the material internal/external distinction drawn by the membrane (the distinction between life and environment, spatial in kind), and the functional internal/external distinction on which given-ness depends (the distinction between the coherent configuration and its dynamic background, relational in kind). The boundary of the "I" discussed in this chapter is precisely where these two layers converge at the level of the subject. This convergence must be clarified, so that the two layers of internal/external do not remain suspended and disconnected.

Recall the claim of Chapter 14: given-ness is internal–external co-evolutionary freezing of a bounded interior, where the "internal" is the coherent configuration currently given, and the "external" is the remaining configuration that co-evolves with it (first and foremost the rest of the brain itself). Now specify the "internal" as the dominant coherent resonance of this chapter—the "I". Thus: the "I", as the internal, continuously co-evolves with the rest of the brain background (and the world extended through the body), as the external, and freezes out given experience. The boundary of the "I" is therefore both the boundary of coherence and the boundary of the "internal" in this internal–external co-evolution. The two are the same line: whatever can lock into the dominant coherence constitutes the "internal" that undergoes internal–external co-evolution; whatever cannot lock into it either falls into the "external" that co-evolves with it (the brain background), or falls outside the co-evolution altogether (the liver and bone marrow, which neither cohere with the dominant coherence nor constitute the external on which its given-ness depends).

The two layers of internal/external therefore converge here without conflict. The material internal/external distinction drawn by the membrane is the precondition of everything: without the boundary of life, there would be no "interior" in which the dominant coherent resonance could reside. But the boundary of the "I" is not identical with the membrane boundary (not identical with the "whole body"). Within the living interior demarcated by the membrane, the dominant coherent resonance draws a more contracted boundary in a functional manner, excluding most of the body (together with part of the brain) from the "I". Thus the boundary of the "I" is nested within the boundary of life and is usually much smaller than it: the boundary of life encloses the whole body, while the boundary of the "I" encloses only the dominant coherent resonance. This nesting relation explains a deep fact of ordinary life: I have a body, but I am not that whole body. I am a dominant coherent resonance nourished and carried by that body; most of the body is "mine" with respect to me, but is not "me".

A distinction that will be fully developed in Chapter 19 should be marked here in advance: when we say that most of the body is "mine" but not "me", the "I" (the dominant coherence itself) is defined by this chapter, while the "mine" (the attribution of body, world, knowledge, and so forth to this dominant coherence) is defined by the $\mathscr{B}$-demarcation mechanism in Chapter 19. This chapter need only point out the difference between the two (I $\neq$ mine). How the "mine" is demarcated one by one at the levels of body, world, knowledge, and judgment will not be developed here, and is left to Chapter 19.

16.4. The Divisibility of the Boundary: Split Brain as a Critical Test

The strictest test for a theory of the boundary of the "I" is how it handles cases in which the boundary becomes anomalous. If the boundary of the "I" is truly the boundary of coherent co-evolution, then when coherence is physically severed, the boundary should split accordingly—one "I" should become two. The split-brain phenomenon provides precisely this test.

Split-brain surgery severs the corpus callosum connecting the two hemispheres in order to prevent epileptic spread. After surgery, in carefully designed experiments, the two hemispheres can display striking independence: information presented to the left visual field (projecting to the right hemisphere) can guide the patient's left hand (controlled by the right hemisphere), while the patient's verbal system (usually in the left hemisphere) reports that it has "seen nothing"—as if there were two mutually unaware cognitive subjects inside the skull. This phenomenon is troublesome for many theories of consciousness, but for this book it is a direct confirmation.

According to this book, the corpus callosum is the main physical bridge that maintains oscillatory phase locking (that is, coherence) between the two hemispheres. When the corpus callosum is intact, oscillations in the two hemispheres can phase-lock across hemispheres and be incorporated into the same dominant coherent resonance; there is therefore only one "I", whose boundary spans the two hemispheres. Once the corpus callosum is severed, cross-hemispheric phase locking is greatly lost, the formerly unified coherent resonance can no longer be maintained as one, and it splits into two separately coherent resonances. Each hemisphere can still maintain its own coherence internally, but the two are no longer coherent as one. According to "the boundary is the boundary of coherence", this should appear as follows: one "I" splits into two subjects, each with its own closed boundary. The experimentally observed independence of the two hemispheres is precisely the behavioral manifestation of this split. Split brain is therefore not an objection to this book, but a critical confirmation of the claim that "boundary is the boundary of coherent co-evolution": it shows that when coherence is physically manipulated (by severing the bridge of phase locking), the boundary of the subject is indeed manipulated (one becomes two).

It should also be noted that in normal circumstances split-brain patients usually behave like a single subject (the split is exposed only in specially designed experiments), and this can also be explained: in ordinary contexts, the two hemispheres still maintain a considerable degree of functional coordination through subcortical structures, a shared sensory environment, and behavioral coordination, so that the two coherent systems are usually synchronized and difficult to distinguish. Only when experiments cut off all such external coordination and force the two hemispheres to receive conflicting information separately does the latent split of coherence become visible as an observable dual subject. This is consistent with this book: boundaries are determined by coherence, and the strength of coherence and degree of coordination determine the extent to which a split can be detected.

16.5. Collapse, Reorganization, and Coexistence of Boundaries: Dissociation and Conjoined Cases

Split brain shows the splitting of boundaries. Other boundary anomalies show the collapse, reorganization, and coexistence of boundaries. This book explains them one by one in terms of the "boundary of coherence", as further tests.

Dissociation: in dissociative disorders, experience that should be unified develops fissures—certain experiential contents, memories, or personalities seem to be separated from the rest of the subject, as if they "do not belong to me". According to this book, this corresponds to a local rupture of coherence within the dominant coherent resonance. Certain configurations that should have locked into the dominant coherence fail to do so for dynamic reasons (often related to intense emotion or trauma), or form secondary coherences that coexist with the dominant coherence without being coherent with it. The phenomenological fissure of dissociation corresponds to a fissure in coherent structure; its degree, from mild "absorption" to severe identity alternation, corresponds to the depth and stability of coherent rupture. Dissociation is therefore not "a foreign object placed inside one I", but "a failure of the dominant coherence to lock together, as one whole, everything that should belong to it". An interface with Chapter 19 should be noted here: in this chapter, dissociation is explained as a local rupture of the coherent boundary (the ontological side), while its phenomenological fissure of attribution—that "certain contents do not belong to me"—corresponds exactly to the failure of $\mathscr{B}$-demarcation in Chapter 19 (the attributional side). The same dissociation, viewed ontologically, is coherent rupture (this chapter); viewed attributionally, it is demarcation failure (Chapter 19). The two are not repetitive, but mutually confirming sides of the same anomaly.

Conjoined twins and rare cases of cranial connection: in very rare cases of conjoined twins with partial cranial connection, some degree of sensory sharing or experiential overlap has reportedly been observed. For many theories that take "one brain, one mind" as their default, this is anomalous. For this book, it is another test: if there is a physical connection between two brains sufficient to support cross-brain phase locking, then in principle a partially coherent resonance spanning the two brains may form—neither two completely independent "I"s, nor one fully fused "I", but an intermediate state that is partially coherent and partially independent. This book does not prejudge the concrete form of such cases (that depends on the actual neural connection), but its framework allows and predicts the existence of such intermediate states: the boundary of the subject is determined by the actual extension of coherence; wherever coherence extends, the "I" extends—even if it crosses what is normally regarded as the boundary between two individuals.

Together with split brain, these two types of phenomena support a unified picture: the boundary of the subject is not something fixed in advance by anatomy or individual identity, but is dynamically demarcated by the actual extension of coherent co-evolution. It can be one (normal), split into two (split brain), locally ruptured (dissociation), or partially extended across two brains (conjoined cases). Each form is the manifestation of the claim that "the boundary is the boundary of coherence" under different physical conditions. A theory capable of accommodating and predicting these forms in a unified way is more credible than one that must treat them as separate exceptions.

16.6. The Body, the World, and "Non-I" Micro-Subjects: What Lies Outside the Boundary

Finally, what lies outside the boundary of the "I" must be clarified, because this directly concerns how the book's anti-panpsychist position (Chapter 3) and its claim about cellular micro-subjects (Chapter 6) can be compatible with the "I" without conflict.

Outside the boundary lies, first, the rest of the body. According to this book, the vast majority of bodily cells—liver cells, muscle cells, immune cells—each has its membrane and its minimal interior, and is therefore a minimal micro-subject (Chapters 3 and 6). But they lack sufficient coherent coupling with the brain's dominant coherent resonance, and therefore fall outside the boundary of the "I". This means a conclusion that may appear strange at first but is in fact coherent: outside the one dominant subject that is the "I", there are many "non-I" micro-subjects distributed within "my" body. They each have their own faint interior, but they are not experienced by "me" and do not constitute "me". This is the precise fulfillment of Chapter 1's claim that "life is a distributed system, while the brain is an integrated system": the body is a distributed system of micro-subjects, and the "I" is the single integrated subject synthesized by the brain's dominant coherence. The "I" sits in the middle of a community of micro-subjects, but becomes coherent as one only with those that can lock into the dominant coherence.

It must be emphasized that this picture does not violate anti-panpsychism. What anti-panpsychism denies is that non-living (unbounded) systems have interiority and experience; it never denies that living systems (bounded cells) have weak interiors. A liver cell has an interior because it has a membrane and a boundary (the gate of Chapter 3 applies to it); a rock has no interior because it has no boundary. The micro-subjects outside the boundary of the "I" are all bounded living units, and therefore their having interiors creates no conflict with anti-panpsychism. Conflict would arise only if one asserted that unbounded things have interiors, and this book never makes that claim.

Outside the boundary lies, second, the world. Through the body's sensorimotor channels, the dominant coherence that is the "I" continuously co-evolves with the external environment (the "connection to the world" of Chapter 14). The world is not a component of the "I", but the extended part of the "external" in the internal–external co-evolution of the "I". It participates in shaping the contents given to the "I" (what I see and hear), but it is not locked into the coherence of the "I", and therefore does not belong to the "I". My relation to the world is an internal–external co-evolutionary relation, not a relation in which the internal contains the external. (The distinction between the ontological side and the attributional side should be marked once more: the world, as "external", does not belong to the "I" [the ontological side, this chapter], but through the "world-oriented demarcation" of $\mathscr{B}$, the world is given as "the world I face" [the attributional side, Chapter 19]. That the world does not belong to the "I" and yet is demarcated as "my world" is not contradictory; it is precisely one example of the claim in Section 16.3 that I $\neq$ mine.)

Thus the place of the "I" in this book is fully settled: it is a single given subject constituted by the dominant coherent resonance in the brain's multicellular space, inside a bounded living system (the body). Its boundary is dynamically demarcated by the actual extension of coherent co-evolution, nested within the material boundary of life and usually much smaller than it. Outside it lie both the micro-subjects in "my" body that are not locked in, and the world that co-evolves with it through sensorimotor channels. This settlement unifies the coherence of Chapter 9, the internal–external co-evolution of Chapter 14, and the boundary and anti-panpsychism of Chapter 3 at the single point of the "I", bringing to a close the first part of Part III (the ontology of given-ness and subjectivity). The "I" whose boundary has been stabilized in this chapter will become, in the later part of Part III (Chapters 17 to 20), the subject around which cognization, comparison, attributional demarcation, and the second compression are organized.

16.7. Preserving the Boundary

In accordance with the discipline of the whole book, this chapter must also preserve its boundary. What this chapter characterizes is what physically determines the boundary of the "I" (the extension of dominant coherent co-evolution), and how this boundary changes with physical conditions (splitting, collapse, reorganization, and co-existence). These are measurable and falsifiable claims (their neural proxies and predictions appear in Chapters 23 and 24). This chapter does not answer, and does not pretend to answer, why this "I" defined by coherent boundary is experienced from the first-person interior—that remains the final cell identified in Chapter 15. This chapter settles the "I" as a coherent resonance with a determinate boundary that can be studied, but it says nothing about why "this coherent resonance has an inner side" or why it is "someone experienced". Nor should it. To characterize the boundary of the "I" clearly and to answer "why this I is experienced" are two different matters; this chapter does only the former.

The division of labor between this chapter and Chapter 19 must also be marked again as part of this boundary: this chapter establishes the ontology of the boundary of the "I" (the boundary of coherent co-evolution), but does not develop how the "mine" of attribution is demarcated by $\mathscr{B}$ (the fivefold demarcation of subjective core, body, world, knowledge, and judgment). That belongs to Chapter 19. This chapter only establishes that I $\neq$ mine and indicates that the two are two sides of the same coherent boundary (the ontological side and the attributional side). The full development of the attributional side is left to Chapter 19. In this way, this chapter and Chapter 19 each take one side, continuing one another without repetition.

16.8. Chapter Summary

This chapter brings the first part of Part III (the ontology of given-ness and subjectivity) to a close by settling the "I" at a determinate point. First, the boundary of the "I" is neither the skin nor the skull, and does not coincide with any anatomical structure (I do not experience the liver, nor the brainstem rhythm). Second, the "I" is the dominant coherent resonance in the brain's multicellular space: whatever can lock into the dominant coherence belongs to me, while what cannot lock into it (unconscious activities of the brainstem and cerebellum) lies outside me; the boundary is drawn by coherent coupling, not anatomy. Third, this boundary is precisely the boundary of the "internal" in internal–external co-evolution, the point where the material internal/external distinction and the functional internal/external distinction converge. The boundary of the "I" is nested within the boundary of life and is usually much smaller than it; therefore "I have a body, but I am not that whole body". The chapter also marks the distinction between "I" (ontology, this chapter) and "mine" (attribution, Chapter 19). Fourth, the chapter tests this through boundary anomalies: split brain as the splitting of coherence (one I becomes two), dissociation as local rupture of coherence (whose attributional fissure corresponds to the failure of $\mathscr{B}$-demarcation in Chapter 19), and conjoined cases as partial cross-brain coherence. The boundary is dynamically demarcated by the actual extension of coherence. Fifth, it clarifies what lies outside the boundary: the rest of the body's cells are "non-I" micro-subjects (they have interiors because they are bounded, so this does not violate anti-panpsychism, but they are not locked into the dominant coherence), and the world is the extended "external" in co-evolution (yet demarcated by $\mathscr{B}$ as "my world").

Sixth, it preserves its boundary: this chapter characterizes what determines the boundary of the "I" and how it changes (measurable), but does not answer "why this I is experienced" (the final cell, Chapter 15), and does not develop the demarcation of the "mine" (Chapter 19).

The failure conditions of this chapter are as follows (summarized in Appendix E).

- F16.1 Boundary and coherence do not correspond. If the boundary of the "I" is shown to have no systematic relation to the extension of the dominant coherent resonance (for example, if coherence remains intact while the subject splits, or coherence splits while the subject remains unchanged), then the claim that "the boundary is the boundary of coherence" fails.
- F16.2 Split brain does not support the splitting prediction. If, when coherence (cross-hemispheric phase locking) is severed, the subjective boundary does not split accordingly (with neither behavioral nor measurable signs of dual subjectivity), then split brain as a critical test falsifies the claim of this chapter.
- F16.3 The two layers of internal/external cannot converge. If the nesting relation between the functional boundary of the "I" and the material boundary of the membrane is internally inconsistent (for example, if there appears a dominant coherent subject without a living material boundary), then the convergence of the two layers of internal/external fails, and the connection between this chapter and Chapter 14 breaks down.
- F16.4 The claim of non-I micro-subjects violates anti-panpsychism. If the inference that "there are non-I micro-subjects in the body" is shown in fact to require acknowledging interiors in unbounded things, then it conflicts with the anti-panpsychist position of Chapter 3 and must be abandoned.

Table 16.1.. The Boundary between What This Chapter Establishes and What It Cannot Derive

What this chapter establishes	What this chapter cannot derive
The "I" is the dominant coherent resonance in the brain's multicellular space, and its boundary is drawn by coherence rather than anatomy.	It cannot answer "why this dominant coherent resonance is experienced" (the final cell, Chapter 15).
The boundary of the "I" is the boundary of the "internal" in internal–external co-evolution, nested within the boundary of life and usually smaller than it; I $\neq$ mine.	It cannot develop in this chapter how the attribution of "mine" is demarcated by $\mathscr{B}$ (this belongs to Chapter 19).
Split brain, dissociation, and conjoined cases as unified tests of boundary splitting/rupture/crossing.	It cannot prejudge the concrete form of rare cases such as conjoined twins (this must depend on actual neural connections).
Outside the boundary lie "non-I" micro-subjects (without violating anti-panpsychism) and the co-evolving world.	It cannot provide in this chapter the neural observable proxies and quantitative predictions of the boundary (Chapters 23 and 24).

17. Language, Symbols, and Experiential Knowledge: The Higher-Order Organization of Memory-Like Gradient Transients

17.1. The Starting Point of a New Line: From Being Given to Being Cognized

The first five chapters of Part III (Chapters 12 to 16) have completed one task: they characterize the given-ness of experience as internal–external co-evolutionary freezing of a bounded interior, and situate the unified "I" as the dominant coherent resonance in the brain's multicellular space. By the end of Chapter 16, we are able to state how experience is given, and where the boundary of the subject to whom it is given (the I) lies.

Yet there remains a level that has not yet been treated, and that is crucial for human experience. When a person opens his eyes and sees the world, what he experiences is not merely a given sensation. He does not see nameless clusters of colors or hear strings of meaningless sounds; he sees "table", "friend", "text", and "danger", and hears "native language", "alarm", and "someone's voice". In other words, what arrives in his experience is not an unorganized density gradient of sensation, but something already organized, recognized, and endowed with meaning by knowledge. Something occurs between a cluster of colors becoming "a familiar face" and a segment of sound waves becoming "a sentence that is understood"—sensation has been cognized.

This "being cognized" is a new layer above given-ness, one not yet touched by the first five chapters of Part III. It opens the new line treated in the later part of this part (Chapters 17 to 20): experience is not only given; it is also cognized (compared with existing experiential knowledge, thereby recognized and endowed with meaning), judged (Chapter 18), and finally demarcated by $\mathscr{B}$ **as "my"** experience (Chapter 19). This line will advance the "given-ness" already established in Part III to the specifically human activity of experience as "cognized and attributed", and on this basis Chapter 20 will compress the residual point of the hard problem one level further.

The nature and limit of this new line must be stated at its starting point (this concerns the C3 prohibition that the book has guarded from the outset, and that must be guarded especially tightly in the later part of this part). What this line advances is the structural and organizational level of experience—how sensation is organized into experience by knowledge, and how experience is attributed as "mine". It does not touch, nor pretend to touch, the irreducible final cell—why the cognized and attributed activity of experience is experienced. With every further step in the later part of this part (knowledge, comparison, demarcation), the articulable structural level is pushed deeper and made more specific, while the cell of "why it is experienced" is left for Chapter 20 to treat with the restraint it deserves. At the beginning of this new line, this book therefore establishes the boundary: every step that follows is a deepening of structure, not an answer to the final cell.

This chapter is the first step of that new line. It treats its most basic material—knowledge: what, within the ρ_f framework, is the experiential knowledge base by which human experience is organized? The answer of this book is: knowledge is the higher-order organization of memory-like gradient transients. This chapter will first connect "memory as gradient transient" with the transient timescale spectrum of the whole book (Section 17.2), then discuss the three features of knowledge as the higher-order organization of memory transients (Sections 17.3 to 17.5), then explain why language and symbols are the transition point of this higher-order organization (Section 17.6), and finally preserve the boundary of this chapter (Section 17.7).

17.2. Memory as Gradient Transient: Connecting with the Timescale Spectrum and the Freezing Scaling Law

To understand knowledge as "the higher-order organization of memory-like gradient transients", one must first secure the premise that "memory is gradient transient"—and this premise has already been prepared by the transient timescale spectrum in Chapter 10. This chapter needs only to connect with it and supplement it with a quantitative scaling law from EQT freezing dynamics.

Chapter 10 established that the retention of experience (from present thickness to long-term memory) is carried by the timescale spectrum of gradient transients—high-frequency transients are short-lived, low-frequency transients are long-lived, and retention duration unfolds continuously along frequency. Memory, then, is not a static item "stored" somewhere, but a stably maintained low-frequency long-duration gradient transient: the retention of a memory is the duration for which the gradient transient carrying it is stably maintained in the brain; the gradual forgetting of memory is the decay and dissipation of that transient (Chapter 10, F10.2).

This picture can be given a quantitative scaling law by EQT freezing dynamics, raising "high frequency short, low frequency long" from a qualitative direction to a computable and testable relation. According to Section 4.3 of *Frozen Cosmology: A Natural Philosophy of Cosmic Evolution in the Perspective of Energy Quantum Theory*, the freezing time of a ρ_f structure follows the scaling relation

$$t_{\text{freeze}}(f) \sim \tau_0(f)/f,$$

where $\tau_0(f)$ is a slowly varying effective background factor—it absorbs corrections from the background, coupling, diffusion, decay, and source terms. It is not an arbitrary free parameter, but an effective quantity that must be slowly varying, calibratable, and testable. Its core tendency is $1/f$: high-frequency structures freeze earlier (short timescale), while low-frequency structures freeze later (long timescale).

This book maintains that memory transients in the brain are directly freezing structures of ρ_f under the brain's background and obey the same scaling law—except that the "background" is no longer cosmic expansion, but the dissipative and competitive background within the brain (neural decay, noise, metabolic constraints, and so forth), absorbed by $\tau_0(f)$. Denoting their retention duration by $\tau_{\text{retain}}(f)$, we have

$$\tau_{\text{retain}}(f) \sim \tau_0(f)/f$$

(freezing under the brain's background).

This is not an analogical transplantation of a cosmological formula, but the same scaling law of the same ρ_f freezing mechanism under different backgrounds: under the cosmic background, it gives the cosmological freezing time of structures; under the brain's background, it gives the retention duration of memory transients. Mind and matter are here once again one: memory retention and the freezing of cosmic structures are manifestations of the same ρ_f freezing mechanism under two different backgrounds (this is one concrete fulfillment, on the problem of memory, of the book's broader ambition of mind–matter unity). From this scaling law: high-frequency transients (τ_{retain} short) carry fleeting and easily vanishing contents (immediate perception, momentary impressions), while low-frequency transients (τ_{retain} long) carry persistent contents—and memory, especially long-term memory, is precisely these low-frequency long-duration transients.

A qualification concerning $\tau_0(f)$ must be stated in accordance with the discipline of the whole book. In the context of consciousness, this book preserves the status of $\tau_0(f)$ as a slowly varying correction factor (consistent with *Frozen Dynamics Spectrum*); it does not treat it as an exact constant, nor does it specifically attach it here to a particular neural frequency band or metabolic rate. It is an effective background quantity that must be slowly varying, calibratable, and testable. What this book claims is the form of the scaling relation ($\tau_{\text{retain}} \sim \tau_0/f$, with $1/f$ dominant and τ_0 as a slowly varying correction) and its direction (high frequency short, low frequency long), not a zero-free-parameter exact law. This restraint preserves the honesty of the whole book in not claiming zero free parameters (Chapter 2). The testable form of the prediction appears in Chapter 25.

17.3. The First Feature of Knowledge: Higher-Order Organization of Memory Transients

Memory is low-frequency long-duration gradient transient—this is the material. But knowledge is not equal to isolated memory transients one by one; knowledge is the higher-order organization of those memory transients. This is the core claim of the chapter, and this section first establishes its general meaning.

An isolated memory transient is a retained gradient configuration—it may correspond to the trace of an experience, a remembered image, or a retained sound. But human knowledge is far more than a pile of such isolated traces. The knowledge of "table" is not a single memory transient of one particular table, but the traces of many experiences involving tables, organized into a structure that can be repeatedly invoked, compared with new experience, and connected with other knowledge. The knowledge of a "native language" is not the memory of a single sentence, but countless experiences of sounds, meanings, and grammar organized into a system capable of generating and understanding endlessly new sentences. Knowledge, then, is the organization of memory transients into a higher-order structure that can be invoked, connected, and generated. It is built on the material of memory transients (low-frequency long-duration transients), but its degree of organization is far higher than that of isolated transients.

The physical counterpart of this "higher-order organization" is clear in the ρ_f framework: it is the connective structure established among low-frequency long-duration transients through same-frequency coherence and cross-frequency resonance (Chapter 9). Isolated memory transients are each retained (low-frequency and long-duration), while knowledge is the organization by which these transients are connected through coherence and resonance into a structure that can be invoked as a whole and can mutually activate itself. To invoke a piece of knowledge (to recall what a "table" is) corresponds to activating the relevant cluster of transients within this connective structure; association among pieces of knowledge (from "table" to "chair" and "study room") corresponds to resonance propagation among neighboring transient clusters in the connective structure. Knowledge as "higher-order organization" is therefore not a metaphor, but has a specific physical counterpart: the coherence–resonance connective structure among memory transients.

From this, the first feature of knowledge can be stated: knowledge is the higher-order organization formed by memory-like gradient transients (low-frequency long-duration transients) through coherence–resonance connections. It takes memory transients as its material, and the connections among transients as its organization. The next two sections discuss two further features of this organization—recoverability/combinability and comparability.

17.4. The Second Feature of Knowledge: Recoverable, Combinable, and Externalizable

As a higher-order organization, knowledge has a first set of functional features: it can be recovered, combined, transmitted, and externalized. This set of features distinguishes knowledge from fleeting present perception and makes it a resource that can be repeatedly used, accumulated, and shared.

Recoverability. As an organization of low-frequency long-duration transients, knowledge can be repeatedly invoked—the same piece of knowledge can be reactivated at different moments, instead of vanishing instantly like a high-frequency short-duration transient. The physical basis of this recoverability is precisely the scaling law in Section 17.2: knowledge is carried by low-frequency long-duration transients; τ_{retain} is long, and therefore it can be maintained for a long time and invoked repeatedly. A person can still recognize a childhood home many years later because the transient organization carrying the knowledge of "home" is low-frequency, long-duration, and has not dissipated over a long interval.

Combinability. Knowledge is not a pile of mutually unrelated items, but can be combined into new structures—existing clusters of knowledge transients can be connected, nested, and reconstructed, generating new knowledge that was not directly experienced before. From the two pieces of knowledge "table" and "metal", one can combine the concept "metal table", which one may not have directly experienced. This combinability corresponds to the reconstructability of the connective structure in the transient organization of knowledge—coherence–resonance connections are not fixed, but can be dynamically reorganized during invocation (this reorganization mechanism is the core of conscious activity in Chapter 18; this chapter first marks its possibility).

Externalizability and transmissibility. This is the feature of knowledge most characteristic of human beings: knowledge can be externalized as language, symbols, writing, and images, thereby leaving the brain of a single individual, being transmitted to others, and being accumulated across generations. The transient organization of knowledge in one person's brain can be externalized through language as transmissible symbols, and reconstructed in another person's brain as a corresponding transient organization of knowledge. This externalizability and transmissibility is the basis for the accumulation of human knowledge and the inheritance of civilization, and is also one of the key features that distinguishes human beings from other forms of life (Chapter 18 will discuss the difference between humans and animals).

This set of features should be connected back to the phenomenological observations of the cellular trilogy in order to make clear its roots. *The Information World of the Cell: Pooling Networks, Continuous-State Computation, and the Architecture of Life Information Processing* already pointed out at the phenomenological level that cells have a dual-state memory system—from rewritable short-term states to long-term solidified stable structures. This book's "high-frequency short-duration transient / low-frequency long-duration transient" is precisely the mechanistic characterization of that dual-state memory within the ρ_f framework: rewritable short-term states correspond to high-frequency short-duration transients, while long-term solidification corresponds to low-frequency long-duration transients. The recoverability, combinability, and externalizability of knowledge are the higher-order forms reached by this memory mechanism in the human brain, a vast integrated system of neural cells, through language and symbols. The earlier work saw the phenomenon that "memory has two states"; this book gives its mechanism through the transient timescale spectrum and the freezing scaling law, and advances it to "knowledge as higher-order organization".

17.5. The Third Feature of Knowledge: Comparability—Toward Conscious Activity

The third feature of knowledge is that it can be compared. It is not only passively retained and invoked, but can actively match, resonate with, conflict with, and revise current perturbations (exogenous perceptions and endogenous thoughts). This feature is the bridge from this chapter to Chapter 18, because it is precisely the mechanistic core of "cognized perception" and "conscious activity".

As an already organized transient structure, knowledge does not stand aside motionless when current experience arrives. When a current exogenous perturbation (a segment of perception) or endogenous perturbation (a thought) enters the brain, it interacts with the existing transient organization of knowledge: if the current perturbation matches a certain knowledge transient (frequency and configuration fitting together), resonance enhancement occurs—recognition accelerates and experience is quickly endowed with meaning ("this is a table"); if the current perturbation does not fit existing knowledge, conflict occurs—attention is heightened, and the system enters a state of searching, judging, explaining, or learning.

This "comparison" is the key that turns knowledge from a static resource into a dynamic participant: knowledge is not merely stored; it actively participates in the organization of current experience. It is this comparison that turns sensation into experience—a segment of sensation that has not undergone comparison is only a density gradient received by a sensory organ; a segment of sensation that has been compared with knowledge and thereby recognized and endowed with meaning becomes what a human being experiences as "seeing a table" or "understanding speech". This insight—that perception without corresponding knowledge is only a density gradient of sensation, and only with the participation of knowledge does it become experience—is the most important proposition of the later part of Part III. This chapter marks its mechanistic basis here (the comparison between knowledge transients and current perturbations), while its full development—conscious activity as the comparison, matching, resonance, conflict, reorganization, and refreezing of exogenous/endogenous perturbations and knowledge transients—is the subject of Chapter 18.

This comparison mechanism must also be connected once more with the perturbation axis (Chapter 8), so that the continuity of the whole book becomes clear. Chapter 8 established that experience is driven by exogenous and endogenous perturbations acting on the evolution of the master equation. The comparison mechanism of this chapter is precisely a refinement of that perturbation-driven process: exogenous/endogenous perturbations drive evolution, while the specific direction of that evolution is constrained and shaped by the existing transient organization of knowledge. Perturbation is compared with knowledge transients; matching leads to resonance enhancement, while conflict leads to reorganization. Thus "comparison with the participation of knowledge" is not a new mechanism, but the specific form of the perturbation-driven evolution of Chapter 8 under the condition that a knowledge base participates. The perturbation axis (drive), the five structural sources (law), and the transient organization of knowledge (the constraint imposed on evolution by existing structure) cooperate here and jointly determine how experience is organized.

17.6. Language and Symbols: The Transition Point of Higher-Order Organization

Knowledge as the higher-order organization of memory transients already has preliminary forms in animals (animals also have learning, memory, and recognition). Yet the reason human knowledge reaches a level far beyond that of animals lies in language and symbols—they are the transition point of this higher-order organization. This section explains why language and symbols produce a qualitative transition in the organization of knowledge.

Animal memory and knowledge are, in large part, tied to direct experience: an animal remembers routes, recognizes food and predators, and forms conditioned reflexes. These are all organizations of memory transients, but they are largely bound to concrete experiences and are difficult to abstract and combine into new structures far removed from direct experience. Language and symbols break this binding: a symbol (a word or a sign) can designate a class of things, an abstract relation, or a possibility not yet experienced. It allows the organization of knowledge transients to transition from "bound to concrete experience" to "capable of abstraction, combination, and reference to what has not been experienced".

In the ρ_f framework, the transitional role of language and symbols can be understood as follows. A symbol is itself a stably maintained and repeatedly invokable memory transient (a low-frequency long-duration transient), but its special feature is that it is organized as a node designating other transient organizations (its referents). The symbol thus becomes a "hub transient" in the organization of knowledge. Through this symbolic hub, transient organizations from many concrete experiences can be abstractly marked, combined into new structures, and externalized for transmission. Language (the systematic organization of symbols) further allows these hub transients to be generated, nested, and inferred according to rules, thereby raising the organization of knowledge to the level at which endlessly new structures can be generated.

The significance of this transition will be developed fully in Chapter 18 (the difference between humans and animals). Here, only its structural basis is established: language and symbols are the hubs and transition points of the transient organization of knowledge—they allow higher-order organization to transition from being bound to concrete experience to becoming abstract, combinable, externalizable, and generative. The reason human experience is experience with a very high degree of "being cognized" (one sees "text", "theory", and "meaning", not merely colors and shapes) is precisely that the human organization of knowledge has reached this post-transition level through language and symbols.

Restraint must again be preserved (consistent with the cellular trilogy). To say that language and symbols are a "transition point" means that they allow the organization of knowledge to reach a higher level of abstraction and generative capacity. This is a claim about organizational complexity; it is not to say that language and symbols "produce" given-ness or being-experienced. Animals have lower-order knowledge organization, while humans have higher-order organization that has undergone the linguistic transition. They are different degrees of complexity within the same kind of mechanism (the higher-order organization of memory transients), not a difference in the presence or absence of given-ness. Whether given-ness is present depends on the internal–external co-evolution of a bounded interior (Chapter 14), not on the level of knowledge organization. A living being with a simple knowledge organization can still have its given-ness; its experience simply has a lower degree of being cognized. To mistake the "higher order of knowledge organization" for the "source of given-ness" would overstep the boundary, and this chapter does not do so.

17.7. The Boundary of This Chapter

This chapter is the first step of the new line in the later part of Part III, establishing "knowledge as the higher-order organization of memory-like gradient transients". Its boundary must be preserved, in accordance with the discipline of the whole book.

What this chapter establishes concerns the structural and organizational level of experience: memory is low-frequency long-duration gradient transient (connected with the scaling law $\tau_{\text{retain}} \sim \tau_0/f$); knowledge is the higher-order organization of memory transients (recoverable, combinable, externalizable, and comparable); and language and symbols are the transition point of this organization. All of this is an articulable structural characterization of "how sensation is organized into experience by knowledge, and how knowledge operates as a higher-order transient structure".

What this chapter does not touch, and does not pretend to touch, is why this cognized and organized activity of experience is experienced. This chapter clarifies the mechanism of "cognization" (the higher-order organization and comparison of transients), but says not one word about "why cognized experience is experienced". That remains the final cell, left for Chapter 20 to treat with restraint. One especially tempting overstep in the later part of Part III must be guarded against: thinking that once "how sensation is organized into experience by knowledge" has been clarified, "why experience is experienced" has also been clarified. The two are separated by the final cell: the former is structure (how knowledge organizes sensation), while the latter is being-experienced (why this organization is felt from within). This chapter does only the former and explicitly does not reach the latter.

Thus the place of this chapter in the whole book is clear: it follows Chapter 16 (the I as dominant coherence), prepares the material for Chapter 18 (cognized perception and conscious activity: knowledge as a comparable higher-order transient organization), and connects through the $\tau(f)$ scaling law with Chapter 10, and through the comparison mechanism with the perturbation axis of Chapter 8. It is the first step of the "second compression line". Next, Chapter 18 will discuss how knowledge participates in conscious activity, Chapter 19 will discuss how $\mathscr{B}$ demarcates all of this as "mine", and Chapter 20 will use these steps to compress the residual point by one more cell. From beginning to end, this line is a deepening of structure, not a crossing of the final cell.

17.8. Chapter Summary

This chapter is the first chapter of the "second compression line" in the later part of Part III, and establishes "knowledge as the higher-order organization of memory-like gradient transients". First, it introduces the new line: experience is not only given, but also cognized, judged, and demarcated by $\mathscr{B}$ as "mine"; it also states that this line is a deepening of structure and does not touch the final cell. Second, it connects with memory as gradient transient: memory is low-frequency long-duration transient, and the chapter cites the scaling law $t_{\text{freeze}}(f) \sim \tau_0(f)/f$ from *Frozen Cosmology: A Natural Philosophy of Cosmic Evolution in the Perspective of Energy Quantum Theory*, maintaining that memory transients in the brain are directly ρ_f freezing under the brain's background and obey $\tau_{\text{retain}}(f) \sim \tau_0(f)/f$ (with τ_0 preserved as a slowly varying correction factor). Memory retention and cosmic freezing are thus manifestations of the same ρ_f mechanism under two backgrounds (another fulfillment of mind–matter unity). Third, it establishes three features of knowledge: higher-order organization formed by memory transients through coherence–resonance connections (first feature); recoverability, combinability, and externalizable transmission (second feature, connected with the dual-state memory in *The Information World of the Cell: Pooling Networks, Continuous-State Computation, and the Architecture of Life Information Processing*); and comparability—matching, resonance, and conflict with exogenous/endogenous perturbations (third feature, leading to Chapter 18 and refining the perturbation axis of Chapter 8). Fourth, it argues that language and symbols are the transition point of higher-order organization: they allow knowledge to transition from being bound to concrete experience to becoming abstract, combinable, externalizable, and generative. This is a claim about organizational complexity, not the source of given-ness (restraint is preserved). Fifth, it preserves the boundary: this chapter establishes the structural mechanism of cognization, but does not reach "why cognized experience is experienced" (the final cell, Chapter 20).

The failure conditions of this chapter are as follows (summarized in Appendix E).

- F17.1 (new) Memory is not gradient transient. If memory is shown in principle to be impossible to understand as low-frequency long-duration gradient transient (that is, if the scaling of $\tau_{\text{retain}} \sim \tau_0/f$ systematically fails for memory in the brain, and retention duration has no systematic inverse relation with carrying frequency), then the premise that "memory is gradient transient" fails, damaging both this chapter and its connection with Chapter 10.

- F17.2 (new) Knowledge is not the higher-order organization of transients. If knowledge is shown to be impossible to understand as the higher-order organization of memory transients through coherence–resonance (for example, if invocation, association, and combination of knowledge have no systematic correspondence with activation, resonance propagation, and reconstruction of transient organizations), then the core claim of this chapter fails.
- F17.3 (new) The comparison mechanism does not hold. If the "comparison between knowledge and current perturbation (matching resonance / conflict reorganization)" is shown to have no systematic correspondence with observable processes of recognition, attention, and learning, then the third feature of knowledge fails, and the bridge to Chapter 18 is broken.
- F17.4 (new) The transition of language and symbols is mistaken for the source of given-ness. If this chapter (or anyone following its logic) uses the "higher-order character of knowledge organization (through language and symbols)" as the source of given-ness or being-experienced, it violates restraint and the C3 prohibition: the level of knowledge organization concerns the degree to which experience is cognized, not whether given-ness is present.

Table 17.1.. The Boundary between What This Chapter Establishes and What It Cannot Derive

What this chapter establishes	What this chapter cannot derive
Memory is low-frequency long-duration gradient transient and obeys $\tau_{\text{retain}} \sim \tau_0/f$ (directly ρ_f freezing in the brain, with τ_0 slowly varying).	It cannot provide in this chapter the exact form of τ_0 or a zero-free-parameter law (τ_0 is a slowly varying calibratable factor).
Knowledge is the higher-order organization of memory transients through coherence–resonance connections (recoverable, combinable, externalizable, and comparable).	It cannot treat in this chapter why cognized experience is experienced (the final cell, Chapter 20).
Language and symbols are the transition point of higher-order organization (abstraction, combination, externalization, generation).	It cannot derive the source of given-ness from the "higher-order character of knowledge organization" (given-ness depends on internal–external co-evolution, Chapter 14).
The comparison mechanism (matching resonance and conflict reorganization between knowledge and exogenous/endogenous perturbations) is a refinement of the perturbation axis of Chapter 8.	It cannot develop in this chapter the full mechanism of conscious activity (Chapter 18) or the demarcation of "mine" (Chapter 19).

18. Experiential Knowledge Base and Conscious Activity: The Comparison Mechanism of Exogenous Perturbations, Endogenous Perturbations, and Knowledge Transients

18.1. The Task of This Chapter: From "What Knowledge Is" to "How Knowledge Participates in Experience"

Chapter 17 established what knowledge is—the higher-order organization of memory-like gradient transients, recoverable, combinable, externalizable, and comparable. This chapter advances one step further and treats what knowledge does—how this higher-order organization participates in the formation of current experience, turning sensation into experience and consciousness into a kind of activity.

The pivot of this advance is the proposition marked at the end of Chapter 17 and fully developed in this chapter. It is the most important insight of the later part of this part:

Perception without corresponding knowledge is merely the density gradient received by the sensory organs (frequency, intensity, threshold, direction, and timescale); perception with the participation of knowledge becomes experience.

This proposition repositions "experience" from "given sensation" to "cognized perception". What a person receives upon opening his eyes is not nameless clusters of colors or meaningless strings of sound waves—those are only density gradients received by the sensory organs. What he experiences is "table", "friend", "native language", and "danger": a meaningful world formed after these density gradients are compared with, recognized by, and endowed with meaning through the existing knowledge base. The difference between sensation and experience therefore does not lie in the strength of sensation, but in whether knowledge participates: sensation is a density gradient that has not been cognized, whereas experience is perception that has been cognized.

The task of this chapter is to articulate this "cognization" as a clear mechanism—conscious activity as the activity of comparison among exogenous perturbations, endogenous perturbations, and memory-like gradient transients in the experiential knowledge base: matching, resonance, conflict, reorganization, and refreezing. This chapter will first establish the proper status of the experiential knowledge base in the ρ_f framework (Section 18.2), then characterize the core mechanism of conscious activity—comparison (Section 18.3), then discuss conscious activity initiated respectively by exogenous and endogenous perturbations (Section 18.4), then analyze the several outcomes of comparison (matching, conflict, reorganization, and refreezing; Section 18.5), and finally preserve the boundary of this chapter (Section 18.6).

The boundary consistently observed in the later part of this part must be restated from the outset (as established in Chapter 17): what this chapter clarifies is the mechanism of "cognization"—how sensation becomes experience through comparison with knowledge. It does not touch "why cognized experience is experienced"—that remains the final cell and is left to Chapter 20. Clarifying "how sensation is organized by knowledge into meaningful experience" and answering "why this meaningful experience is felt from within" are separated by the final cell; this chapter does only the former.

18.2. The Experiential Knowledge Base: A Distributed Transient Organization, Not a Substantive Region

To explain how conscious activity invokes knowledge, one must first place the "experiential knowledge base" properly within the ρ_f framework—especially avoiding the mistaken view that it is some specialized "knowledge warehouse" region in the brain.

The experiential knowledge base is the totality of knowledge about oneself, others, and the world that an individual establishes through all his experience and life history. In the ρ_f framework, it is not a substantive region in the brain dedicated to storing knowledge (such as some "memory center"), but the higher-order organization of low-frequency long-duration memory transients distributed across the entire dominant coherent system (Chapter 17). Knowledge is not stored centrally in one place, but is distributedly embodied in a large number of low-frequency long-duration transients and their coherence–resonance connective structures. This is consistent with the distributed position of the book from the beginning (Chapter 1: "life is a distributed system, while the brain is an integrated system", and the distributed consciousness of the cellular trilogy).

Understanding the experiential knowledge base as a distributed transient organization rather than a substantive region has two important meanings. First (consistent with neuroscience): contemporary neuroscience does not support the view that "knowledge is centrally stored in a single brain region", but supports the view that knowledge is distributedly embodied in broad neural connections and activity patterns. This book's "distributed organization of low-frequency long-duration transients" accords with that picture rather than positing a suspicious "knowledge center". Second (consistent with this book's distributed line): since knowledge is a distributed transient organization, invoking and comparing knowledge is not a matter of "retrieving items from a warehouse", but of activating and resonating relevant transient clusters within a distributed coherent system. This is consistent with Chapter 16's "I as dominant coherence" and Chapter 9's "unity carried by coherence". The experiential knowledge base is therefore the broad field of low-frequency long-duration transients within the dominant coherent system, organized into invokable structures—the totality of knowledge carried and invokable by the coherent resonance called the "I".

A restraint must be preserved here (consistent with the cellular trilogy): this book does not describe the experiential knowledge base as a sharply bounded, functionally specialized "region" or "organ". It is a distributed and dynamic transient organization whose boundary is dynamically demarcated by the extension of coherence (Chapter 16), and whose content is continuously accumulated, updated, and reorganized through experience. Reifying it as a fixed region would be inconsistent with both neuroscience and this book's distributed line; this chapter avoids that reification.

18.3. The Core Mechanism of Conscious Activity: Comparison

Since the experiential knowledge base is a distributed, invokable, and comparable transient organization, the core mechanism of conscious activity can be summarized in one word: comparison—the interaction between the current perturbation (exogenous perception or endogenous thought) and the relevant knowledge transients in the experiential knowledge base. This section characterizes that core mechanism.

When a perturbation enters the dominant coherent system (a segment of perception or a thought), it is not given in isolation, but immediately interacts with the existing organization of knowledge transients. This interaction is what this book calls comparison. The physical counterpart of comparison is the resonance relation between the gradient configuration excited by the current perturbation and existing knowledge transients (low-frequency long-duration transients) (Chapter 9, resonance kernel A4):

If there is matching, resonance is enhanced. If the configuration of the current perturbation matches that of a knowledge transient (frequency and gradient structure match), the two undergo resonance enhancement: the knowledge transient is activated by the current perturbation, while the current perturbation is recognized and endowed with meaning by the knowledge transient. This resonance enhancement corresponds to "recognition" in experience: a visual density gradient resonates with the knowledge transient "table" and is therefore experienced as "seeing a table"; a segment of sound waves resonates with the knowledge transient "native language" and is therefore experienced as "understood speech". The speed of recognition (recognizing a familiar person at a glance) corresponds to the rapid enhancement of resonance (strong matching).

If there is no matching, conflict occurs. If the current perturbation does not fit existing knowledge transients (there is no match, or it conflicts with an expected knowledge transient), conflict occurs: resonance fails to form, or destructive interference appears. This conflict corresponds to "strangeness", "surprise", and "incomprehension" in experience: an object never seen before, a sentence one cannot understand, or a scene that violates expectations all correspond to a current perturbation that cannot find a matching knowledge transient or conflicts with an expected transient. The consequence of conflict is important: it raises attention and drives the system into a state of searching, judging, explaining, or learning (Section 18.5). Conflict is not the failure of comparison, but the engine of higher-order conscious activity (thinking and learning).

This comparison mechanism is precisely the continuation and refinement of the perturbation axis in Chapter 8 and the end of Chapter 17. Chapter 8 established that perturbation drives the evolution of the master equation. This chapter refines that claim as follows: the specific direction of perturbation-driven evolution is shaped by comparison between the perturbation and existing knowledge transients. Matching quickly evolves along a path of resonance enhancement into recognition, whereas conflict evolves into heightened attention and search-reorganization. Conscious activity is therefore not the free evolution of perturbation in an empty brain, but evolution of perturbation under the constraints of an existing organization of knowledge transients. The knowledge base is the "terrain" of evolution, and perturbation is the water flowing across it; the terrain (existing knowledge) shapes the direction of the water (current experience). This picture of the "knowledge base as evolutionary terrain" is key to understanding why human experience is highly cognized: the knowledge terrain of the human brain is extremely rich (higher-order organized through language and symbols, Chapter 17), and therefore every segment of perception flowing across it is deeply shaped by this rich terrain into meaningful experience rather than nameless sensation.

18.4. Two Kinds of Initiation: Conscious Activity Initiated by Exogenous and Endogenous Perturbations

Since conscious activity is comparison between perturbation and knowledge transients, there are two ways in which conscious activity is initiated, according to the source of perturbation (Chapter 8, the perturbation axis: exogenous/endogenous). This section discusses them separately in order to show the full scope of conscious activity—it includes both perception of the world and thinking apart from the world.

Conscious activity initiated by exogenous perturbation: cognized perception. When conscious activity is initiated by exogenous perturbation (the environment entering through the senses), it appears as cognized perception of the world: external changes in density gradients (light, sound, touch) enter through the sensory organs, are compared with the experiential knowledge base, are recognized and endowed with meaning, and become meaningful experience about the world. This is precisely what the proposition in Section 18.1 characterizes: external stimuli do not directly become a "meaningful world", but become experience only after comparison with the knowledge base. Seeing, hearing, and touching, whenever they are caused by external objects, are all forms of cognized perception initiated by exogenous perturbation. The degree to which such experience is "cognized" depends on the richness of the knowledge participating in comparison. A domain expert looking at a professional chart (with rich relevant knowledge participating) and a layperson looking at the same chart (with impoverished relevant knowledge participating) have markedly different experiences: the same exogenous perturbation is cognized into different experiences because the knowledge bases compared with it differ.

Conscious activity initiated by endogenous perturbation: thinking, remembering, and imagining. Conscious activity need not be initiated by external objects. It may also be initiated by endogenous perturbation (Chapter 8). Remembering, imagining, reasoning, planning, and even aimless wandering thoughts are all conscious activities initiated by endogenous perturbation. What enters is not an external object; rather, existing knowledge transients within the brain perturb and compare with one another: one knowledge transient activates adjacent transients, producing association; a group of knowledge transients compare with one another, producing reasoning; knowledge transients are reorganized, producing imagination. Conscious activity initiated by endogenous perturbation is conscious activity "apart from the present world"—it does not depend on sensory input at the moment, but unfolds within the knowledge base. Human beings can think with their eyes closed, conceive things in imagination, and relive the past in memory precisely because conscious activity can be initiated by endogenous perturbation within the knowledge base, without external objects.

The key commonality of the two kinds of initiation (connected with Chapter 8) is that they share the same comparison mechanism and the same knowledge base. In exogenous initiation, an external perturbation is compared with the knowledge base; in endogenous initiation, knowledge transients compare with one another (the endogenous perturbation itself is an activation of knowledge transients). The mechanism is the same, differing only in the source of perturbation (external/internal). This is fully consistent with Chapter 8's thesis that "exogenous/endogenous perturbations share the same evolutionary law", and is its specific form under the condition that the knowledge base participates. The full scope of conscious activity is thereby unified: whether initiated by external objects or by internal thoughts, conscious activity is comparison between perturbations and knowledge transients within the dominant coherent system. Perception of the world and thinking apart from the world are two kinds of initiation of the same mechanism.

The significance of "thinking need not come from exogenous perturbation" should be stated explicitly. It refutes a naive view that consciousness is always "a response to the external world". This book shows that a substantial portion of conscious activity, and indeed the most characteristic human forms of it (reasoning, imagining, reflection), are initiated by endogenous perturbations and unfold within the knowledge base, without depending on present external objects at all. This is consistent with Chapter 14's claim that "given-ness continues as usual in pure thinking and meditation" (exogenous channels are closed while co-evolution is maintained): pure thinking is possible and vivid precisely because conscious activity can be initiated by endogenous perturbation within the knowledge base, while its givenness depends on the maintenance of internal–external co-evolution (Chapter 14), not on exogenous input. Chapter 14 established this from the angle of "given-ness"; this chapter establishes it again from the angle of "the initiation of conscious activity". The two chapters mutually confirm one another at this point.

18.5. The Outcomes of Comparison: Matching, Conflict, Reorganization, and Refreezing

Once comparison occurs, it has several outcomes, corresponding to several directions of conscious activity. This section characterizes these outcomes, with particular emphasis on the path of "conflict–reorganization–refreezing", because it is the basis of learning, understanding, and creativity.

Matching and recognition (resonance enhancement). This is the most common outcome: the current perturbation matches a knowledge transient, resonance is enhanced, and experience is quickly recognized and endowed with meaning. This is smooth cognization—a familiar object is recognized at a glance, and familiar speech is understood immediately. Its experiential character is fluency, absence of obstruction, and lack of effort.

Conflict and heightened attention. The current perturbation conflicts with existing knowledge (no match or violation of expectation), resonance fails to form, and the system enters a state of heightened attention. This corresponds to surprise, confusion, and alertness in experience. Conflict is a signal: current experience exceeds the organizational capacity of existing knowledge, and more resources must be mobilized to process it.

Search and reorganization. After conflict, the system enters search: it mobilizes more knowledge transients to compare with the current perturbation and tries different matches. If no existing knowledge transient organization matches, reorganization occurs: existing knowledge transients are reconnected and reconstructed to accommodate the current experience. This reorganization corresponds to "understanding", "insight", and "learning" in experience: a phenomenon that was initially incomprehensible becomes understood through reorganization of knowledge; a new concept is mastered through the reorganization of existing knowledge. Reorganization is the core mechanism of learning. Learning is not the addition of items into a warehouse, but the reconstruction of the transient organization of knowledge so that it can accommodate and organize new experience.

Refreezing: the stabilization of new knowledge. If reorganization stabilizes, refreezing occurs: the reorganized transient organization is stabilized as a new low-frequency long-duration transient and becomes new content in the knowledge base. This refreezing is precisely the manifestation of the ρ_f freezing mechanism (Section 17.2) at the level of knowledge: a transient organization stabilized through reorganization is frozen into a low-frequency long-duration transient (τ_{retain} long), and from then on becomes recoverable and invokable new knowledge. To learn something new is to stabilize a new low-frequency long-duration transient organization in the knowledge base through "conflict–search–reorganization–refreezing". Forgetting something newly learned is the decay and dissipation of that new transient because it has not undergone sufficient refreezing (remaining in a relatively high-frequency short-duration band) (Chapter 10, F10.2).

This cycle of "matching–conflict–search–reorganization–refreezing" characterizes the dynamic scope of conscious activity: when things proceed smoothly, there is matching and recognition; when blocked, there is conflict; after blockage, there is search, reorganization, and refreezing into learning. Human consciousness can constantly learn, understand, and create because this cycle continuously reorganizes and refreezes the knowledge base through new experience. The knowledge base is not fixed, but grows through the continuous reorganization and refreezing of conscious activity. This must again be connected with the perturbation axis and freezing scaling law to show the continuity of the whole book: initiation is perturbation (Chapter 8), comparison is constrained by the knowledge terrain (Section 18.3), the stabilization of reorganization is refreezing (Section 17.2), and new knowledge is low-frequency long-duration transient (the timescale spectrum of Chapter 10). The later part of Part III and the earlier parts of the book are unified in this cycle.

18.6. The Boundary of This Chapter: Clarifying Cognization, Not Being-Experienced

This chapter must preserve the boundary consistently observed in the later part of Part III, and because this chapter comes closest to "seeming to explain experience", that boundary must be guarded especially tightly.

What this chapter clarifies is the mechanism of cognization: how sensation (density gradients received by the sensory organs) becomes experience (meaningful, recognized, and endowed with meaning as cognized perception) through comparison with the experiential knowledge base (matching, resonance, conflict, reorganization, and refreezing). This mechanism answers "how sensation becomes experience"—that is, why the structure and organization of experience are so rich and so full of meaning. It is articulable, comparable with neuroscience, and in principle testable (its measurable predictions appear in Chapter 25).

But this chapter does not answer, nor does it pretend to answer, why this cognized experience is experienced. This boundary must be drawn with utmost clarity because this chapter can easily tempt one into overstepping. This chapter explains "why what is seen is 'a table' rather than nameless color blocks" (the result of knowledge comparison)—this is "why the content of experience is cognized". But the question "why this cognized experience (seeing a table) is experienced, felt from within, rather than the same comparison and recognition being completed without experience" cannot be answered by this chapter, and this chapter does not pretend to answer it. Whether a system that performs the same comparison (matching input with a knowledge base, recognition, endowment of meaning) is accompanied by being-experienced is not determined by the comparison mechanism itself (as established in Chapters 14 and 15: the being-experienced of given-ness is not entailed by mechanism). This chapter clarifies the comparison mechanism (the structure of cognization), not the being-experienced of this mechanism (the final cell).

One particularly dangerous overstep in the later part of Part III must be guarded against: thinking that once "experience is cognized perception and the result of comparison" has been clarified, "why experience is experienced" has also been clarified. These two are separated by the final cell: the former concerns how the content of experience is organized by knowledge (the structural level), while the latter concerns why this organized experience is felt from within (being-experienced). Mistaking "experience is cognized perception" for "the being-experienced of experience has been explained" is precisely the topic substitution strictly prohibited by C3. This chapter fully clarifies the mechanism of cognization and fully leaves open "why it is experienced", to be treated with restraint in Chapter 20. What this chapter achieves is to advance experience from "given sensation" to "cognized, comparison-organized experiential activity". This is a deepening at the structural level, an approach toward the final cell (pushing the residual point deeper), not a crossing of the final cell.

18.7. Conclusion of This Chapter

This chapter advances knowledge, as established in Chapter 17 (the higher-order organization of memory transients), into the mechanism of conscious activity. Its main points are as follows. First, it establishes the core proposition: perception without corresponding knowledge is merely a density gradient received by the sensory organs, while perception with the participation of knowledge becomes experience (the difference between sensation and experience lies in whether knowledge participates). Second, it properly situates the experiential knowledge base: it is the higher-order organization of low-frequency long-duration memory transients distributed in the dominant coherent system, not a substantive region in the brain (consistent with neuroscience and this book's distributed line). Third, it establishes the core mechanism of conscious activity—comparison: the resonance relation between the current perturbation and knowledge transients. Matching produces resonance enhancement (recognition and endowment of meaning), while conflict heightens attention (driving thinking and learning). This is a refinement of the perturbation axis of Chapter 8 (the knowledge base is the terrain of evolution, shaping the direction of perturbation-driven evolution). Fourth, it distinguishes two kinds of initiation: exogenous initiation as cognized perception (the same exogenous perturbation becomes different experiences because the knowledge bases compared with it differ), and endogenous initiation as thinking, remembering, and imagining (unfolding inside the knowledge base and independent of external objects). The two share the same comparison mechanism and knowledge base, confirming Chapter 14's account of the given-ness of pure thinking. Fifth, it characterizes the cycle of comparison outcomes: matching and recognition–conflict–search–reorganization–refreezing (learning is the reorganization of the transient organization of knowledge and its refreezing into new low-frequency long-duration transients), thereby connecting the perturbation axis, freezing mechanism, and timescale spectrum. Sixth, it preserves the boundary: this chapter clarifies the mechanism of cognization (how sensation becomes experience), but does not reach "why cognized experience is experienced" (the final cell, Chapter 20).

The failure conditions of this chapter are as follows (summarized in Appendix E).

- F18.1 (new) The cognization proposition does not hold. If "experience = cognized perception" is falsified—that is, if it is shown that sensation is already complete experience without the participation of knowledge (recognition and endowment of meaning do not depend on comparison with existing knowledge)—then the core proposition of this chapter fails.
- F18.2 (new) The experiential knowledge base is in fact a sub-

stantive region. If the experiential knowledge base is shown to have to be understood as a specialized and sharply bounded substantive storage region in the brain, rather than as a distributed transient organization, then this chapter's characterization of its status fails (and also conflicts with this book's distributed line).

- F18.3 (new) The comparison mechanism has no counterpart. If conscious activities (recognition, surprise, search, learning) have no systematic correspondence with "comparison between perturbations and knowledge transients (resonance enhancement / conflict / reorganization / refreezing)", then the core mechanism fails.

- F18.4 (new) The two kinds of initiation cannot be unified. If exogenous initiation (perception) and endogenous initiation (thinking) are shown to require two different mechanisms, rather than being two kinds of initiation of the same comparison mechanism, then this chapter's unification of conscious activity fails (linked with the perturbation axis of Chapter 8).

- F18.5 (new) Cognization is made to masquerade as an explanation of being-experienced. If this chapter (or anyone following its logic) uses "experience is cognized perception / the result of comparison" as the answer to "why experience is experienced", it violates the C3 prohibition and constitutes topic substitution (the failure against which this chapter must guard most carefully).

Table 18.1.. The Boundary between What This Chapter Establishes and What It Cannot Derive

What this chapter establishes	What this chapter cannot derive
Core proposition: perception without the participation of knowledge is merely density gradient; only with the participation of knowledge does it become experience.	It cannot derive why experience is experienced from "experience is cognized perception" (the final cell, Chapter 20).
The experiential knowledge base is the higher-order organization of distributed low-frequency long-duration transients, not a substantive region.	It cannot locate the knowledge base in this chapter as a fixed brain region (which would violate the distributed line and neuroscience).
The core mechanism of conscious activity: comparison between perturbations and knowledge transients (matching resonance / conflict / reorganization / refreezing), as a refinement of the perturbation axis of Chapter 8.	It cannot be decided by the comparison mechanism itself whether it is accompanied by being-experienced (being-experienced is not entailed by mechanism).
The two kinds of initiation (cognized perception through exogenous perturbation, and thinking/imagination through endogenous perturbation) share the same mechanism, confirming Chapter 14's account of the given-ness of pure thinking.	It cannot demarcate in this chapter why these activities are attributed as "mine" (this belongs to the $\mathscr{B}$ demarcation of Chapter 19).
Learning is the reorganization and refreezing of the transient organization of knowledge (connecting the perturbation axis, freezing mechanism, and timescale spectrum).	It cannot provide in this chapter the measurable proxies and predictions of cognization (these belong to Chapter 25).

19. Subjective Boundary and the Formation of "Mine": The Demarcation Mechanism of $\mathscr{B}$

19.1. The Task of This Chapter: From "Experience" to "My Experience"

Chapters 17 and 18 advanced experience from "given sensation" to "cognized experiential activity"—sensation becomes meaningful, recognized, and endowed experience through comparison with the experiential knowledge base (matching, resonance, conflict, reorganization, and refreezing). Yet one final step remains, and this step is crucial for human experience: why are those cognized experiential activities—this perception, that judgment, this bodily feeling, that world-orientation—not scattered events occurring separately, but uniformly attributed to one and the same "I"? What I see is "what I see"; what I think is "what I think"; my body, the world I face, my knowledge, my judgments—all of these carry the attribution of "mine". The step from "experience" to "my experience" is the step this chapter addresses.

This step must be distinguished from the task of Chapter 16, otherwise the two would overlap. Chapter 16 established that the "I" is the dominant coherent resonance in the brain's multicellular space, and that its boundary is dynamically demarcated by the actual extension of coherent co-evolution. This concerns the ontological and boundary level of the "I" (what I physically am, and where I end). This chapter addresses another level: how the manifold and distributed activities within that dominant coherent system (perception, knowledge, judgment, bodily sense, and world-orientation) are demarcated as "mine". This concerns the attributional level of the "I" (how all these become mine as one and the same subject). Chapter 16 concerns "what the I is and where its boundary lies"; this chapter concerns "how manifold distributed activities are demarcated and attributed as belonging to one and the same mine". The two continue one another without repetition: the demarcation treated in this chapter occurs precisely within the dominant coherent system established in Chapter 16 and is defined by that coherent boundary.

This book denotes this subjective boundary and demarcation mechanism by $\mathscr{B}$. Its core thesis can be condensed into one sentence, drawing on the discussion in *The Final Attempt*; this chapter will develop it fully:

Open detection relations allow content to be given (Chapters 13 and 14); $\mathscr{B}$ makes given content belong to one and the same "I".

Given-ness and attribution are two different matters: given-ness (internal–external co-evolutionary freezing, Chapter 14) makes content experienced; $\mathscr{B}$-demarcation makes this experienced content **"mine"**—uniformly attributed to a subject. This chapter will first place $\mathscr{B}$ properly within the ρ_f framework (Section 19.2), then unfold the five layers of $\mathscr{B}$-demarcation (Section 19.3), then use the **failure modes** of demarcation to confirm, in reverse, what $\mathscr{B}$ does (Section 19.4), maintain the restraint that "$\mathscr{B}$ should not bear too much" (Section 19.5), and finally preserve the boundary of this chapter (Section 19.6).

The boundary consistently observed in the later part of this part must be restated from the outset: the $\mathscr{B}$-demarcation established in this chapter is the attributional structure of experience—how manifold activities are uniformly attributed as "mine". It does not touch, nor pretend to touch, "why this experiential activity demarcated as 'mine' is experienced"—that remains the final cell and is left to Chapter 20.

Clarifying "how content is demarcated as mine" and answering "why this 'my' experience is felt from within" are separated by the final cell; this chapter does only the former.

19.2. The Status of $\mathscr{B}$ in the ρ_f Framework: The Attributional Projection of Dominant Coherence, Not a New Entity

To avoid misplacing $\mathscr{B}$ as a mysterious "self-center" or an added "subject entity", one must first place its status correctly within the ρ_f framework.

$\mathscr{B}$ is not a new entity, not a special "self-organ" in the brain, and not something added outside the dominant coherence. It is a functional aspect of the dominant coherent resonance established in Chapter 16. More specifically, when the dominant coherence demarcates its boundary (which activities lock into the dominant coherence and become part of this unified coherence), it simultaneously realizes an attributional projection: whatever locks into the dominant coherence is thereby marked by this coherence as "belonging to this same coherent whole"; and "belonging to this same coherent whole", seen from within, is precisely "belonging to the same I". $\mathscr{B}$, then, is the manifestation of the boundary of dominant coherence (Chapter 16) at the attributional level—the same coherent boundary, viewed ontologically, is "the boundary of the I" (Chapter 16), and viewed attributionally, is the demarcation of "mine" (this chapter).

This placement has two important meanings. First, it prevents $\mathscr{B}$ from being suspended outside the ontology: $\mathscr{B}$ is not an added entity, but the attributional aspect of dominant coherence (Chapter 16, and also the coherence that carries unity in Chapter 9). To lock into the same coherence is to be demarcated as "mine". The unity of attribution corresponds to the unity of coherence: precisely because manifold activities lock into one and the same coherent whole, they are given as "belonging to the same mine"; what cannot lock in (brainstem rhythms and the activities of the remaining bodily organs) is not demarcated as "mine" (Chapter 16 has already established this distinction). Second, it gives $\mathscr{B}$-demarcation a physical counterpart rather than leaving it as a mysterious assertion: demarcation is not a "self" looking down from above and claiming ownership, but the physical fact of whether something locks into coherence. What can lock into dominant coherence is demarcated as "mine"; what cannot lock in is not so demarcated. The boundary of "mine" is therefore the same line as the boundary of the "I" (Chapter 16): as far as coherence reaches, so far the "I" reaches, and so far "mine" reaches.

Thus $\mathscr{B}$ is continuous with the mechanisms already established: given-ness is provided by internal–external co-evolution (Chapter 14), its content is organized by comparison with the knowledge base (Chapters 17 and 18), and all of this, when locked into the same dominant coherence (Chapters 9 and 16), is demarcated by $\mathscr{B}$ as "mine". $\mathscr{B}$ is not a new link outside this chain, but the name of the attributional aspect of the fact of "locking into dominant coherence".

19.3. The Fivefold Demarcation of $\mathscr{B}$

The way $\mathscr{B}$ demarcates "mine" is not a single vague act, but operates across several layers, each with its own target of demarcation. Following the line developed in *The Final Attempt*, this section unfolds five layers of $\mathscr{B}$-demarcation. Together, they demarcate scattered activities as the experience of one unified subject.

First, demarcation of the subjective core. $\mathscr{B}$ first demarcates the core of experience—the "I who is now experiencing". Manifold perceptions, thoughts, and judgments are attributed to this core: not "a seeing occurs" or "a thought appears", but "I see" and "I think". In the ρ_f framework, this corresponds to the core mode of dominant coherence—the strongest, most stable coherent core running through the various activities, around which manifold activities are organized and into which they lock. The demarcation of the subjective core is the axis of the other four layers: body, world, knowledge, and judgment are all demarcated as "mine" relative to this core.

Second, demarcation of bodily attribution. $\mathscr{B}$ demarcates a body as "my body"—the sensations of this body (proprioception, interoception, pain, itch, touch, pressure) are attributed to me, whereas the states of external objects are not attributed in the same way. I feel the position of "my" hand and the posture of "my" body, but I do not feel external objects in the same manner. In the ρ_f framework, this corresponds to body-related sensory configurations being locked into the dominant coherence and demarcated as "mine". It must be noted (following Chapter 16) that what is demarcated as "my body" is the layer of bodily activity whose sensations lock into dominant coherence, not the activity of every cell in the entire body. The activities of the liver and kidneys, although belonging to "my body", do not enter the experience of the "I" (Chapter 16). Thus bodily attribution demarcates the layer that is "experienced as my body".

Third, demarcation of world-orientation. $\mathscr{B}$ demarcates a "world that I face"—experience is always oriented from the perspective of "I" toward a world: the table I see is "in front of me", and the sound I hear is "around me". The world does not present itself without ownership; it is demarcated as the world "toward which I am oriented and which I face". In the ρ_f framework, this corresponds to the configurations excited by exogenous perturbations (Chapter 8) being locked into dominant coherence and organized relative to the subjective core as an "orientation" structure. The world is given through internal–external co-evolution (Chapter 14), and is further demarcated by $\mathscr{B}$ as "what I face". This directedness is precisely consistent with phenomenological intentionality (to be discussed in detail in Chapter 27): consciousness is always "about" something and oriented toward something, and the origin of this orientation is precisely the subjective core demarcated by $\mathscr{B}$.

Fourth, demarcation of knowledge attribution. $\mathscr{B}$ demarcates the experiential knowledge base as "my knowledge"—the knowledge I invoke, my memories, and what I have learned and know are attributed to me: "this is what I remember", "this is what I understand". Chapters 17 and 18 established knowledge as a transient organization and as a participant in conscious activity; this layer further demarcates this knowledge as "mine". In the ρ_f framework, this corresponds to the invoked and compared knowledge transients (Chapter 18) being locked into dominant coherence and demarcated as belonging to the same subject. The significance of knowledge attribution is that it makes knowledge not ownerless information, but part of "my" experience—my memories constitute my past, and my knowledge constitutes my way of understanding the world. Only when they are demarcated as "mine" does the "I" acquire diachronic thickness and individual uniqueness (different individuals have different experiential knowledge bases, and therefore different "I"s).

Fifth, demarcation of judgment and action. $\mathscr{B}$ demarcates judgments and actions as "mine"—"this is my judgment", "this is what I did". Chapter 18 established conscious activity (comparison, judgment, reasoning); this layer demarcates those activities as "initiated and governed by me". In the ρ_f framework, this corresponds to the pre-initiating class of endogenous perturbations (the sense of agency in Chapter 11): the endogenous perturbations that initiate judgment and action are locked into dominant coherence and demarcated as "initiated by me". The demarcation of judgment and action is precisely the attributional aspect of Chapter 11's account of the sense of agency as a pre-initiating endogenous perturbation: the sense of agency (that it is I who do this) is judgment and action being demarcated by $\mathscr{B}$ as "mine".

Taken together, the fivefold demarcation unifies scattered activities into the experience of a subject: there is a core of experience (subjective core); it has a body (bodily attribution), faces a world (world-orientation), relies on its knowledge (knowledge attribution), and makes its judgments and actions (judgment and action). "My experience" is therefore not a mysterious unity, but the attributional structure jointly realized by these five layers of demarcation—manifold activities lock into one and the same dominant coherence and are demarcated across these five layers as "mine".

19.4. Failure Modes of $\mathscr{B}$-Demarcation: Confirming What It Does in Reverse

The strongest confirmation of the reality of a demarcation mechanism is what happens when it fails. If $\mathscr{B}$-demarcation really performs the task of "attributing activities as mine", then when one layer of demarcation fails, corresponding observable abnormalities of attribution should appear. Clinically, there is precisely such a range of abnormalities, each corresponding to the failure of one layer of $\mathscr{B}$-demarcation. This both confirms the reality of $\mathscr{B}$ and makes it a mechanism with failure conditions and falsifiability, rather than an empty assertion.

Failure of subjective-core demarcation: depersonalization. In depersonalization, experience still occurs and content is still given, but patients report that these experiences "do not feel like mine", that they feel "like an observer", or that experience has "lost its mineness". This corresponds precisely to a weakening of subjective-core demarcation: activities still lock into coherence and are still given, but their demarcation as belonging to the subjective core is weakened, so that experience occurs but does "not belong to me". Depersonalization shows that given-ness and attribution are separable matters (the core thesis of Section 19.1): content can be given without being demarcated as "mine". This is exactly a problem of $\mathscr{B}$-demarcation, not of given-ness.

Failure of knowledge and thought attribution: thought insertion. In certain psychotic states, patients report that some thoughts are "not thought by me", "inserted", or "someone else's thoughts". The thought still appears in experience (it is given), but its demarcation as "initiated by me and belonging to me" fails, so it is experienced as coming from outside. This corresponds to the failure of knowledge attribution and judgment demarcation: the thought initiated by endogenous perturbation is not demarcated as "mine" and is therefore experienced as non-mine or inserted.

Failure of bodily attribution: somatoparaphrenia and alien hand. After certain brain injuries, patients may report that one of their limbs is "not mine" and deny that it belongs to them (somatoparaphrenia), or that one hand "acts on its own" and is not under the control of the "I" (alien hand syndrome). This corresponds to the failure of bodily attribution: the sensation or activity of that limb is not demarcated as "my body", and is therefore experienced as an external object or autonomous other.

Failure of agency attribution: made experiences. In certain states, patients report that their actions are "not done by me" or are "being made". The action still occurs, but the demarcation of "initiated by me" (judgment and action demarcation) fails, so it is experienced as passive and not governed by the self. This is precisely the reverse side of the sense of agency in Chapter 11: when the sense of agency (demarcated as done by me) fails, action is experienced as made.

This series of failure modes corresponds, one by one, to the failure of specific layers of the fivefold $\mathscr{B}$-demarcation, and all of them display the same structure: content is still given, but its attribution as "mine" fails. This strongly confirms the core thesis of Section 19.1—given-ness and attribution are separable matters, and $\mathscr{B}$-demarcation is responsible specifically for attribution. A mechanism that can unify and explain this series of attributional abnormalities (depersonalization, thought insertion, somatoparaphrenia, and made experiences) is much stronger than a claim that merely asserts "there is an I". It specifies what $\mathscr{B}$ demarcates, how it demarcates, and what happens when it fails. This also makes $\mathscr{B}$ a falsifiable mechanism: if these attributional abnormalities are shown to have no systematic correspondence with failures of the corresponding demarcations, then the characterization of $\mathscr{B}$ fails (the failure conditions of this chapter).

19.5. $\mathscr{B}$ Should Not Bear Too Much: A Necessary Restraint

Precisely because $\mathscr{B}$-demarcation can explain so many attributional phenomena, a danger arises: attributing too much to $\mathscr{B}$, and making it bear roles that it should not bear. This section establishes a necessary restraint: $\mathscr{B}$ demarcates attribution only; it does not generate given-ness, and still less does it explain being-experienced.

$\mathscr{B}$ does not generate given-ness. The distinction must be strict: given-ness (content becoming experienced) is provided by internal–external co-evolution (Chapter 14), not by $\mathscr{B}$. $\mathscr{B}$ demarcates the attribution of already-given content (making it "mine"). The evidence is precisely the failure modes in Section 19.4: in depersonalization, given-ness remains (experience still occurs), while attribution fails (it does not belong to me). This shows that given-ness does not depend on $\mathscr{B}$ ($\mathscr{B}$ can fail while given-ness remains), and therefore $\mathscr{B}$ does not generate given-ness, but only demarcates its attribution. To attribute given-ness to $\mathscr{B}$ would be an overreach—given-ness has another source (co-evolution).

$\mathscr{B}$ does not explain being-experienced. This must be guarded even more strictly: $\mathscr{B}$-demarcation makes content belong as "mine", but "why this experience attributed as 'mine' is experienced" is not explained by $\mathscr{B}$. $\mathscr{B}$ is an attribution mechanism—it explains how content is demarcated as "mine" (the structural level), but "why this 'my' experience is felt from within" (being-experienced) remains the final cell. To mistake "$\mathscr{B}$ demarcates my experience" for "$\mathscr{B}$ explains why my experience is experienced" is precisely the overstep prohibited by C3 (Chapter 20 will preserve this boundary directly).

The boundary of $\mathscr{B}$ is the boundary of coherence, and no additional standard is introduced. A further restraint is also needed: the scope of $\mathscr{B}$-demarcation is determined by the actual extension of coherence (Chapter 16), not by an independent "attribution criterion" introduced by $\mathscr{B}$. What can lock into dominant coherence is demarcated as "mine"; what cannot lock in is not demarcated as mine. Whether demarcation occurs depends on coherence locking, not on some autonomous judgment made by $\mathscr{B}$. This prevents $\mathscr{B}$ from becoming an autonomous "reviewer" (which would reintroduce the homunculus problem), and makes it only the attributional aspect of the physical fact of coherence locking.

Taken together, these three restraints place $\mathscr{B}$ where it belongs: it demarcates attribution (making content "mine"), does not generate given-ness (which belongs to co-evolution), does not explain being-experienced (which belongs to the final cell), and does not introduce an independent standard (demarcation is coherence locking). If these restraints are preserved, $\mathscr{B}$ is a powerful but non-overreaching attribution mechanism. If any one of them is crossed, $\mathscr{B}$ becomes a pseudo-mechanism that bears too much and may even masquerade as a solution to the final cell. This chapter chooses the former.

19.6. The Boundary of This Chapter

This chapter must preserve the boundary consistently observed in the later part of Part III. What this chapter establishes, $\mathscr{B}$-demarcation, is the attributional structure of experience—how manifold, given, and cognized activities are uniformly attributed as "my" experience through fivefold demarcation, and how the failure modes of attributional abnormalities confirm this structure. All of this is an articulable structural characterization of "how content becomes mine", and is in principle comparable with clinical abnormalities of attribution and falsifiable (its measurable side appears in Chapters 24 and 25).

This chapter does not answer, nor does it pretend to answer, why the experiential activity demarcated as "mine" is experienced. This boundary must be drawn clearly: this chapter explains "how experience becomes my experience" (the structure of attribution), but the question "why this my experience is experienced and felt from within, rather than the same demarcation being completed without experience" cannot be answered by this chapter, and this chapter does not pretend to answer it. Whether a system that performs the same demarcation (attributing activities to a unified core) is accompanied by being-experienced is not determined by the demarcation mechanism itself (just as the being-experienced of given-ness is not entailed by mechanism, Chapters 14 and 15). This chapter clarifies the demarcation mechanism (the structure of attribution), not the being-experienced of this mechanism (the final cell).

At this point, the three steps in the later part of this part have been completed: Chapter 17 established knowledge (the higher-order organization of memory transients), Chapter 18 established cognized conscious activity (comparison between perturbations and knowledge), and this chapter establishes $\mathscr{B}$-demarcation (the attribution of activity as "mine"). Together, the three steps advance experience from "given sensation" all the way to "cognized experiential activity demarcated as mine". This is a gradual deepening at the structural level and a gradual approach toward the final cell (pushing the residual point from "general given-ness" toward "subjective attribution"), not a crossing of the final cell. Chapter 20 will use these three steps to compress the residual point of the hard problem once more, more deeply yet still honestly.

19.7. Chapter Summary

This chapter establishes the demarcation mechanism of $\mathscr{B}$, advancing experience into "my" experience. First, it clarifies the task and division of labor: this chapter concerns how manifold activities are demarcated and attributed as "mine" (the attributional level), distinct from Chapter 16's concern with "what the I is and where its boundary lies" (the ontological/boundary level). The core thesis is that open detection makes content given, while $\mathscr{B}$ makes content belong to the same I. Second, it places $\mathscr{B}$ properly: it is not a new entity or self-organ, but the attributional aspect of dominant coherence (Chapter 16). To lock into the same coherence is to be demarcated as "mine", so the boundary of "mine" is the boundary of the "I". Third, it unfolds the fivefold demarcation: subjective core (the experiencing I, the axis of the rest), bodily attribution (my body), world-orientation (the world I face, consistent with intentionality), knowledge attribution (my knowledge, giving the I diachronic thickness and individual uniqueness), and judgment and action (what I initiate, the attributional side of the sense of agency in Chapter 11). Fourth, it confirms the mechanism in reverse through failure modes: depersonalization (failure of subjective-core demarcation), thought insertion (failure of knowledge/judgment attribution), somatoparaphrenia and alien hand (failure of bodily attribution), and made experiences (failure of agency attribution). These all display the structure of "content still given, while attribution fails", confirming that given-ness and attribution are separable, that $\mathscr{B}$ is responsible specifically for attribution, and that $\mathscr{B}$ is a falsifiable mechanism. Fifth, it establishes three restraints ($\mathscr{B}$ should not bear too much): it does not generate given-ness (depersonalization as evidence), it does not explain being-experienced (the final cell), and it does not introduce an independent standard (demarcation is coherence locking, avoiding the homunculus problem). Sixth, it preserves the boundary: this chapter establishes the structure of attribution, not "why my experience is experienced" (the final cell, Chapter 20).

The failure conditions of this chapter are as follows (summarized in Appendix E).

- F19.1 (new) Given-ness and attribution are inseparable. If given-ness and attribution are shown to be inseparable (that is, if there are no cases in which "content is given but not demarcated as mine", and attributional abnormalities such as depersonalization are reinterpreted as abnormalities of givenness itself), then the core thesis of this chapter ($\mathscr{B}$ is responsible specifically for attribution and is distinct from given-ness) fails.

- F19.2 (new) $\mathscr{B}$-demarcation has no correspondence with attributional phenomena. If the fivefold demarcation has no systematic correspondence with observable structures of attribution (subjective core, bodily attribution, world-orientation, knowledge attribution, and agency), then the characterization of $\mathscr{B}$ fails.

- F19.3 (new) Failure modes do not correspond. If attributional abnormalities (depersonalization, thought insertion, somatoparaphrenia, made experiences) cannot correspond to failures of the corresponding demarcations, then the status of $\mathscr{B}$ as a falsifiable mechanism is weakened.

- F19.4 (new) $\mathscr{B}$ bears too much. If this book actually uses $\mathscr{B}$ to generate given-ness, or to explain being-experienced, or makes $\mathscr{B}$ an autonomous reviewer with an independently introduced standard (thereby reintroducing the homunculus problem), then it violates the restraint of Section 19.5. Among these, "using $\mathscr{B}$ to explain being-experienced" violates the C3 prohibition and is the failure against which this chapter must guard most carefully.

Table 19.1.. The Boundary between What This Chapter Establishes and What It Cannot Derive

What this chapter establishes	What this chapter cannot derive
$\mathscr{B}$-demarcation: open detection makes content given, and $\mathscr{B}$ makes content belong to the same I (the distinction between given-ness and attribution).	It cannot explain from $\mathscr{B}$-demarcation "why my experience is experienced" (the final cell, Chapter 20).
$\mathscr{B}$ is the attributional aspect of dominant coherence (locking in is demarcation as mine), and the boundary of "mine" is the boundary of the "I".	It cannot generate given-ness from $\mathscr{B}$ (given-ness depends on internal–external co-evolution, Chapter 14; depersonalization serves as evidence).
Fivefold demarcation (subjective core, bodily attribution, world-orientation, knowledge attribution, judgment and action).	It cannot re-establish in this chapter the ontology and boundary of the "I" (this belongs to Chapter 16; this chapter treats its attributional aspect).

What this chapter establishes	What this chapter cannot derive
Failure modes as confirmation (depersonalization / thought insertion / somatoparaphrenia / made experiences corresponding to failures of different layers of demarcation).	It cannot make $\mathscr{B}$ an autonomous reviewer with an independently introduced standard (demarcation is coherence locking, avoiding the homunculus problem).

20. The Second Compression of the Final Cell: Interface, Interior, Information Processing, and Subjective Demarcation

20.1. The Position of This Chapter: The Moment of the Second Compression

Chapter 15 already compressed the final cell once. After stripping away all the parts of the hard problem that can be addressed (structure and the mechanism of given-ness), it compressed the irreducible residual point into one cell, which this book called the given-ness residual point: "given all the structures and co-evolutionary facts of a bounded interior, why is all this experienced?" That was the result of the first compression: the fog diffused across the whole hard problem was compressed into the single cell of "why given-ness is experienced".

But Chapters 16 to 19 added a layer not yet reached by Chapter 15. Chapter 16 placed the "I" as dominant coherence; Chapters 17 and 18 showed that experience is not only given, but also cognized (recognized and endowed with meaning through comparison with the experiential knowledge base); Chapter 19 showed that cognized experiential activity is further demarcated by $\mathscr{B}$ as "mine". The addition of this layer means that the "given-ness residual point" of Chapter 15 can be compressed one cell further: the "given-ness" spoken of there was general given-ness, whereas we now see that the given-ness of human experience is a given-ness that is cognized and demarcated as "mine". The residual point can therefore be further compressed and made more precise, from "why general given-ness is experienced" to "why this subjective-attribution activity, cognized and demarcated by $\mathscr{B}$ as 'mine', is experienced". This book calls the residual point after this second compression the subject-attribution residual point.

This chapter is the moment of this second compression. But a boundary must be drawn at the very beginning, one that is most critical for this chapter and even for the entire book—because this chapter is precisely where Version A and Version B diverge, and it is the final touchstone of whether the book remains honest: the second compression makes the residual point smaller and locates it more precisely; it absolutely does not eliminate the residual point. What this chapter does is relocate that cell, moving it from "general given-ness" to "subjective attribution"; this chapter never claims that, through this compression, the hard problem has "completely ceased to exist" or has "been solved". To mistake "further compression" for "elimination" is precisely the topic substitution this book has rejected from the beginning—it would cause the honesty preserved through the first nineteen chapters to collapse at the final moment. The entire measure of this chapter depends on preserving these six words: compression, not elimination.

20.2. The Interface Lineage: From "Why an Interior Arises" to "How Interfaces Are Upgraded"

The second compression can be carried out through a reframing: correcting the very formulation of the traditional hard problem. This correction is this book's response to the "mistaken premise" of the hard problem, but it must be used with restraint; otherwise it will slide into the overreach of Version B.

The traditional hard problem is posed in this way: on one side there are external, third-person physical processes without an interior; on the other side there is internal, first-person experience. It therefore asks, "How can physical processes without an interior produce an interior?" This book points out that this formulation contains an unexamined premise: it assumes that physics has only one kind of interface, namely the "external interaction interface". But Chapter 3 has already established that physics does not have only external interaction interfaces—enclosure (the lipid membrane) produced a genuine internal interface (the inside and outside of life). Once it is acknowledged that physical systems can form different levels of interface through organization, the formulation of the hard problem can be corrected: the problem is no longer "why physics without an interior produces an interior", but "what kind of interface, under what conditions, becomes the interior of experience".

This correction can be unfolded through an interface lineage (its phenomenological basis lies in the cellular trilogy, while its mechanism is provided by the preceding parts of this book):

- Non-living external interaction interface: structures formed by the passive aggregation of the four forces, possessing only external interaction surfaces and no interior (Chapter 3).

- Membranous life interface: enclosure first produces a topological interior, demarcating the inside and outside of life (Chapter 3)—this is the precondition of given-ness (only with an interior is there the possibility of "being given to it").

- Cellular information-processing interface: pooling networks and the PFR loop inside the membrane (the cellular trilogy) turn the interior into an information-processing system—elementary internal–external co-evolution (Chapter 14) occurs here.

- Human-brain subjective-experience interface: massive neural-cell integration, the experiential knowledge base, comparison between internal and external perturbations (Chapter 18), and demarcation by $\mathscr{B}$ as a unified subject (Chapter 19)—a higher-order experiential interface that is cognized and attributed as "mine".

The main line of the hard problem is thus restated from "why an interior arises" to "how interfaces are upgraded along this lineage": from no interior, to interior, to internal information processing, to a subjective experiential interface. Each level of upgrading has articulable physical–organizational conditions (enclosure, pooling networks, the knowledge base, and $\mathscr{B}$-demarcation), rather than a mysterious production. This restatement transforms the hard problem from a "mysterious generation from nothing" into a "conditional problem of interface upgrading along a lineage"—and a conditional problem can be addressed, mechanized, and given failure conditions.

But the C3 boundary must be raised immediately here, because this is precisely the entry point of Version B's overreach. To restate the hard problem as "what kind of interface becomes an experiential interior" answers the question of "what kind of interface" (the conditions and upgrading of interfaces); it does not answer "why that interface is an experienced experiential interior". The interface lineage clarifies "which level of interface, under which conditions" becomes the bearer of experience—this is addressable; but the question "why this highest-level subjective interface is experienced from within rather than completing the same information processing and demarcation without experience" cannot be answered by the interface lineage. To say that "the interface lineage has clarified this, so the hard problem no longer exists" is precisely to answer "what kind of interface" and then pretend that "why it is experienced" has also been answered. That is topic substitution, the trap of Version B. The interface lineage is a powerful framework for restatement and compression, but it compresses the residual point; it does not eliminate it.

20.3. The Second Compression: From the Given-ness Residual Point to the Subject-Attribution Residual Point

With the interface lineage and Chapters 16 to 19 in place, the second compression can now be completed precisely—the residual point can be relocated from the "given-ness residual point" to the "subject-attribution residual point".

The residual point obtained by the first compression (Chapter 15) was: "given the structure and co-evolution of a bounded interior, why is it experienced?" The "being experienced" here is general and has not yet been further specified.

Chapters 16 to 19 show that experiential activity at the highest interface of the human brain has a more specific characterization: it is the activity of a massive integrated neural-cell system (Chapter 16: I = dominant coherence), with the participation of the experiential knowledge base (Chapter 17), comparing, judging, and reasoning over exogenous and endogenous perturbations (Chapter 18), and being demarcated by $\mathscr{B}$ as the unified field of "mine" belonging to one subject (Chapter 19). All of this—integration, cognization, comparison, and demarcation—belongs to the articulable, mechanizable structural level, and Chapters 16 to 19 have clarified it fully.

The residual point is therefore compressed one cell further and relocated once more. It is no longer the general question of "why structure and co-evolution are experienced", but is made precise as follows:

Subject-attribution residual point: why is the experiential activity of the human brain, a system with boundary, a cellular information-processing basis, comparison with the participation of a knowledge base, and demarcation by $\mathscr{B}$ as a unified subject—an activity that is cognized and demarcated as "mine"—experienced from the first-person interior, rather than completing the same integration, comparison, and demarcation without experience?

This residual point is smaller and more precise than the "givenness residual point" of the first compression: it no longer contains "what process given-ness corresponds to" (answered in Chapter 14), "how experience is cognized" (answered in Chapter 18), or "how experience is demarcated as mine" (answered in Chapter 19). Everything addressable has been stripped away. What remains is only the final point: why is the full activity that has been integrated, cognized, and demarcated as "mine" accompanied by first-person being-experienced, rather than being completed in the absence of experience?

This is the result of the second compression: the residual point is moved and located from "general given-ness" to "subjective attribution". It remains irreducible (for the same reason as in Chapter 15: being-experienced is not entailed by structure and mechanism—here, by the more specific mechanisms of integration, cognization, and demarcation), but it has been made smaller and located more precisely. Every layer clarified in Chapters 16 to 19 (integration, cognization, demarcation) removes another addressable component from the residual point and pushes it deeper and more specifically—from the "given-ness residual point" to the "subject-attribution residual point".

20.4. Preserving the Boundary: Further Compression Is Not Elimination

This section is the final guardrail of honesty for this chapter and for the entire book. With maximum clarity, one must decisively separate "second compression" from "eliminating the hard problem"—because only a thin line separates them, and crossing that line is the topic substitution of Version B, the very "cheating" this book has rejected from the beginning.

The residual point remains; it has only been made smaller and more precise. The second compression has stripped away integration, cognization, and demarcation as addressable components, but it has not, and cannot, strip away the final point: "why all this is experienced". The subject-attribution residual point remains a residual point not entailed by structure and mechanism: even if all the physical–organizational facts of human-brain integration, cognization, comparison, and $\mathscr{B}$-demarcation are fully given, the fact that "this activity demarcated as 'mine' is experienced from within" is still not logically entailed by those facts (the reason is the same as in Chapter 15: being-experienced is not contained within structural–relational language). Whether a system executing the same integration, the same cognized comparison, and the same $\mathscr{B}$-demarcation is accompanied by first-person being-experienced is not determined by those mechanisms themselves. Thus the residual point has not disappeared; it has only been made precise as the cell of subjective attribution.

This book decisively rejects the claim that "the hard problem has completely ceased to exist". Here this book explicitly rejects a tempting but overreaching conclusion: that because the hard problem has been restated as "what kind of interface becomes an experiential interior", and because the conditions of interface upgrading have been clarified, the hard problem has "completely ceased to exist" or has "been solved". This conclusion is the topic substitution of Version B: it passes off the answered question "what kind of interface, under what conditions, becomes the bearer of experience" as the unanswered and unanswerable question "why that interface is experienced". The entire honesty of this book depends on not taking this step. The hard problem has not "completely ceased to exist"—it has been compressed twice and precisely located as the subject-attribution residual point, but that cell remains irreducible and remains the boundary of physical–structural language. To call "restatement and compression" "elimination" would cause the honesty preserved through the first nineteen chapters to collapse at the final moment, and would confirm the criticism that the project "looks like cheating". This book would rather acknowledge in the final chapter that the residual point remains than cover it with the attractive phrase "solved".

Interface and demarcation are tools of compression, not spells of elimination. One must be especially vigilant here: the interface lineage (Section 20.2) and $\mathscr{B}$-demarcation (Chapter 19) are powerful tools, but they are easily misused as spells that "eliminate" the residual point—as if, once one has clarified how interfaces are upgraded and how activity is demarcated as "mine", one has clarified why it is experienced. But the interface lineage compresses the addressable part of "why an interior arises" (the conditions and upgrading of interfaces), and $\mathscr{B}$-demarcation compresses "how given-ness becomes mine" (the structure of attribution). Both make the residual point smaller, but neither touches the final point (why the upgraded and demarcated activity is experienced). To use interface or demarcation to "eliminate" the residual point is precisely the overreach that Chapter 27 will analyze directly and prohibit through failure conditions (F23.4 / F20.4 of this chapter): using identity, interface, or demarcation to masquerade as the dissolution of being-experienced. This chapter uses these tools to compress the residual point, while explicitly stating that they cannot and have not eliminated it.

20.5. Why Further Compression Is Still an Achievement, Not Cheating

If the residual point has not disappeared, is the second compression still an achievement? Does it instead make the book seem to have taken a long detour (adding four chapters) while still stopping at "one irreducible cell"—would that not be futile, or even a case of using the term "compression" to conceal the fact that the problem has not really been solved? This section responds to that concern and explains why the second compression is a substantive achievement, and why precisely because it "does not eliminate" the residual point, it is not cheating.

First, locating the residual point more precisely is progress in understanding. The first compression located the residual point at "general given-ness"; the second compression locates it precisely at "subjective attribution". This is not moving in circles. It marks more precisely where the "irreducible" lies. It shows that the addressable range has again advanced a long way: how given-ness is cognized and how it is demarcated as "mine" were themselves addressable and mechanizable. The irreducible cell is therefore pushed deeper, made smaller, and made more specific. The more precisely a boundary is marked, the more understanding advances. The second compression moves forward the line of "how far we can explain" and marks it more precisely.

Second, "not eliminating" is exactly what distinguishes it from cheating. A theory that claims "through this compression, the hard problem has completely ceased to exist" (Version B) is precisely cheating—it answers what can be addressed and then declares the whole problem solved. By contrast, this book's second compression explicitly says that "the residual point remains, only smaller and more precise". Precisely because it does not claim elimination, it is not cheating. The reason earlier drafts of this book dissatisfied the author (they "looked like cheating") lay exactly in the fact that restatement was used to obscure the residual point. The second compression of this chapter is the remedy for that problem: it compresses everything that can be compressed to the limit, and honestly leaves what cannot be eliminated. "Not eliminating" is not a defect of this chapter; it is the mark that distinguishes it from cheating. It shows that this book can distinguish "compression" from "elimination", and does not pass off the former as the latter.

Third, it prepares the most precise object for the philosophical closure. The second compression locates the residual point as the "subject-attribution residual point"—a single point with an extremely precise position. This prepares the most precise object for Chapter 27 (the philosophical closure and final analysis of the final cell): philosophy will no longer face a diffuse hard problem, but this single point, compressed twice and located at subjective attribution. A residual point located with such precision is far more suitable for the philosophical treatment it deserves (recognition of boundary, convergence with phenomenology, what identity can and cannot do; Chapter 27) than a diffuse difficulty. The achievement of the second compression therefore also lies in this: it polishes the object handed over to philosophy until it is minimal and precise.

20.6. The Boundary of This Chapter and the Handoff

This chapter has completed the second compression; it must preserve its boundary and hand the issue onward.

What this chapter does is compress and relocate the final cell from the "given-ness residual point" to the "subject-attribution residual point". Through the restatement of the interface lineage and through the integration/cognization/demarcation of Chapters 16 to 19, it strips away one more layer of what can be addressed and pushes the residual point deeper, smaller, and more precise. What this chapter explicitly does not do is eliminate this residual point: the residual point remains irreducible, and this chapter decisively rejects the conclusion that "the hard problem has completely ceased to exist", preserving the distinction between "compression and not elimination".

At this point, the "second compression line" of the later part of this part (Chapters 17 to 20) is complete: knowledge (Chapter 17), cognized conscious activity (Chapter 18), $\mathscr{B}$-demarcation (Chapter 19), and the second compression (this chapter). It advances experience from "given sensation" all the way to "subjective-attribution activity that is cognized and demarcated as 'mine'", and compresses and locates the residual point of the hard problem from the "given-ness residual point" to the "subject-attribution residual point". Everything addressable has been addressed; what is not addressable has been left with precision.

This chapter hands this twice-compressed residual point, now located at subjective attribution, to Part V (the philosophical boundary):

Chapter 26 will directly fulfill "mind–matter unity" (placing this unity without eliminating the residual point), Chapter 27 will provide the final philosophical analysis of this subject-attribution residual point (as boundary rather than object, convergence with phenomenology, what identity can and cannot do, and a direct rejection of the temptation to "dissolve the residual point through identity/interface/demarcation"), and Chapter 28 will conclude the whole book. This chapter only compresses and locates the residual point with the greatest precision; its philosophical treatment is handed to Chapter 27.

One sentence should close this chapter and the entire second-compression line: this book compresses the hard problem twice—first to the given-ness residual point, and then to the subject-attribution residual point; in both compressions, everything addressable is fully addressed, while the one cell that cannot be addressed is located with increasing precision and yet honestly remains there, never eliminated and never falsely presented as eliminated.

20.7. Chapter Summary

This chapter crowns the four newly added chapters, completes the second compression of the final cell, and sets the final tone of Version A (restraint) for the whole book. First, it clarifies its position: Chapter 15 first compressed the problem into the "given-ness residual point"; Chapters 16 to 19 added the layer in which "experience is cognized and demarcated by $\mathscr{B}$ as mine", so the residual point can be compressed one cell further. Second, it restates the hard problem through the interface lineage: correcting its premise that "physics has only external interaction interfaces", and changing "why an interior arises" into the conditional problem of "how interfaces are upgraded along a lineage" (non-living external interaction interface $\rightarrow$ membranous life interface $\rightarrow$ cellular information-processing interface $\rightarrow$ human-brain subjective-experience interface). But it immediately preserves C3: answering "what kind of interface" does not entail "why that interface is experienced". Third, it completes the second compression: the residual point is precisely located from "why general given-ness is experienced" to the "subject-attribution residual point" (why activity integrated, cognized, and demarcated by $\mathscr{B}$ as mine is experienced), smaller and more precise, yet still irreducible. Fourth, it preserves the boundary (the final guardrail of honesty for the whole book): further compression is not elimination, and it decisively rejects the Version B conclusion that "the hard problem has completely ceased to exist"; interface and demarcation are tools of compression, not spells of elimination. Fifth, it argues that further compression is still an achievement rather than cheating: locating the residual point more precisely is progress in understanding, and "not eliminating" is the mark that distinguishes it from cheating (the remedy for the earlier drafts' problem of "looking like cheating"); it also prepares the most precise object for Chapter 27. Sixth, it preserves its boundary and hands the issue to Part V (Chapter 26's placement of unity, Chapter 27's philosophical analysis of the residual point, and Chapter 28's conclusion).

The failure conditions of this chapter are as follows (summarized in Appendix E).

- F20.1 (new) The second compression does not hold. If what Chapters 16 to 19 strip away (integration, cognization, and $\mathscr{B}$-demarcation) is not in fact addressable structural content, but itself contains a residual of being-experienced, then the residual point cannot be compressed from "given-ness" to "subjective attribution", and the second compression fails.

- F20.2 (new) The subject-attribution residual point is in fact reducible. If there exists a valid derivation that, without appealing to any first-person premise and starting purely from the structural facts of integration/cognization/demarcation, infers that "this activity is experienced", then the subject-attribution residual point is reducible (this is a failure the book would welcome, because it would mean that the hard problem has truly been solved, connected with F15.1 and F23.1).

- F20.3 (new) The restatement by the interface lineage is mistaken. If the hard problem cannot be consistently restated as a "conditional problem of interface upgrading along a lineage" (for example, if the upgrading conditions of some level in the interface lineage cannot be articulated, or if the lineage is incoherent), then the restatement framework fails.

- F20.4 (new) Further compression is made to masquerade as elimination (Version B overreach). If this book (or anyone following its logic) uses the second compression, the interface restatement, or $\mathscr{B}$-demarcation to claim that the hard problem has "completely ceased to exist / been solved", then it violates the C3 prohibition, confirms topic substitution, and causes the honesty of the whole book to collapse (the failure against which this chapter, and the whole book, must guard most carefully).

Table 20.1.. The Boundary between What This Chapter Establishes and What It Cannot Derive

What this chapter establishes	What this chapter cannot derive
The second compression: the residual point is precisely located from the given-ness residual point to the subject-attribution residual point (smaller, more precise, and still irreducible).	It cannot eliminate the residual point through the second compression (further compression is not elimination; the residual point remains).
Interface-lineage restatement: the hard problem is changed from "why an interior arises" into the conditional problem of "how interfaces are upgraded".	It cannot infer from "answering what kind of interface" to "why that interface is experienced" (interface is a tool of compression, not a spell of elimination).
The second compression as an achievement: more precise location is progress in understanding, and "not eliminating" is the mark distinguishing it from cheating.	It cannot claim that the hard problem has "completely ceased to exist / been solved" (Version B overreach, the most important danger for the whole book, F20.4).

What this chapter establishes	What this chapter cannot derive
It prepares the most precise object for Chapter 27 (the subject-attribution residual point) and hands the issue to Part V's philosophical closure.	It cannot provide in this chapter the philosophical analysis of the residual point (boundary / phenomenology / what identity can and cannot do belong to Chapter 27).

21. Measurable Projections on the Isomorphic Structural Space: Ψ

21.1. The Turn of Part IV and the Pivotal Status of This Chapter

The first three parts stand at different levels. Part II treats the structure of experience—it is isomorphic to the gradient configurations of ρ_f (at the identity level); Part III treats the given-ness of experience—it is the internal–external co-evolutionary freezing of a bounded interior. Both proceed at the level of ontology and mechanism, answering the question "what experience is". Part IV turns to a new level: measurement. It no longer asks "what experience is", but "given that experience is so, what can we measure, how can we measure it, and what falsifiable predictions can be made on that basis". This turn is the step by which this book moves from ontology–mechanism toward empirical science, and it is also where the meta-norm M (Chapter 2) is implemented on the measurable side.

This chapter is the opening and pivot of Part IV, because it must place an object that runs through the whole book but has not yet been formally defined—the consciousness-state tensor $\Psi = (\tilde{A}, C, \mathscr{B})$. In the preceding chapters, $\tilde{A}$, C, and $\mathscr{B}$ have occasionally been mentioned (for example, Chapter 16's boundary of the "I" involves $\mathscr{B}$), but their exact status has remained suspended. The task of this chapter is to give Ψ its proper name—and this "proper naming" must be carried out within the reconstructed framework of this book, while drawing a clear line from several popular approaches to "quantifying consciousness"; otherwise it can easily be misunderstood as yet another "definition of consciousness" or "generator of consciousness".

This chapter first establishes the exact status of Ψ—what it is and what it is not (21.2, 21.3), then defines one by one the three components $\tilde{A}$, C, and $\mathscr{B}$, together with what each projects (21.4 to 21.6), then explains the relation between Ψ, the five sources of Part II, and the given-ness of Part III (21.7), and finally draws the limits of Ψ (21.8).

21.2. Ψ as the Projection of the Isomorphic Structural Space onto the Observable Plane

To give Ψ its proper name, one must first return to the three-level architecture of this book, because the status of Ψ can be accurately placed only within that architecture.

This book contains three levels that must be distinguished. The first level is experience itself—given, first-person experience. The second level is the ρ_f gradient configuration identical with experience—according to the identity level (Chapter 5), experience is what this configuration is from within; this configuration has its complete structure (carried by the five sources, Part II). The third level is observable projection—we cannot directly measure first-person experience (it is given), nor can we directly measure the complete ρ_f gradient configuration (it belongs to the foundational ρ_f level and is not directly accessible to existing instruments); what we can measure is the projection of this configuration onto currently available neural measurement methods—the part captured by observables such as EEG, MEG, metabolism, and connectivity.

Ψ is precisely the object at this third level: $\Psi = (\tilde{A}, C, \mathscr{B})$ is the projection, onto the currently observable plane, of the ρ_f gradient configuration that is identical with experience. It is not experience itself (the first level, the given), nor the complete gradient configuration (the second level, the ρ_f level), but the set of summary quantities that become available after this configuration is projected onto the measurable plane, and on the basis of which we can infer and test.

The term "projection" must be taken in its strict sense. When a high-dimensional object is projected into a lower dimension, information is necessarily lost—the projection captures certain summary features of the original object, not the whole of it. The same is true of Ψ as a projection: the complete ρ_f gradient configuration is fivefold-rich (Chapters 1 and 5), while the three components of Ψ ($\tilde{A}$, C, $\mathscr{B}$) are far from sufficient to exhaust it. They are several measurable summary dimensions of this rich configuration, capturing "how well it is maintained, how well it is organized, and where its boundary lies", rather than all the details of the configuration. This information loss is not a defect of Ψ, but its nature as a measurable projection: measurement always uses finite observables to summarize a richer reality.

Placing Ψ as a projection clarifies its epistemological status: Ψ is an observable window through which we approach that reality that cannot be directly measured (experience / gradient configuration), not that reality itself. To study Ψ is to study the projection laws of experience on the measurable plane, and thereby to infer and test conclusions about experience—just as astronomy infers the interior of a star from measurable spectral projections, rather than directly touching the stellar core.

21.3. What Ψ Is Not: Neither a Definition of Consciousness nor a Generator of Experience

To give Ψ its proper name is, in one half, to clarify what it is; but the other and more important half is to clarify what it is not—because in popular approaches to "quantifying consciousness", there are two misunderstandings that can easily be imposed on Ψ, and both must be explicitly excluded.

First, Ψ is not a definition of consciousness. A common approach is to define consciousness directly by some quantity ("consciousness is Φ", "consciousness is some index of global availability")—consciousness is equated with that quantity, and that quantity is taken to be what consciousness is. This book explicitly rejects using Ψ in this way. Ψ is not a definition in the sense that "consciousness is Ψ"; consciousness (experience) is at the first level, while Ψ is a projection at the third level, with two levels between them. To say "consciousness is Ψ" is equivalent to saying "a star is its spectrum"—it confuses a reality with its projection. This book never claims that experience is Ψ; it claims that experience is the interior side of a gradient configuration (the second level), while Ψ is the projection of that configuration onto the measurable plane (the third level). Ψ is a projection quantity describing conscious states, not a definition that defines consciousness.

Second, Ψ is not a generator of experience. A deeper misunderstanding is to treat such a quantity as the producer of experience—as if experience were "generated" once Ψ (or Φ) reaches a certain value. This book firmly rejects this misunderstanding, for reasons that run through the whole book: the given-ness of experience is not produced by any structural quantity; it is the internal–external co-evolutionary freezing of a bounded interior (Chapter 14), and its being-experienced is ultimately the irreducible final cell (Chapter 15). As a projection quantity, Ψ describes how strong a conscious state is, how it is organized, and where its boundary lies, but it never produces given-ness. To mistake a descriptive quantity for a generator is precisely the overreach warned against in Chapter 15, in which a mechanistic statement masquerades as an explanation of being-experienced. A high Ψ means that the gradient configuration summarized by this projection is well maintained, well organized, and clearly bounded; it does not mean that "Ψ manufactures experience".

These two negatives can be combined in one sentence: Ψ is a

descriptive quantity, not a definition and not a generator. It describes the projection, on the measurable plane, of the gradient configuration identical with experience; it does not usurp the definitional status of "consciousness is this quantity", nor does it usurp the causal status of "this quantity produces experience". Only by preserving this point can Ψ avoid repeating the errors of single-scalar identity theories (defining consciousness by one quantity) or mechanism-masquerade theories (making one quantity produce consciousness).

21.4. $\tilde{A}$: The Dissipative Power Required to Maintain Co-Evolution

We now define the three components one by one and explain what each projects. First, $\tilde{A}$—consciousness intensity.

Experience has degrees of strength: it is strong during wakefulness, weak during drowsiness, and tends toward absence in deep sleep or anesthesia. What does this strength correspond to at the ρ_f level? According to this book, experience is given internal–external co-evolution (Chapter 14), and maintaining an active internal–external co-evolution—maintaining a coherent configuration, its continuous coupling with the background, and its freezing—requires continuous energy dissipation (the dissipative term of the master equation, Chapter 2; non-equilibrium structures must be maintained by dissipation). The strength of experience therefore corresponds to the dissipative power required to maintain this co-evolution: the more active the co-evolution, the stronger the coherence, and the more frequent the freezing, the greater the dissipative power required; when power decays, co-evolution weakens, and experience tends toward drowsiness or disappearance.

Thus $\tilde{A}$ projects the dissipative power required to maintain internal–external co-evolution. Its observable proxy is the brain's energy metabolism (for example, cerebral glucose metabolic rate measured by PET, or neural energy consumption estimated by other methods)—metabolic rate is the projection of dissipative power on the measurable plane.

Two qualifications concerning $\tilde{A}$ must be established, in order to preserve the discipline of the whole book. First, the dimensional distinction (already anticipated in Chapter 2): the dimensional physical dissipative power is denoted by A (with units of power), while what enters Ψ is its dimensionless relative quantity $\tilde{A}$ (dimensionless, used for comparisons across systems and states). The two must not be conflated; throughout this chapter and the following chapters, whenever it enters Ψ, the quantity is $\tilde{A}$. Second, necessity without sufficiency: dissipative power is a necessary condition for maintaining co-evolution, not a sufficient condition. A typical counterexample is a generalized epileptic seizure: cerebral metabolic power is extremely high ($\tilde{A}$ high), yet consciousness is lost. This precisely shows that $\tilde{A}$ alone is insufficient to characterize a conscious state—high power without proper organization (low C) and without a proper boundary (abnormal $\mathscr{B}$) does not yield strong experience. The necessary-but-not-sufficient status of $\tilde{A}$ is one reason why Ψ must be three-dimensional rather than merely one-dimensional in $\tilde{A}$.

21.5. *C*: Differentiation and Integration of the Configuration

The second component, C—consciousness complexity—projects the organization of the gradient configuration identical with experience: it is both highly differentiated and highly integrated.

At any given moment, experience is both extremely differentiated (rich in content, with many discernible objects and features) and highly integrated (all of this is unified into one experiential field, Chapter 9). This organization in which "differentiation and integration coexist" corresponds, at the ρ_f level, to the corresponding organization of the gradient configuration: the configuration contains rich and distinguishable local structures (differentiation), while it is also linked into a whole by coherence and cross-frequency resonance (integration, Chapter 9). C projects precisely the degree of this "differentiation × integration" organization—neither chaotic high differentiation without integration (as in random noise: differentiation extremely high, integration zero), nor high integration without differentiation (as in whole-brain synchronous discharge: integration extremely high, differentiation tending toward zero, as seen precisely in epilepsy and deep sleep), but a high product of both: richness together with unity.

Observable proxies for C are quantities that characterize the coexistence of differentiation and integration in neural activity—for example, complexity measures of responses to perturbation (a typical case is the perturbational complexity index, PCI, obtained by TMS stimulation with EEG recording; it measures exactly the extent to which the response is both differentiated and integrated), as well as various measures of effective connectivity and information integration. These quantities capture, on the measurable plane, the organizational degree projected by C.

The connection between C and Chapter 9 must be made explicit: the "integration" projected by C is precisely the measurable projection of the unity carried by same-frequency coherence in Chapter 9; the "differentiation × integration coexistence" projected by C corresponds to the coexistence of unity (coherence) and rich contents (multiple configurations) in Chapter 9. Therefore, C is not an isolated quantity introduced ad hoc, but the projected summary, on the measurable plane, of the structure of unity and binding developed in Chapter 9.

21.6. $\mathscr{B}$: The Boundary of Coherent Co-Evolution

The third component, $\mathscr{B}$—consciousness boundary—projects the boundary of the coherent co-evolution that carries experience: how far it extends and how well it is closed.

According to Chapter 16, the "I" is the dominant coherent resonance in the brain's multicellular space, and its boundary is dynamically demarcated by the actual extension of coherent co-evolution—what can lock into dominant coherence is inside, and what cannot lock in is outside. $\mathscr{B}$ projects precisely this boundary: the range over which coherent co-evolution extends and the degree to which it is closed (whether the boundary is clear and complete). A high $\mathscr{B}$ means that there is a well-extended and clearly closed dominant coherence (a unified subject with a definite boundary); a low or fragmented $\mathscr{B}$ means that coherent extension is insufficient, closure is incomplete, or the coherence splits into several parts (as in split brain and dissociation in Chapter 16).

Observable proxies for $\mathscr{B}$ are quantities that characterize the extension and closure of neural coherence—for example, the spatial extension of cross-regional phase locking/coherence, the range of causal closure defined by effective connectivity, and whether coherence splits in special states (such as the rupture of cross-hemispheric coherence in split brain). These quantities capture, on the measurable plane, the boundary projected by $\mathscr{B}$.

The relation between $\mathscr{B}$ and the two layers of "boundary" in Chapters 3 and 16 must be clarified, so as to avoid confusion. The boundary of Chapter 3 is the material boundary of life drawn by the membrane (the precondition of given-ness); the boundary of the "I" in Chapter 16 is the boundary of coherent co-evolution (nested within the boundary of life and usually smaller). What $\mathscr{B}$ projects is the latter—the boundary of coherent co-evolution (the boundary of the "I"), not the material boundary of the membrane. Within a living system (which has a membrane and therefore the precondition of given-ness), $\mathscr{B}$ projects where its dominant coherence extends and how well it is closed. $\mathscr{B}$ is therefore the projected summary, on the measurable plane, of the boundary of the "I" in Chapter 16.

21.7. The Relation between Ψ, the Five Sources, and Given-ness

Since Ψ is a projection quantity, it must be clarified: what is it a projection of? What is its relation to the five sources of Part II and to the given-ness of Part III? Only by clarifying this can one see the exact place of Ψ in the whole book, as well as its inevitable incompleteness.

The relation between Ψ and the five sources: Ψ is a three-dimensional measurable summary of a fivefold-rich structure. Part II established the five sources of experiential structure (single-frequency configurations, the frequency dimension, master-equation evolution, resonance modes, and transient timescales), which together constitute the fivefold-rich gradient-configuration space. The three components of Ψ are not another parallel list alongside these five sources, but a summary projection of this fivefold-rich configuration onto the measurable plane: $\tilde{A}$ summarizes the power required to maintain this configuration; C summarizes the differentiation and integration of the configuration (especially in relation to the integration of resonance modes, Chapter 9); $\mathscr{B}$ summarizes the boundary of the coherent configuration (especially in relation to the extension of dominant coherence, Chapter 16). The five sources are the complete structure of the configuration (the ontological level, the second level), while Ψ is its measurable summary (the projection level, the third level). Therefore Ψ is necessarily incomplete—three summary dimensions cannot capture all the details of the fivefold-rich configuration (for example, it cannot capture the specific qualitative kinds, which require direct characterization of frequency families and far exceed what $\tilde{A}$, C, and $\mathscr{B}$ can summarize). This incompleteness is the nature of projection, as already stated (21.2).

The relation between Ψ and given-ness: Ψ describes the measurable features of co-evolution, but does not touch given-ness itself. Part III established given-ness—the internal–external co-evolutionary freezing of a bounded interior. The three components of Ψ precisely summarize three measurable features of this co-evolution: $\tilde{A}$ summarizes the power required to maintain it (whether co-evolution is active), C summarizes its organization (whether the configuration linked by co-evolution is differentiated and integrated), and $\mathscr{B}$ summarizes its boundary (whether co-evolution extends and closes properly). Ψ is therefore the projected summary of internal–external co-evolution on the measurable plane. But the boundary must be preserved: Ψ summarizes the measurable features of co-evolution (power, organization, boundary), and does not touch "why this co-evolution is given / experienced"—the latter is given-ness itself, and ultimately the final cell (Chapter 15), which is not contained in Ψ. A high Ψ describes an internal–external co-evolution that is well maintained, well organized, and clearly bounded; it does not describe or explain the being-experienced of this co-evolution. This is the inevitable limit of Ψ as a descriptive quantity (21.3).

From these two relations, the place of Ψ in the whole book becomes clear: it is the bridge connecting ontology–mechanism (Parts II and III) with the measurable plane (Part IV). It projects the gradient configuration / internal–external co-evolution that is identical with experience but not directly measurable into three measurable summary quantities, thereby allowing the ontological–mechanistic claims of Parts II and III to become observable and falsifiable research in Part IV. Ψ as bridge is precisely the key that makes Part IV possible.

21.8. Conclusion of This Chapter and the Development of Part IV

This chapter has given Ψ its proper name and established its exact status in the whole book. $\Psi = (\tilde{A}, C, \mathscr{B})$ is the projection, onto the observable plane, of the ρ_f gradient configuration identical with experience—it is not experience itself (the first level), not the complete gradient configuration (the second level), but its measurable projection (the third level). It is not a definition of consciousness (it does not claim "consciousness is Ψ"), nor a generator of experience (it does not claim "Ψ produces given-ness"), but a projection quantity describing conscious states. Its three components each have what they project: $\tilde{A}$ projects the dissipative power required to maintain internal–external co-evolution (proxy: cerebral metabolism; necessary but not sufficient, with epilepsy as a counterexample); C projects the differentiation $\times$ integration of the configuration (proxy: complexity measures such as PCI; connected with the integration of Chapter 9); $\mathscr{B}$ projects the boundary of coherent co-evolution (proxy: the extension and closure of cross-regional coherence; the projection of the boundary of the "I" in Chapter 16). Ψ is a three-dimensional measurable summary of a fivefold-rich structure (and therefore necessarily incomplete), and it is a summary of the measurable features of internal–external co-evolution (and therefore does not touch given-ness itself). It is the bridge connecting ontology–mechanism with the measurable plane, allowing the falsifiable research of Part IV to become possible.

The remaining chapters of Part IV unfold along this bridge: Chapter 22 gives the details of the physical characterization of $\tilde{A}$, C, and $\mathscr{B}$; Chapter 23 gives the interface from ρ_f to concrete measurement methods (EEG, MEG, metabolism, connectivity, and perturbation); Chapter 24 uses Ψ and internal–external co-evolution to provide a unified explanation of transitions in conscious states (sleep, anesthesia, split brain, and so on); Chapter 25 gives falsifiable predictions and a roadmap for testing. The status established in this chapter—that Ψ is a projection quantity, descriptive rather than definitional or generative—is the precondition that prevents all the work of Part IV from overstepping: Part IV studies the projection laws of experience on the measurable plane, infers and tests conclusions about experience on that basis, and never mistakes the projection quantity for experience itself, still less for the generator of given-ness.

21.9. Chapter Summary

This chapter is the opening and pivot of Part IV, and gives the consciousness-state tensor Ψ its proper name. First, it establishes the three-level architecture: experience itself (the first level, given) / the ρ_f gradient configuration identical with it (the second level, ontological) / observable projection (the third level); $\Psi = (\tilde{A}, C, \mathscr{B})$

is the projection quantity at the third level, an observable window onto a reality that cannot be directly measured. Second, it clarifies what Ψ is not: it is not a definition of consciousness (it does not claim "consciousness is Ψ", which would confuse reality with projection, as in saying "a star is its spectrum"), nor a generator of experience (it does not claim "Ψ produces given-ness", which would overstep by masquerading as an explanation of being-experienced); Ψ is a descriptive quantity. Third, it defines the three components and what they project: $\tilde{A}$ projects the dissipative power maintaining internal–external co-evolution (proxy: cerebral metabolism; dimensionally, A is dimensional and $\tilde{A}$ is dimensionless; necessary but not sufficient, with high-power unconscious epilepsy as a counterexample); C projects the differentiation $\times$ integration of the configuration (proxy: PCI and related measures, connected with Chapter 9's integration, excluding the two extremes of random noise and whole-brain synchrony); $\mathscr{B}$ projects the boundary of coherent co-evolution (proxy: the extension and closure of cross-regional coherence; the projection of Chapter 16's boundary of the "I", distinct from the material membrane boundary of Chapter 3). Fourth, it clarifies the relation between Ψ, the five sources (as their three-dimensional measurable summary, and therefore necessarily incomplete), and given-ness (as the summary of its measurable features, not touching given-ness itself); Ψ is the bridge connecting ontology–mechanism with the measurable plane.

The failure conditions of this chapter are as follows (summarized in Appendix E).

- F21.1 Ψ is mistaken for a definition or generator. If this book, in what follows, actually uses Ψ as a definition of consciousness ("consciousness is Ψ") or as a generator of experience ("Ψ produces given-ness"), then it violates the status established in this chapter, and the relevant arguments must be corrected (this is the failure against which this chapter must guard most carefully).

- F21.2 A component has no observable proxy. If any of $\tilde{A}$, C, or $\mathscr{B}$ is shown in principle to have no observable proxy (that is, it cannot be projected onto anything measurable), then that component fails to qualify as a "measurable projection quantity", and the status of Ψ as a measurable bridge is impaired.

- F21.3 The necessity of $\tilde{A}$ does not hold. If maintaining internal–external co-evolution is shown not to require dissipative power (that is, if strong experience can occur at zero power), then the basis for $\tilde{A}$ as a projection fails.

- F21.4 The three-dimensional projection omits a core measurable feature. If some observable feature systematically relevant to conscious states can neither be summarized by $\tilde{A}$, C, or $\mathscr{B}$, nor dispensed with in characterizing conscious states, then Ψ is insufficient as a three-dimensional projection and must be supplemented by an additional projection dimension (note: this is an expansion at the projection level, different from the issue of a sixth source on the structural axis in Chapter 11).

Table 21.1.. The Boundary between What This Chapter Establishes and What It Cannot Derive

What this chapter establishes	What this chapter cannot derive
Ψ as the projection of the isomorphic structural space onto the observable plane (the third level in the three-level architecture).	It cannot claim in this chapter that "consciousness is Ψ" (Ψ is not a definition; confusing reality with projection is an error).
Ψ as a descriptive quantity, neither a definition of consciousness nor a generator of experience.	It cannot produce or explain given-ness through Ψ (Ψ does not touch given-ness itself or the final cell).
The three components $\tilde{A}/C/\mathscr{B}$, what each projects, and their observable proxies (metabolism / complexity / coherent boundary).	It cannot provide in this chapter the detailed physical characterization and measurement interface of the three quantities (these belong to Chapters 22 and 23).

What this chapter establishes	What this chapter cannot derive
Ψ as a three-dimensional measurable summary of a fivefold structure (necessarily incomplete), and as a summary of the measurable features of internal–external co-evolution.	It cannot provide in this chapter the unified explanation and predictions concerning conscious states (these belong to Chapters 24 and 25).

22. The Physical Characterization of Consciousness Intensity, Complexity, and Boundary

22.1. The Task of This Chapter: From "What Is Projected" to "What It Is Physically"

Chapter 21 gave $\Psi = (\tilde{A}, C, \mathscr{B})$ its proper status, establishing its identity as a projection of the isomorphic structural space onto the observable plane, a descriptive quantity rather than a definition of consciousness or a generator of experience; it also specified what each of the three components projects: $\tilde{A}$ projects the dissipative power required to maintain internal–external co-evolution, C projects the differentiation and integration of the configuration, and $\mathscr{B}$ projects the boundary of coherent co-evolution. But Chapter 21 stopped at "what is projected" and did not yet explain "what, physically, these three quantities are, what their physical basis is, and how they are related to one another". This chapter supplies that level of physical characterization.

The division of labor between this chapter and Chapters 21 and 23 must first be made clear in order to avoid overlap. Chapter 21 establishes the status of Ψ (it is a projection quantity, not a definition/-

generator); this chapter characterizes the physical content of the three quantities (what each is at the ρ_f level, what its physical basis is, and how they are related); Chapter 23 then provides the interface from these physical quantities to concrete measurement methods (how they can be measured through EEG, metabolism, and connectivity). This chapter stands between the two: it is more concrete than Chapter 21 (moving down to physical content), but more foundational than Chapter 23 (not yet involving specific instruments or measurement paradigms).

This chapter must also observe two disciplines from the outset. First, citation without re-proof (Chapter 2): the ρ_f physics used in this chapter (the dissipative and nonlinear terms of the master equation, eigenmodes, resonance kernel, criticality, and so on) is established in the foundational volume; this chapter cites its conclusions in order to characterize the three quantities, without re-deriving the underlying physics. Second, the three quantities remain descriptive projections throughout (Chapter 21): this chapter characterizes the physical content of the three quantities, but from beginning to end they are measurable features of the gradient configuration / internal–external co-evolution identical with experience. They do not touch given-ness itself (Part III), still less the final cell (Chapter 15). Every step this chapter takes into physics must preserve this boundary.

This chapter characterizes $\tilde{A}$ (22.2), C (22.3), and $\mathscr{B}$ (22.4) in sequence, then discusses their interrelations and why three dimensions are required (22.5), and finally clarifies how the three quantities connect with the five sources and with given-ness (22.6).

22.2. Consciousness Intensity $\tilde{A}$: The Dissipative Power Required to Maintain Non-Equilibrium Co-Evolution

Chapter 21 established that $\tilde{A}$ projects the dissipative power required to maintain internal–external co-evolution. This section characterizes its physical content.

Physical basis: a non-equilibrium structure must be maintained by dissipation. An active internal–external co-evolution—the maintenance of a coherent configuration, its continuous coupling with the background, and the repeated formation of freezing (Chapter 14)—is an organized dynamic process far from equilibrium. According to thermodynamics and the dissipative term of the master equation (Chapter 2), any organized structure far from equilibrium must continuously dissipate energy in order to be maintained; once energy supply is interrupted, the structure relaxes toward equilibrium and its organization collapses. The internal–external co-evolution of consciousness is precisely such a non-equilibrium structure: it must maintain its coherence, coupling, and renewed freezing through continuous dissipative power. Physically, $\tilde{A}$ is the dissipative power required to maintain this non-equilibrium co-evolution—the more active the co-evolution (the stronger the coherence, the more frequent the freezing, and the more configurations linked), the greater the dissipative power required.

Relation to metabolism: metabolism is the supply-side proxy of dissipative power. The brain's energy metabolism (consumption of glucose and oxygen) supplies precisely the energy required to maintain this non-equilibrium co-evolution; therefore the cerebral metabolic rate is the supply-side proxy, on the observable plane, of the dissipative power projected by $\tilde{A}$ (the measurement interface belongs to Chapter 23). This explains a robust empirical fact: the strong correlation between level of consciousness and cerebral metabolic rate—metabolism is high during wakefulness and markedly decreases during deep sleep and anesthesia. In the framework of this book, this is not a coincidence, but the direct consequence of the claim that "maintaining active co-evolution requires high dissipative power, and when co-evolution collapses, dissipative demand drops sharply".

Dimensional distinction (disciplinary constraint). Chapters 2 and 21 have already established that dimensional physical dissipative power is denoted by A (with units of power, such as watts or a neural-energy-consumption equivalent), whereas what enters Ψ is its dimensionless relative quantity $\tilde{A}$. The purpose of nondimensionalization is to make systems of different scales, species, and states comparable (for example, comparing the "consciousness intensity" of an awake human brain and that of an animal brain requires a dimensionless relative quantity rather than absolute power). Throughout this chapter and the whole book, whenever the quantity enters Ψ, it is $\tilde{A}$; whenever absolute physical power is discussed, it is A. The two must not be conflated.

Necessary but not sufficient (a key qualification). $\tilde{A}$ is a necessary condition for maintaining co-evolution, but it is far from sufficient—this is the primary reason why $\tilde{A}$ alone is insufficient to characterize consciousness, and why Ψ must be three-dimensional. The strongest counterexample is a generalized epileptic seizure: during the seizure, neurons discharge in a highly synchronized and high-frequency manner, and cerebral metabolic power is extremely high (both A and $\tilde{A}$ are high), yet consciousness is completely lost. Why is there high power but no consciousness? According to this book: during the seizure, although dissipative power is extremely high, neural activity becomes highly synchronized—differentiation is lost (C tends toward zero, 22.3), and this global synchrony often disrupts the normal coherent boundary ($\mathscr{B}$ becomes abnormal, 22.4). A generalized epileptic seizure is therefore a state of "high $\tilde{A}$ but failed C and $\mathscr{B}$": there is abundant power, but no proper organization and boundary, and therefore no consciousness. This counterexample shows precisely that power is the fuel that maintains co-evolution, but abundant fuel does not equal a well-organized, clearly bounded co-evolution—the latter also requires C and $\mathscr{B}$.

22.3. Consciousness Complexity C: Richness and Integration of Eigenmodes, and Criticality

Chapter 21 established that C projects the differentiation and integration of the configuration (differentiation $\times$ integration). This section characterizes its physical content and introduces a key physical concept—criticality.

Physical basis: richness of eigenmodes (differentiation) and their coherent integration (integration). The "organization" of a ρ_f gradient configuration can be characterized by its eigenmodes (the eigen-oscillatory modes of the configuration under the master equation). Differentiation corresponds to the richness of eigenmodes—the more numerous the distinguishable eigenmodes a configuration can support, the higher its differentiation (and the richer the distinguishable experiential contents it can carry). Integration corresponds to the coherent linkage among these eigenmodes—the more strongly the eigenmodes are linked into a whole through same-frequency coherence and cross-frequency resonance (Chapter 9), the higher their integration. Physically, C is the high value reached by the product of these two: there are both rich eigenmodes (high differentiation) and coherent linkage into a whole (high integration).

This characterization by "product" excludes two extremes, exactly in accordance with experience. The first is high differentiation without integration—eigenmodes are extremely rich but have no coherent linkage with one another (as in random noise: many modes, but no overall organization), so differentiation is high but integration is zero, and the product is low; this corresponds to a non-conscious state of "fragmentation without unity". The second is high integration without differentiation—all activity is locked into a single mode (as in generalized epileptic seizure or global synchronous discharge in deep sleep: integration is extremely high, while distinguishable modes tend toward one), so integration is high but differentiation is zero, and the product is low; this corresponds to a non-conscious state of "unity without content". What consciousness requires is a high product of the two—richness together with unity, neither fragmentation nor impoverished synchrony.

Criticality: the dynamical state required by a high product of differentiation and integration. Physically, when can a system possess both high differentiation and high integration? The answer points to criticality—when the system lies in the critical region between order (high integration, low differentiation) and disorder (high differentiation, low integration), it can maintain both rich and distinguishable modes (differentiation) and cross-scale coherent linkage (integration). This book cites the concept of criticality (its physics belongs to the foundational volume and statistical physics, and is not re-proved here) to characterize C: the high C required by consciousness corresponds to the gradient configuration being in a critical or near-critical dynamical state. Far from criticality (whether in over-ordered synchrony or over-disordered noise), C is low. This characterization is consistent with the direction of neuroscientific findings that "consciousness is associated with criticality" (for example, wakefulness being near-critical, and anesthesia and deep sleep deviating from criticality), and this book explains that association through the projection C.

Observable proxies. Proxies for C are complexity measures that characterize the coexistence of differentiation and integration—especially perturbation-response complexity indices (such as PCI: perturbing with TMS and recording the response with EEG, measuring the degree to which the response is both differentiated and integrated), as well as various indices of criticality (such as avalanche statistics and long-range temporal correlations). These quantities capture, on the measurable plane, the degree of organization projected by C (the measurement interface belongs to Chapter 23).

Connection between C and Chapter 9. The "integration" projected by C is precisely the measurable projection of the unity carried by same-frequency coherence in Chapter 9; the "coexistence of differentiation $\times$ integration" projected by C corresponds to the coexistence of unity (coherence) and rich content (multiple eigenmodes / multiple configurations) in Chapter 9. Therefore, C is not introduced in isolation, but is the summary projection, on the measurable plane, of the structure of unity and binding developed in Chapter 9.

22.4. Consciousness Boundary $\mathscr{B}$: The Closure Range and Topology of Coherent Co-Evolution

Chapter 21 established that $\mathscr{B}$ projects the boundary of coherent co-evolution. This section characterizes its physical content.

Physical basis: the extension and closure of dominant coherence. According to Chapter 16, the "I" is the dominant coherent resonance in the brain's multicellular space, and its boundary is dynamically demarcated by the actual extension of coherent co-evolution—what can lock into dominant coherence is inside, and what cannot lock in is outside. Physically, $\mathscr{B}$ characterizes the extension range of this dominant coherent co-evolution (which configurations/regions it covers) and the degree of its closure (whether its boundary is clear and whether closure is complete). A high $\mathscr{B}$ means that there is a well-extended and clearly closed dominant coherence (a unified subject with a clear boundary); a low $\mathscr{B}$ means insufficient coherent extension; a fragmented or split $\mathscr{B}$ means incomplete closure or division into several parts (as in coherence splitting in split brain and local coherent rupture in dissociation, Chapter 16).

Two meanings of closure: coherent closure and causal closure. The "closure" characterized by $\mathscr{B}$ must be understood from two related angles. The first is coherent closure—the range within which dominant coherence maintains phase locking as one (the unity of Chapter 9); this defines which configurations are locked into the same coherent whole. The second is causal closure—the range within which this coherent whole constitutes a relatively self-contained causal loop (its internal causal connections are significantly stronger than cross-boundary causal connections). The two usually coincide: a whole with good coherent closure usually also has good causal closure. What $\mathscr{B}$ projects is the range and degree of this coherent–causal closure. It should be noted that this book adopts relative and dynamic closure (the boundary is demarcated by the actual extension of coherence and can change with state), not absolute and fixed closure—the boundary of the "I" is not a fixed anatomical line, but is dynamically demarcated by the extension of coherence (split brain and dissociation in Chapter 16 are precisely dynamic changes of the boundary).

Distinguishing the two layers of boundary in Chapters 3 and 16 (disciplinary constraint). Chapter 21 already anticipated this, and it is stated here explicitly: this book contains two layers of "boundary" that must be distinguished. The first is the material boundary of life drawn by the membrane in Chapter 3—it is the precondition of given-ness (only with a membrane is there an interior). The second is the subject boundary of coherent co-evolution in Chapter 16—the boundary of the "I", nested within the boundary of life and usually much smaller than it. What $\mathscr{B}$ projects is the latter—the coherent boundary of the subject, not the material boundary of the membrane. Within a living system (which has a membrane and therefore the precondition of given-ness), $\mathscr{B}$ projects how far its dominant coherence extends and how well it is closed. Confusing these two layers of boundary would misplace the boundary of the "I" onto the whole body (the boundary of the membrane), whereas this book explicitly states that "I have a body, but I am not that whole body" (Chapter 16)—$\mathscr{B}$ projects precisely that subject boundary, smaller than the body, drawn by coherence.

Observable proxies. Proxies for $\mathscr{B}$ are quantities that characterize the extension and closure of neural coherence—the spatial extension of cross-regional phase locking/coherence, the range of causal closure defined by effective connectivity, and whether coherence splits in special states (the rupture of cross-hemispheric coherence in split brain). These quantities capture, on the measurable plane, the boundary projected by $\mathscr{B}$ (the measurement interface belongs to Chapter 23).

22.5. The Relation among the Three: Why Three Dimensions Are Required Rather than Any Single Dimension

After characterizing the three quantities, one must explain why all three must stand together, and why they cannot be replaced by any one dimension or by some combination. This is the physical reason why Ψ is three-dimensional (rather than one-dimensional like single-scalar theories), and it is also the measurable-side expression of this book's critique of single-scalar identity theories (Chapter 5).

The three characterize three different matters and cannot replace one another. $\tilde{A}$ characterizes the fuel of co-evolution (the power required for maintenance), C characterizes its organization (differentiation $\times$ integration), and $\mathscr{B}$ characterizes its boundary (extension and closure). These three matters are physically orthogonal: a system can have abundant power but failed organization (epilepsy: high $\tilde{A}$, low C); it can have relatively intact organization but a fragmented boundary (dissociation: C may remain relatively intact, while $\mathscr{B}$ is broken); it can have a clear boundary but insufficient power (a certain low-arousal state: $\mathscr{B}$ may remain relatively intact, while $\tilde{A}$ is low). Precisely because the three are orthogonal and do not entail one another, no single dimension is sufficient to characterize a conscious state. This is the physical basis for the inevitable inadequacy of single-scalar theories (which characterize consciousness by a single quantity): conscious states must be jointly characterized by "fuel, organization, and boundary", and none of the three can be omitted.

The joint space of the three quantities gives the state space of consciousness. The three quantities span a three-dimensional state space; each conscious state corresponds to a point (or a region) in that space, and transitions in conscious states correspond to trajectories in that space. Wakefulness corresponds to a region in which all three dimensions are high (abundant power, rich and integrated organization, clear boundary); deep sleep and anesthesia correspond to a reduction in all three dimensions (especially a marked reduction in C and $\mathscr{B}$); generalized epileptic seizure corresponds to a special point with high $\tilde{A}$ but failed C and $\mathscr{B}$; dissociation corresponds to a region in which $\mathscr{B}$ is broken, while $\tilde{A}$ and C need not be low. This three-dimensional state space is the geometric framework used in Chapter 24 to provide a unified explanation of transitions in conscious states: state transitions are trajectories in this space, not vague one-dimensional changes in "strength of consciousness".

The joint three-dimensional representation is still a projection and remains incomplete. The qualification from Chapter 21 must be reiterated: although the three-dimensional joint representation is far better than any single dimension, it is still a projection of the fivefold-rich gradient configuration. It captures three measurable summary dimensions—"fuel, organization, and boundary"—but not the whole configuration. For example, it cannot capture specific qualitative kinds, which require direct characterization of frequency families and exceed what $\tilde{A}$, C, and $\mathscr{B}$ can provide. Predictions concerning qualitative content must therefore appeal to the frequency dimension itself, that is, to the ρ_f-specific predictions in Chapter 25, rather than to the three components of Ψ. The joint three-dimensional representation is the best measurable summary of conscious states on the observable plane, but a summary is still not the whole. This incompleteness is the nature of projection, not a defect of Ψ.

22.6. The Connection of the Three Quantities with the Five Sources and with Given-ness

Finally, the three quantities must be connected back to Part II (structure) and Part III (given-ness), in order to see their exact place in the whole book and to preserve their boundary as descriptive projections.

Connection with the five sources. The three quantities are a three-dimensional measurable summary of a fivefold-rich structure (Chapter 21): $\tilde{A}$ summarizes the power required to maintain the configuration (related to the activity level of the configuration); C summarizes the differentiation $\times$ integration of the configuration (especially in relation to the integration of resonance modes, Chapter 9, and the richness of transient eigenmodes); $\mathscr{B}$ summarizes the boundary of dominant coherence (related to the extension of the unified coherence formed by resonance modes, Chapters 9 and 16). It must be noted that some dimensions among the five sources (such as the frequency families on which qualitative kinds depend, and the transient timescales on which memory depends) are not directly summarized by $\tilde{A}$, C, and $\mathscr{B}$. They must be characterized by the frequency dimension and timescale spectrum themselves (Chapters 7 and 10), and their measurable predictions appear in Chapter 25. Therefore, the three components of Ψ and the five sources do not stand in one-to-one correspondence, but in a relation of "three-dimensional summary vs. fivefold complete structure": Ψ summarizes the overall state features of the configuration (fuel / organization / boundary), whereas the content features of the configuration (qualitative kinds and memory timescales) must go beyond Ψ and appeal directly to the frequency dimension and timescale spectrum.

Connection with given-ness. The three quantities summarize three measurable features of internal–external co-evolution: $\tilde{A}$ summarizes its maintenance (whether co-evolution is active), C summarizes its organization (the differentiation and integration of the configurations it links), and $\mathscr{B}$ summarizes its boundary (the extension and closure of co-evolution). But the boundary from Chapter 21 must be preserved: the three quantities summarize the measurable features of co-evolution, and do not touch "why this co-evolution is given / experienced". When $\tilde{A}$, C, and $\mathscr{B}$ are all high, they describe an internal–external co-evolution that is well maintained, well organized, and clearly bounded; they do not describe or explain the being-experienced of this co-evolution. This book asserts that a state with all three dimensions of Ψ high corresponds to a strong, rich, unified, and clearly bounded given experience—but "why it is experienced as such" is not contained in the three quantities; that is given-ness itself, and ultimately the final cell (Chapter 15). The physical characterization of the three quantities in this chapter has preserved this boundary throughout: what this chapter characterizes is the physical content of measurable features of the gradient configuration / internal–external co-evolution identical with experience, not the being-experienced of that experience.

Thus the place of the three quantities in the whole book is complete: they are the three piers of the bridge connecting ontology–mechanism (Parts II and III) with the measurable plane (Part IV). The pier of $\tilde{A}$ rests on "non-equilibrium must be maintained by dissipation"; the pier of C rests on "differentiation $\times$ integration and criticality"; the pier of $\mathscr{B}$ rests on "the boundary of coherent–causal closure". Together, the three piers support one bridge, allowing the ontological–mechanistic claims of Parts II and III to become observable and falsifiable research in Chapters 23, 24, and 25. The far end of the bridge—the being-experienced of given-ness—always remains beyond the bridge's reach. That is the philosophical boundary (Chapter 27).

22.7. Conclusion of This Chapter

This chapter has supplied the physical characterization of the three projection quantities whose proper status was established in Chapter 21. Physically, $\tilde{A}$ (consciousness intensity) is the dissipative power required to maintain non-equilibrium internal–external co-evolution, with cerebral metabolism as its supply-side proxy; A is dimensional, while $\tilde{A}$ is dimensionless (for cross-system comparison); it is necessary but not sufficient (in generalized epileptic seizure, $\tilde{A}$ is high while consciousness is lost, because C and $\mathscr{B}$ fail). Physically, C (consciousness complexity) is the high product of eigenmode richness (differentiation) and coherent integration (integration) of the gradient configuration; it excludes the two extremes of "fragmentation without unity" (high differentiation, zero integration) and "unity without content" (high integration, zero differentiation), and the dynamical state it requires is critical or near-critical. Its proxies include PCI and other measures of complexity and criticality, and it connects with the unity of Chapter 9. Physically, $\mathscr{B}$ (consciousness boundary) is the extension range and closure degree of dominant coherent co-evolution (including coherent closure and causal closure, understood as relative and dynamic closure), projecting the subject boundary of the "I" in Chapter 16 (distinct from the material membrane boundary of Chapter 3). The three quantities characterize three orthogonal and mutually irreducible matters—"fuel, organization, and boundary"—and therefore Ψ must be three-dimensional (the physical basis for the inadequacy of single-scalar theories); together they span the state space of consciousness (the geometry of state transitions in Chapter 24). Yet their joint three-dimensional form remains a projection and remains incomplete (it cannot capture content features such as qualitative kinds, which require the frequency dimension itself, Chapter 25). The three quantities are the three piers of the bridge connecting ontology–mechanism with the measurable plane, and they remain descriptive projections throughout; they do not touch the being-experienced of given-ness (the final cell, Chapter 15).

The failure conditions of this chapter are as follows (summarized in Appendix E).

- F22.1 $\tilde{A}$ has no systematic relation with dissipative power/metabolism. If the dissipative power required to maintain internal–external co-evolution has no systematic correspondence with consciousness intensity or with cerebral metabolism (for example, if strong experience can occur with zero dissipation, or if metabolism is systematically unrelated to level of consciousness), then the physical characterization of $\tilde{A}$ fails.

- F22.2 C has no systematic relation with differentiation $\times$ integration / criticality. If level of consciousness has no systematic correspondence with the high product of "eigenmode differentiation $\times$ coherent integration" and with criticality (for example, if a highly synchronized state with no differentiation corresponds instead to strong consciousness), then the physical characterization of C fails.

- F22.3 $\mathscr{B}$ has no systematic relation with coherent–causal closure. If the subject boundary has no systematic correspondence with the extension of coherent–causal closure (for example, if coherent closure remains intact while the subject splits, or coherence splits while the subject remains unchanged), then the physical characterization of $\mathscr{B}$ fails (linked to F16.1 in Chapter 16).

- F22.4 The three dimensions are reducible to one. If conscious states are shown to be completely characterizable by one of $\tilde{A}$, C, or $\mathscr{B}$, or by some scalar combination, such that the separation of the three dimensions has no empirical consequence, then the claim that three dimensions are required fails, and this book retreats to single-scalar theory on the measurable plane.

- F22.5 The three quantities overstep into given-ness. If this book, in what follows, uses the three quantities (or their combination) to explain or produce the being-experienced of givenness, then it violates the identity of the three quantities as descriptive projections (Chapter 21) and the prohibition of the whole book; the relevant arguments must be corrected.

Table 22.1.. The Boundary between What This Chapter Establishes and What It Cannot Derive

What this chapter establishes	What this chapter cannot derive
The physics of $\tilde{A}$: dissipative power maintaining non-equilibrium co-evolution (metabolism as proxy; necessary but not sufficient, with epilepsy as counterexample).	It cannot produce or explain the being-experienced of given-ness through $\tilde{A}$ ($\tilde{A}$ is a descriptive projection).

What this chapter establishes	What this chapter cannot derive
The physics of C: high product of eigenmode differentiation $\times$ coherent integration, requiring criticality (excluding the two extremes of noise and synchrony).	It cannot provide in this chapter the concrete measurement paradigm for C (this belongs to Chapter 23).
The physics of $\mathscr{B}$: extension of dominant coherence and coherent–causal closure (projecting the subject boundary, distinct from the material boundary of the membrane).	It cannot define in this chapter the precise form of boundaries in rare states (this belongs to Chapters 16 and 24 and the roadmap).
The three quantities are orthogonal and mutually irreducible, so three dimensions are required (the physical basis for the insufficiency of single-scalar theories), and they span a state space.	It cannot summarize content features such as qualitative kinds through the three dimensions (these require the frequency dimension itself, Chapter 25).
The three quantities are the three piers of the bridge connecting ontology–mechanism with the measurable plane, and they remain descriptive projections throughout.	It cannot use the three quantities to touch the being-experienced of given-ness (the final cell, Chapter 15).

23. The Measurement Interface: From ρ_f to EEG/MEG/fMRI/PET/TMS

23.1. The Task of This Chapter: Building a Bridge from Theory to Instruments

Chapter 21 placed Ψ as a measurable projection, and Chapter 22 characterized the physical contents of the three quantities $\tilde{A}$, C, and $\mathscr{B}$. But between "physical content" and "actual measurement", there remains a gap: we do not possess a "ρ_f meter" or a "$\tilde{A}$ meter" that can directly read off these quantities. What we do possess are existing neural measurement methods, such as EEG, MEG, metabolic imaging, connectivity analysis, and transcranial magnetic stimulation. The task of this chapter is to build the interface from theoretical quantities ($\rho_f, \tilde{A}, C$, and $\mathscr{B}$) to these existing methods—to clarify which part of ρ_f each method measures, which of the three quantities it can proxy, where its blind spots and uncertainties lie, and why multiple methods must be complementary.

A core perspective runs throughout this chapter and prevents it from becoming merely a list of instruments: every measurement method is a projection of ρ_f through a specific "frequency–gradient window". ρ_f is a rich field spanning extremely wide frequencies and very broad spatial scales (Chapter 2), whereas any existing method opens only one window onto a certain frequency range, a certain spatial scale, and a certain type of signal—EEG opens a window onto rapidly changing electrical signals, metabolic imaging opens a window onto slowly changing energy consumption, and so forth. Each window projects only one part of ρ_f, and each has its own distortions and blind spots. To understand measurement is to understand what each window projects and what it misses. Precisely because no single window can project the whole, multimodal complementarity is necessary rather than optional.

This chapter first establishes the overall perspective that "measurement is windowed projection" (23.2), then characterizes one by one the window opened by each method and the quantities it proxies (23.3 to 23.6), then specifies the measurement implications of the "consciousness frequency-band" hypothesis (23.7), and finally discusses multimodal complementarity and the reporting of uncertainty (23.8). From the outset, the boundary must be preserved: everything measured in this chapter is a projection of ρ_f onto the observable plane, and all of it consists of third-person observable quantities. However precise the measurement becomes, what it reaches remains a projection (the third level, Chapter 21); it does not touch given-ness itself (Part III), still less the final cell (Chapter 15).

23.2. Measurement as Windowed Projection: A Unified Perspective

To understand why multiple methods are needed, and why each method has both capacities and blind spots, one must first establish the perspective that "measurement is windowed projection". This is the unified framework by which this chapter understands all concrete methods.

According to Chapter 2, ρ_f is an energy quantum density field partitioned by frequency families, and its structure spans extremely wide frequencies (from very high to very low frequencies) and extremely broad spatial scales. From this book's perspective, a "measurement" is the use of a certain physical device to couple to a certain part of ρ_f, and to record certain features of that part as readable signals. This is precisely the manifestation, at the level of measurement, of the detection mechanism in Chapter 13: measurement is co-evolutionary transient freezing between the detector and ρ_f. But any existing device can couple only to a certain frequency range, a certain spatial scale, and a certain physical quantity of ρ_f: it has its response-frequency window (sensitive only to a certain frequency range), its spatial-resolution window (able to resolve only a certain scale), and its signal-type window (measuring only electricity, or only metabolism, or only blood flow). These three windows define the "window" opened by that method; through this window, ρ_f is projected into the signal recorded by that method.

This perspective immediately yields several implications that run through the whole chapter. First, every window has blind spots: the frequencies, scales, and signal types outside the window cannot be measured by that method. EEG cannot measure slowly changing metabolism (outside its frequency window), and metabolic imaging cannot measure rapidly changing oscillations (outside its temporal-resolution window). Second, every window has distortion: projection necessarily involves information loss and distortion (Chapter 21), as in the spatial blurring of EEG signals caused by conduction through the skull and scalp (a distortion of the spatial-resolution window). Third, no window can project the whole: because the richness of ρ_f far exceeds what any single window can capture, the Ψ projection obtained from any single method must be partial. This is why multimodal complementarity is necessary (23.8).

Understanding measurement as windowed projection also clarifies a common confusion: when two methods yield inconsistent results for the same conscious state, this does not necessarily mean that the measurement is wrong or that the theory is wrong. It may mean that the two windows project different parts of ρ_f—each correctly projects the part inside its own window, and those two parts are themselves different. Once this point is understood, the integration of multimodal data will not become a matter of "taking the average" or "choosing one and trusting it", but of assembling the projections from multiple windows into a more complete inference concerning the configuration that cannot be measured directly.

23.3. Electroencephalography and Magnetoencephalography (EEG/MEG): The Window onto Fast Oscillations

Electroencephalography (EEG) and magnetoencephalography (MEG) record rapid fluctuations in the electric and magnetic fields generated by neural activity. Their window opens onto the rapidly changing oscillatory range of ρ_f—with high temporal resolution (on the millisecond scale) and coverage of a relatively wide range from low to high frequencies, but with relatively coarse spatial resolution (especially EEG, because of blurring by skull conduction).

This window is especially important for this book because it is the only window that can project "frequency structure" relatively well—and frequency structure is precisely the most distinctive resource of this book (the frequency dimension carries qualitative kinds, Chapter 7; the timescale spectrum carries memory, Chapter 10). Spectral analysis of EEG/MEG can distinguish the distribution and changes of neural activity across different frequency bands, and therefore becomes the main window for testing this book's "consciousness frequency-band" hypothesis (23.7), the frequency–quality correspondence, and the frequency–timescale correspondence (the two ρ_f-specific predictions in Chapter 25). In this sense, EEG/MEG is not merely a method for measuring "state", but a key window for measuring "frequency structure"—that is, for approaching the distinctive content of this book.

With respect to the three quantities of Ψ, EEG/MEG mainly proxy C (through spectral complexity, the complexity of responses to perturbation—see TMS-EEG, 23.6—and criticality indices such as long-range temporal correlations) and $\mathscr{B}$ (through phase locking across electrodes/sources and the spatial extension of coherence, projecting the coherent boundary). Their proxy for $\tilde{A}$ is more indirect (the total power of high-frequency activity is related to metabolism, but it is not a direct measurement of energy consumption).

Their blind spots must be marked honestly: spatial resolution is coarse (especially for EEG, where precise source localization is difficult; MEG is somewhat better but still limited), so the fine spatial boundary of $\mathscr{B}$ and the activity of deep structures are poorly projected by this window. Moreover, what they measure is mainly the summed synchronous electrical activity of large populations of neurons, so they are insensitive to asynchronous processes or non-electrical processes.

23.4. Metabolic and Hemodynamic Imaging (PET/fMRI): The Window onto Slow Energy Consumption

Positron emission tomography (PET, especially measurements of glucose metabolism) and functional magnetic resonance imaging (fMRI, measuring blood-oxygen-level-dependent signals) record the energy metabolism and blood flow of neural activity. Their window opens onto the slowly changing energy-consumption range of ρ_f—with relatively high spatial resolution (localizable to smaller brain regions, including deep structures), but slow temporal resolution (on the scale of seconds, constrained by metabolic and hemodynamic delay).

This window is the main window for proxying $\tilde{A}$. As Chapter 22 established, $\tilde{A}$ projects the dissipative power required to maintain non-equilibrium co-evolution, while metabolic imaging (especially glucose metabolic rate measured by PET) directly measures the brain's energy consumption—a relatively direct supply-side measurement of the dissipative power projected by $\tilde{A}$. The strong correlation between level of consciousness and cerebral metabolism (high in wakefulness, low in deep sleep and anesthesia) is measured through this window.

fMRI, with its relatively high spatial resolution, also contributes to the characterization of C (the spatial pattern of differentiation and integration among regions) and $\mathscr{B}$ (the spatial extension of activity).

Its blind spot is slow temporal resolution, which means that it completely misses rapidly changing frequency structure—precisely where this book's most distinctive resource (the frequency dimension) lies. The metabolic/hemodynamic window projects "state" well (strength and spatial distribution), but is almost blind to "frequency content" (the frequency correspondence of qualitative kinds). This confirms the point of 23.2: the metabolic window and the EEG window project different parts of ρ_f (slow energy consumption versus fast frequency), and the two are complementary rather than mutually replaceable. Relying only on the metabolic window, this book's frequency-specific predictions cannot be tested; they require supplementation by the EEG/MEG window.

23.5. Effective Connectivity Analysis: The Window onto Causal Coupling

Effective connectivity analysis (for example, based on dynamic causal modeling or various causal inference methods) is not an independent measurement device, but an analytic method for inferring causal coupling among neural regions from EEG, MEG, or fMRI data. The window it opens is onto the causal coupling structure of ρ_f—the direction and strength of mutual influence among regions.

This window primarily proxies $\mathscr{B}$, and is especially important. As Chapter 22 established, $\mathscr{B}$ projects the boundary of dominant coherent co-evolution, and its "closure" includes both coherent closure and causal closure. Effective connectivity analysis is precisely a method for approaching "causal closure": it infers which regions constitute a relatively self-contained causal loop (whose internal causal links are significantly stronger than cross-boundary links), thereby defining the causal boundary of dominant coherence. In states such as split brain, effective connectivity analysis can test the rupture of cross-hemispheric causal coupling (the causal counterpart of subject splitting in Chapter 16). It also contributes to C (the causal structure of integration).

Its blind spots and limits must be marked honestly: effective connectivity is inference rather than direct measurement—it depends on model assumptions and data quality, and different methods may yield inconsistent causal structures. Moreover, it usually characterizes the strength of causal coupling among regions, while it remains difficult to characterize the topology of coherent closure in detail (the geometric form of boundary closure, not merely coupling strength). The latter is precisely one of the developmental tasks listed in the roadmap of Chapter 25 for directly measuring $\mathscr{B}$ as "subject boundary".

23.6. Transcranial Magnetic Stimulation–EEG (TMS-EEG): The Window onto Active Perturbation

Transcranial magnetic stimulation combined with EEG (TMS-EEG) actively perturbs a brain region by transcranial magnetic stimulation and records, through EEG, how this perturbation propagates and dissipates across the whole brain. The window it opens differs from the passive recording windows discussed above—it is a window of active perturbation, corresponding exactly to the element in the detection mechanism of Chapter 13 in which "perturbation is actively applied so that the object enters a recordable state".

This window is the strongest proxy for C, and it resonates closely with the framework of this book. As Chapter 13 established, detection requires active perturbation, threshold crossing, and a recordable response through nonlinear feedback. TMS-EEG does exactly this: it actively perturbs and observes the complexity of the response. The perturbational complexity index (PCI) derived from it measures the degree to which the response is "both differentiated and integrated", precisely what C projects (Chapter 22). If the response is both differentiated (propagating to multiple regions with rich patterns) and integrated (the responses of different regions are coherently linked), then C is high. If the response is either restricted to the stimulation site and immediately dissipates (low in both differentiation and integration, as in deep sleep and anesthesia), or globally stereotyped and synchronous (high integration but low differentiation), then C is low. The ability of PCI to distinguish conscious states (high in wakefulness/REM, low in deep sleep/anesthesia/vegetative state) is explained in this book through the projection C.

More importantly, within this book's framework, TMS-EEG has a significance beyond "measuring C": it is the direct manifestation, at the level of measurement, of the detection mechanism (Chapter 13)—active perturbation and observation of the co-evolutionary response and freezing. This makes it not merely a measurement method, but a paradigm for testing this book's claim that "given-ness is tied to internal–external co-evolution": the complexity of perturbation–response projects how well internal–external co-evolution is maintained.

Its blind spots: the PCI of TMS-EEG mainly projects C (organization), while its projection of frequency structure (the distinctive content of this book) is limited; its spatial precision is also constrained by the limits of EEG source localization.

23.7. The Measurement Meaning of the "Consciousness Frequency-Band" Hypothesis

The most distinctive resource of this book is the frequency dimension, and the measurement meaning of that frequency dimension is concentrated in a proposition this book calls the **"consciousness frequency-band" hypothesis**. This section clarifies its measurement meaning because it is the core to be tested by the EEG/MEG window (23.3), and it is also the measurement basis for the two ρ_f-specific predictions in Chapter 25.

The "consciousness frequency-band" hypothesis must first be stated accurately, so as not to be misunderstood as an overly strong claim. It does not say that "consciousness is located in one fixed frequency band" (such as certain theories claiming "gamma equals consciousness")—Chapter 2 has already established the discipline that "consciousness is not locked to one frequency band". Rather, it says two more structured things. First (corresponding to frequency–quality, Chapters 7 and 25): different qualitative kinds correspond to different and distinguishable frequency structures in neural activity, and the discontinuities among qualitative kinds correspond to intervals among frequency families. Thus the "content" of consciousness has a measurable structure on the frequency axis, rather than being diffuse and undifferentiated. Second (corresponding to frequency–timescale, Chapters 10 and 25): the retention duration of experiential content has a systematic inverse correspondence with the frequency that carries it—short-term retention corresponds to higher frequency, and long-term retention corresponds to lower frequency.

The measurement implication of this hypothesis is that it hands the distinctive content of this book to the EEG/MEG window for testing—because only this fast-changing frequency window can distinguish the frequency structure of neural activity. More specifically: high-frequency-resolution MEG/EEG combined with content decoding (inferring the qualitative kind of the subject's current experience) can test the systematic correspondence between qualitative kinds and frequency structures (frequency–quality); time-resolved spectral analysis can test the inverse correspondence between retention duration and frequency (frequency–timescale). The metabolic window (23.4) is almost entirely blind to this; the connectivity window (23.5) and perturbation window (23.6) are not its main instruments either. Therefore, testing the "consciousness frequency-band" hypothesis depends almost entirely on the frequency window of EEG/MEG. This again shows that the distinctive content of this book does not lie in the three quantities of Ψ (which are more often proxied by metabolic, connectivity, and perturbation windows), but in frequency structure (proxied by the EEG/MEG window). The division of labor stated in Chapter 22—"Ψ remains modest, while the frequency dimension bears the main distinctive load"—is confirmed at the measurement level.

Its current measurement status must be marked honestly (consistent with Chapter 25): the directional correspondences of frequency–quality and frequency–timescale can already be preliminarily tested with current EEG/MEG. However, a precise "qualitative kind–frequency family" map and the fine frequency–timescale correspondence at the end of present thickness require further development in frequency resolution and content-decoding capacity, and belong to the roadmap of Chapter 25.

23.8. Multimodal Complementarity and the Reporting of Uncertainty

The foregoing shows that no single window can project the whole of ρ_f: EEG/MEG opens onto fast-changing frequency (projecting C, $\mathscr{B}$, and frequency structure, but blind to slow energy consumption); metabolic/hemodynamic imaging opens onto slow energy consumption (projecting $\tilde{A}$ and spatial distribution, but blind to frequency structure); connectivity analysis opens onto causal coupling (projecting the causal closure of $\mathscr{B}$, but limited to inference); the perturbation window opens onto active response (projecting C, but blind to frequency). Each window projects one part of ρ_f, each with its own blind spots and distortions. This yields two concluding methodological points for this chapter.

First, multimodal complementarity is necessary, not optional. To approach the gradient configuration identical with experience, which cannot be measured directly, the projections from multiple windows must be assembled: the metabolic window to measure $\tilde{A}$; the perturbation window and spectral complexity to measure C; connectivity and coherence to measure $\mathscr{B}$; and the frequency structure of EEG/MEG to measure frequency–quality and frequency–timescale. Any single window is partial; only by assembling multiple windows can one make a relatively complete inference about the three quantities of Ψ and the frequency structure. Moreover, assembly is not simple addition: when the results of two windows differ, one must judge, according to the perspective of "measurement as windowed projection" (23.2), whether they project different parts of ρ_f, and then integrate them, rather than hastily averaging them or choosing one over the other.

Second, Ψ should be reported as an uncertainty region, not as an exact point. Since every windowed projection has distortion and blind spots, the Ψ inferred from them must carry uncertainty. This book therefore maintains that Ψ should be reported as an uncertainty region in the state space (Chapter 22), rather than as an exact point. The size of the region reflects how strongly the multimodal data constrain that state (the more abundant and consistent the data, the smaller the region; the fewer or more conflicting the data, the larger the region). This practice of "reporting as a region" is required by measurement honesty: to report, as an exact point, a quantity inferred through multiple distorted windows and through model-based inference would be false precision; reporting it as a region more truthfully reflects the limits of what we know about the configuration that cannot be measured directly. This also resonates with the view of objectivity in Chapter 13—a measurement result is the projection of a condition-constrained co-evolutionary transient; its objectivity lies in "repeatability", not in "absolute precision".

Finally, the boundary of this chapter must be preserved (consistent with Chapters 21 and 22). The interface built in this chapter leads, from beginning to end, to observable projections: what EEG, metabolism, connectivity, and perturbation measure are all third-person observable projections of ρ_f through their respective windows; the Ψ and frequency structures inferred from them are also quantities at the projection level (the third level, Chapter 21). However precise measurement becomes, however complete multimodality becomes, and however small the uncertainty region becomes, what is reached remains projection. It enables us to infer with increasing precision the measurable features of the configuration identical with experience, but it never touches "why this configuration is experienced". The being-experienced of given-ness (the final cell, Chapter 15) lies within no window and is not touched by any improvement in measurement precision. This chapter builds the bridge between theory and instruments; the far cell beyond that bridge remains outside all windows.

23.9. Conclusion of This Chapter

This chapter has built the interface from theoretical quantities (ρ_f, $\tilde{A}$, C, and $\mathscr{B}$) to existing measurement methods. Its unified perspective is "measurement as windowed projection": each method is a projection of ρ_f through a specific frequency–gradient window, each projecting something and each having blind spots, with no method able to project the whole. EEG/MEG opens the window onto fast-changing frequency, mainly proxying C, $\mathscr{B}$, and frequency structure, and it is especially the only window capable of projecting frequency structure (the distinctive content of this book). Metabolic/hemodynamic imaging (PET/fMRI) opens the window onto slow-changing energy consumption, mainly proxying $\tilde{A}$, while being blind to frequency. Effective connectivity analysis opens the window onto causal coupling, mainly proxying the causal closure of $\mathscr{B}$, while remaining inferential. TMS-EEG opens the window onto active perturbation, mainly proxying C, and is the direct manifestation of the detection mechanism (Chapter 13) at the level of measurement. Testing the "consciousness frequency-band" hypothesis (not "consciousness is located in a fixed frequency band", but rather that qualitative kinds correspond to frequency structures and retention duration inversely corresponds to frequency) depends almost entirely on the frequency window of EEG/MEG, confirming the division of labor that "Ψ remains modest, while the frequency dimension bears the main distinctive load". Because no window can project the whole, multimodal complementarity is necessary, and Ψ should be reported as an uncertainty region in the state space rather than as an exact point (as required by measurement honesty). The bridge built in this chapter leads, from beginning to end, to observable projections (third-person, third-level quantities). However precise measurement becomes, it remains projection and does not touch the being-experienced of given-ness (the final cell, Chapter 15).

The failure conditions of this chapter are as follows (summarized in Appendix E).

- F23.1 The proxy relation between window and quantity does not hold. If a given method is shown not to project the quantity it is claimed to proxy (for example, if metabolism has no systematic relation with the dissipative power projected by $\tilde{A}$, or if EEG frequency structure has no systematic relation with frequency–quality / timescale), then that interface fails.
- F23.2 Frequency structure cannot be projected by EEG/MEG. If the frequency structure on which this book's distinctive content depends (frequency–quality and frequency–timescale) is shown to be in principle unprojectable by the frequency window of EEG/MEG (that is, if no frequency window can reach it), then the "consciousness frequency-band" hypothesis loses its measurement basis.
- F23.3 Multimodal projections cannot be assembled. If projections from multiple windows are shown to be in principle impossible to assemble into a consistent inference concerning the same configuration (that is, if the projections from different windows are mutually incompatible and cannot be integrated), then the methodology of multimodal complementarity fails.
- F23.4 Measurement oversteps into given-ness. If this book uses any measurement result to explain or support the being-experienced of given-ness (rather than only its measurable features), then it violates the boundary of this chapter and the prohibition of the whole book.

Table 23.1.. The Boundary between What This Chapter Establishes and What It Cannot Derive

What this chapter establishes	What this chapter cannot derive
The unified perspective that "measurement is windowed projection": each method is a frequency–gradient windowed projection of ρ_f, with its own blind spots.	It cannot project the whole of ρ_f through any single window (therefore multimodal complementarity is necessary).
The quantities proxied by each method: EEG/MEG (C, $\mathscr{B}$, frequency structure), PET/fMRI ($\tilde{A}$), connectivity ($\mathscr{B}$ causal closure), and TMS-EEG (C).	It cannot provide in this chapter a unified explanation of transitions in conscious states (this belongs to Chapter 24) or the complete list of falsifiable predictions (this belongs to Chapter 25).
The measurement meaning of the "consciousness frequency-band" hypothesis depends almost entirely on the frequency window of EEG/MEG.	It cannot provide in this chapter a precise frequency–quality map (this belongs to the roadmap of Chapter 25).
Multimodal complementarity is necessary; Ψ should be reported as an uncertainty region rather than an exact point.	It cannot touch the being-experienced of given-ness through improved measurement (the final cell lies outside all windows).

24. The Dynamics of Conscious States: Sleep, Anesthesia, Split Brain, Dissociation, and Disorders of Consciousness

24.1. The Task of This Chapter: Bringing Disparate States into a Unified Geometry

Consciousness has many different states: wakefulness, light sleep, deep sleep, dreaming, various forms of anesthesia, vegetative state and minimally conscious state, split brain, dissociation, and so forth. Traditionally, these states have often been described separately, without a unified framework among them. Why deep sleep is unconscious, why anesthesia makes consciousness disappear, why split brain divides the subject, and why dissociation introduces fissures into experience are often explained by mechanisms unrelated to one another. The task of this chapter is to show that these apparently disparate states can be understood uniformly as different configurations within one and the same framework: they are different points or trajectories in the three-dimensional state space of Ψ (Chapter 22), and at the same time different configurations of internal–external co-evolution (Chapter 14) and different forms of coherent boundary (Chapter 16).

This unification does not force all states into a single mold. Rather, it shows that several elements already established in the preceding parts of this book—the three-dimensional state space ($\tilde{A}$ as fuel, C as organization, and $\mathscr{B}$ as boundary, Chapter 22), the configuration of internal–external co-evolution (whether co-evolution is maintained, and whether the exogenous channel is connected, Chapter 14), and the form of the coherent boundary (extension, splitting, rupture, Chapter 16)—jointly provide a coordinate system sufficient to locate all conscious states. Each state is a locatable point or describable trajectory in this coordinate system. The explanatory power of unification lies precisely here: not one mechanism for each state, but one coordinate system accommodating many states.

The boundary of this chapter must first be restated (consistent with the chapters of Part IV): what this chapter explains is how transitions among conscious states are characterized and predicted on the measurable plane. A state's position in Ψ-space, its internal–external co-evolutionary configuration, and the form of its boundary are all observable projections. This chapter does not explain, and does not touch, "why these states are each experienced as they are" (that belongs to given-ness itself and ultimately to the final cell, Chapter 15). This chapter brings the dynamics of conscious states into a measurable geometry, while the "being-experienced" of each point in that geometry always remains outside the geometry.

This chapter first establishes the state space and its three coordinate axes (24.2), then locates the individual states one by one (24.3 to 24.6), then discusses the trajectories of state transitions and their falsifiable implications (24.7), and finally preserves the boundary (24.8).

24.2. Three Coordinate Axes: The Geometry of State Space

To locate the various states, the coordinate system must first be made clear. The framework this book uses to locate conscious states is not a one-dimensional scale of "strength of consciousness", but a space jointly spanned by three groups of coordinates.

First group of coordinates: the three dimensions of Ψ ($\tilde{A}$, C, $\mathscr{B}$). This is the main body of the state space (Chapter 22). $\tilde{A}$ is fuel: the dissipative power that maintains co-evolution, proxied by cerebral metabolism. C is organization: eigenmode differentiation $\times$ coherent integration, proxied by PCI and related measures. $\mathscr{B}$ is boundary: the extension and closure of dominant coherence, proxied by cross-regional coherence and effective connectivity. Each state has its position in these three dimensions.

Second group of coordinates: the configuration of internal–external co-evolution (Chapter 14). This coordinate characterizes two matters: whether the two poles of internal–external co-evolution are maintained (which determines whether given-ness exists), and whether the exogenous perturbation channel is connected (which determines whether given-ness is anchored in the real world). The former is closely related to $\tilde{A}$, C, and $\mathscr{B}$ (maintaining co-evolution requires sufficient fuel, good organization, and a clear boundary); the latter is relatively independent (the exogenous channel may be connected or disconnected while co-evolution remains maintained).

Third group of coordinates: the form of the coherent boundary (Chapter 16). This coordinate characterizes whether dominant coherence is single and complete, split into multiple parts, or locally ruptured. It refines $\mathscr{B}$: $\mathscr{B}$ is the degree of boundary extension and closure (quantity), while boundary form is the topology of the boundary (one, two, or fragmented).

These three groups of coordinates are not three mutually independent systems, but interconnected coordinates that characterize the same state of affairs from different angles: the three dimensions of Ψ are the state's "summary quantities", the internal–external configuration is the state's "co-evolutionary mechanism", and the boundary form is the state's "boundary topology". A complete localization of a state must provide all three. In what follows, each state is located from the standpoint of these three groups of coordinates: where it lies in the three dimensions of Ψ, what its co-evolutionary configuration is, and what form its boundary takes.

24.3. Sleep and Anesthesia: The Collapse and Reconstruction of Co-Evolution

Wakefulness (baseline). All three dimensions are high: $\tilde{A}$ is high (vigorous metabolism), C is high (near-critical, with differentiation and integration coexisting), and $\mathscr{B}$ is high (dominant coherence is well extended and well closed). Internal–external co-evolution is fully maintained, the exogenous channel is connected to the real world, and the boundary form is single and complete. This is the region in the state space where all three dimensions are high, with given-ness strong, rich, unified, and anchored in the world. It is the baseline against which the other states are compared.

Deep sleep (dreamless slow-wave sleep). The three dimensions decline markedly, especially C and $\mathscr{B}$: during slow-wave sleep, neural activity tends toward large-amplitude synchronized low-frequency slow waves. This is a typical case of low C (high synchrony = high integration but low differentiation = low C, Chapter 22), and this global slow wave also disrupts the coherent boundary of wakefulness ($\mathscr{B}$ decreases). Because of the decreases in C and $\mathscr{B}$, the two poles of internal–external co-evolution become difficult to maintain: there is no stable "internal" (coherent configuration), nor a well-formed "external" co-evolving with it. Given-ness therefore largely recedes or disappears. In this book, the "unconsciousness" of deep sleep is not "fuel exhaustion" ($\tilde{A}$ decreases but does not reach zero), but the collapse of organization and boundary that makes co-evolution unsustainable. It is a state of low C and $\mathscr{B}$, rather than a state of merely low $\tilde{A}$.

Dreaming (REM sleep). This is sharply different from deep sleep. During REM, neural activity again exhibits a pattern close to wakefulness, in which differentiation and integration coexist: C rises again, $\mathscr{B}$ is reconstructed, and the two poles of internal–external co-evolution are maintained again (so given-ness reappears and dreams are vivid). However, the exogenous channel is largely closed (sensory input is blocked and motor output inhibited), so experience is driven by endogenous perturbation (Chapter 8) and is not anchored in the real world. In the state space, REM is the configuration in which "C and $\mathscr{B}$ rise again, co-evolution is maintained, but the exogenous channel is disconnected". This explains two features of dreams: they are vividly given (co-evolution is maintained), yet detached from reality (exogenous input is cut off and endogenous generation dominates). The difference between deep sleep and dreaming therefore does not lie in $\tilde{A}$ (both are sleep states), but in C, $\mathscr{B}$, and whether co-evolution is maintained. This contrast can be clearly distinguished by Ψ and by co-evolutionary configuration.

Anesthesia. Different anesthetic agents have different mechanisms, but with respect to the disappearance of consciousness, this book characterizes them uniformly as follows: anesthesia collapses the two poles of internal–external co-evolution—either by suppressing C (disrupting differentiation and integration, and making responses stereotyped or localized, as seen in the marked decline of PCI under anesthesia), or by disrupting $\mathscr{B}$ (collapsing cross-regional coherent closure), so that dominant coherence cannot be maintained. The unconsciousness of anesthesia belongs to the same class as that of deep sleep: "the collapse of C and $\mathscr{B}$ makes co-evolution unsustainable", rather than "$\tilde{A}$ is exhausted". A key falsifiable implication follows: the disappearance of consciousness should primarily track the collapse of C and $\mathscr{B}$, rather than the decrease of $\tilde{A}$ (metabolism). Therefore, there should be anesthetic states in which $\tilde{A}$ has not declined markedly, yet consciousness disappears because C and $\mathscr{B}$

collapse (under some anesthetics, metabolism may decrease only moderately while consciousness is already lost), and there should also be states in which $\tilde{A}$ is extremely high yet consciousness is entirely absent (generalized epileptic seizure, 24.5). This implication that "consciousness depends on C and $\mathscr{B}$, not solely on $\tilde{A}$" is one of the important falsifiable predictions of this chapter (Chapter 25).

24.4. Split Brain and Conjoined Cases: Boundary Splitting and Boundary Crossing

Split brain. Chapter 16 already established its ontological explanation; this chapter locates it in the state space. Split-brain surgery severs the corpus callosum, the main physical bridge maintaining cross-hemispheric coherence (the cross-hemispheric extension of $\mathscr{B}$). In the state space, split brain is not a major change in $\tilde{A}$ or C (the fuel and organization within each hemisphere are largely preserved), but a change in the boundary form of $\mathscr{B}$: the originally single dominant coherence spanning both hemispheres splits into two separately closed coherences because the cross-hemispheric coherence bridge is severed. Thus the split-brain state is characterized, across the three coordinate groups, as follows: $\tilde{A}$ and C are largely preserved; the extension and closure of $\mathscr{B}$ remain possible within each hemisphere; but the boundary form changes from "one" to "two". This corresponds exactly to the experimentally observed dual-subject phenomena (Chapter 16), and yields a falsifiable implication: after split-brain surgery, the degree of rupture in cross-hemispheric coherence should systematically correlate with the observable degree of dual subjectivity—the more thoroughly coherence is severed, the more readily dual subjectivity should appear in experiments.

The fact that split-brain patients usually behave as a single subject in daily life also needs to be characterized in state-space terms (as already discussed in Chapter 16): in everyday contexts, the two hemispheres maintain considerable functional coordination through subcortical structures, a shared environment, and behavioral coordination. The two coherences therefore usually move in step. In the state space, this is a configuration in which "two coherences are separately closed in boundary, but highly synchronized", making them difficult to distinguish as two. Only when experiments cut off this coordination and force the two hemispheres to receive conflicting information does the latent boundary split become visible as observable dual subjectivity.

Conjoined cases (cases of partial cranial connection). Chapter 16 already touched on these cases; here their state-space localization is given. If there is a physical connection between two brains sufficient to support cross-brain coherence, then a coherence whose boundary partially crosses two brains may form. In the state space, this is an intermediate boundary form of $\mathscr{B}$, neither "two completely independent subjects" nor "one fully fused subject". This book does not prejudge its concrete form (that must depend on actual neural connections), but its framework allows and predicts the existence of such intermediate forms: the boundary is drawn by the actual extension of coherence; wherever coherence extends, the boundary extends.

24.5. Generalized Epileptic Seizure: High Fuel with Failed Organization and Boundary

Generalized epileptic seizure is a particularly instructive point in this book's state space, because it displays most clearly that "one dimension is insufficient, and the three dimensions must be considered jointly" (Chapter 22).

During a seizure, $\tilde{A}$ is extremely high (neurons discharge in a highly synchronized, high-frequency manner, and metabolic power surges). Yet C is extremely low (high synchrony = extremely high integration but differentiation tending toward zero = low C, the excluded extreme of "unity without content" in Chapter 22). $\mathscr{B}$ is also impaired: global synchronized discharge destroys the normal, differentiated coherent boundary, and the normal closure of dominant coherence is overwhelmed by stereotyped global synchrony. Internal–external co-evolution cannot be maintained. Although fuel is extremely abundant, there is no well-organized, clearly bounded co-evolution because both C and $\mathscr{B}$ fail. Consciousness is therefore entirely absent.

The significance of generalized epileptic seizure lies in its role as a decisive counterexample to "high $\tilde{A}$ with consciousness": by itself, this example falsifies the claim that "consciousness is fuel (metabolism/power)" or any attempt to characterize consciousness by a single $\tilde{A}$-like quantity. Here $\tilde{A}$ is extremely high, yet consciousness is entirely absent. Its location in the state space (high $\tilde{A}$, low C, failed $\mathscr{B}$) makes clear at a glance that consciousness requires the three dimensions jointly (sufficient fuel + good organization + clear boundary), and that without one of them (here, without C and $\mathscr{B}$) consciousness is absent. Through this point, the claim in Chapter 22 that "three dimensions are required" becomes a concrete, vivid, and observable state.

24.6. Dissociation and Disorders of Consciousness: Local Boundary Rupture and Global Depression

Dissociation. Chapter 16 already established its ontological explanation (local rupture within dominant coherence); this chapter locates it in the state space. Dissociation is not a global depression of $\tilde{A}$ or C (dissociated persons are usually awake, and fuel and organization are largely preserved), but a local fragmentation in the boundary form of $\mathscr{B}$: fissures appear within dominant coherence, so that certain configurations that should have locked into the dominant coherence fail to do so, or form secondary coherences that coexist with the dominant coherence without being coherent with it. In the state space, dissociation is a configuration in which "$\tilde{A}$ and C are largely preserved, while the boundary form of $\mathscr{B}$ has local ruptures". Its degree (from mild absorption to identity alternation) corresponds to the depth and stability of coherent rupture. Dissociation therefore belongs to the same class as split brain, that is, "change in boundary form" (split brain is global splitting into two; dissociation is local rupture), and is sharply different from deep sleep and anesthesia (global reduction of $\tilde{A}$, C, and $\mathscr{B}$, with collapse of co-evolution). This distinction is a clear prediction of the present framework: dissociation should appear as "largely preserved global state quantities with local abnormality of coherent boundary", as distinct from the "global depression of state quantities" seen in sleep and anesthesia.

Disorders of consciousness (coma, vegetative state, minimally conscious state). This group of states corresponds to abnormalities of consciousness level after severe brain injury. This book characterizes their spectrum through the three-dimensional state space. Coma: all three dimensions are severely depressed ($\tilde{A}$, C, and $\mathscr{B}$ are all low; co-evolution is nearly collapsed), and given-ness is nearly absent. Vegetative state (unresponsive wakefulness syndrome): $\tilde{A}$ may be relatively high (there is wakefulness, sleep–wake cycles, and partial metabolic recovery), but C and $\mathscr{B}$ are severely depressed (no organization of differentiation and integration, and no proper coherent boundary), so there is the appearance of wakefulness without given-ness. This is another case of "relatively high $\tilde{A}$, but low C and $\mathscr{B}$ leading to unconsciousness", following the same logic as epilepsy and deep sleep. Minimally conscious state: C and $\mathscr{B}$ are partially restored (there is fluctuating and limited differentiation/integration and coherence), so there is fluctuating and limited given-ness. In the state space, this spectrum is a path extending from "all three dimensions low" (coma) toward "partial restoration of the three dimensions" (minimally conscious state).

Covert consciousness (a key falsifiable implication). From the above, this book yields a significant implication: in some behaviorally unresponsive patients (for example, some patients diagnosed as vegetative), if C and $\mathscr{B}$ are in fact partially maintained (as measurable by PCI, coherence, and related indices), then this book predicts that they are actually given, even though they show no behavioral expression. This is because given-ness depends on the maintenance of internal–external co-evolution (C and $\mathscr{B}$), not on behavioral output (the presence or absence of motor output, as in locked-in syndrome, Chapter 14). This implication of "covert consciousness" is falsifiable and is consistent with current research directions that use PCI and related measures to detect residual consciousness in unresponsive patients: those who are behaviorally unresponsive but maintain C and $\mathscr{B}$ should be judged to have given-ness; only those who are behaviorally unresponsive and whose C and $\mathscr{B}$ both fail should be judged truly without given-ness. The basis of this implication is precisely this book's criterion that given-ness depends on C and $\mathscr{B}$ (maintenance of co-evolution), rather than on behavioral output.

24.7. State Transitions as Trajectories, and the Fulfillment of Unified Explanatory Power

Once the states have been located, transitions among states naturally appear as trajectories in the state space. This is where the unified explanatory power of this chapter is most concentrated.

Falling asleep is a trajectory descending from wakefulness (all three dimensions high), through light sleep, toward deep sleep (decline of C and $\mathscr{B}$, collapse of co-evolution), and then oscillating in REM toward the configuration in which "C and $\mathscr{B}$ rise again while the exogenous channel remains disconnected". Anesthetic induction is a trajectory descending from wakefulness toward "collapse of C and $\mathscr{B}$" (the path differs across anesthetic agents, but the endpoint is collapse of co-evolution). Emergence from anesthesia, recovery from coma, and recovery from minimally conscious state toward consciousness are upward trajectories in the reverse direction, from low-dimensional depression toward restoration of the three dimensions. Every transition of conscious state is a describable, and in principle measurable, trajectory in this three-dimensional space (with internal–external configuration and boundary form superimposed).

This picture of "state as point, transition as trajectory" is the fulfillment of the unified explanatory power of this chapter. Deep sleep, dreaming, anesthesia, split brain, dissociation, epilepsy, and vegetative state—phenomena traditionally described separately and without connection—are unified in this book as different points and trajectories within one coordinate system. Their differences are no longer explained by mutually unrelated mechanisms, but by different values on the same set of coordinates ($\tilde{A}$, C, $\mathscr{B}$, internal–external configuration, and boundary form): deep sleep is a global depression of C and $\mathscr{B}$; dreaming is a recovery of C and $\mathscr{B}$ with the exogenous channel disconnected; epilepsy is high $\tilde{A}$ with failed C and $\mathscr{B}$; split brain is a boundary split into two; dissociation is a local boundary rupture; vegetative state is relatively high $\tilde{A}$ but low C and $\mathscr{B}$. One coordinate system accommodates many states. This unification is simpler, more predictive, and easier to falsify consistently than a scattered description with one mechanism for each state.

Several core falsifiable implications can therefore be extracted from this chapter (see Chapter 25 for details). First, the presence or absence of consciousness varies primarily with C and $\mathscr{B}$ (the maintenance of co-evolution), rather than primarily with $\tilde{A}$ (fuel). Therefore, states with high $\tilde{A}$ but no consciousness should exist (epilepsy and vegetative state). Second, given-ness depends on the maintenance of co-evolution, not on behavioral output. Therefore, covert consciousness should exist in behaviorally unresponsive patients who maintain C and $\mathscr{B}$. Third, changes in boundary form (splitting, rupture) are independent of changes in global state quantities. Therefore, split brain and dissociation should appear as "largely preserved global quantities with boundary abnormality", distinct from the "global depression of quantities" in sleep and anesthesia. All three implications can be tested by measuring the three dimensions of Ψ and the internal–external configuration.

24.8. Preserving the Boundary: What Is Explained Is the Measurable State, Not Being-Experienced

This chapter must finally preserve the boundary consistently maintained throughout Part IV. By means of the three-dimensional state space, internal–external configuration, and boundary form, this chapter has provided a unified explanation of transitions among a wide range of conscious states. But what it explains, from beginning to end, is how these states are characterized, located, and predicted on the measurable plane: where the state lies in the three dimensions of Ψ, whether its co-evolution is maintained, whether its exogenous channel is connected, and what form its boundary takes. All these are observable projections (the third level, Chapter 21), and all are falsifiable.

This chapter does not explain, and does not touch, why these states are each experienced as they are: the vividness of wakefulness, the strangeness of dreams, the "nothingness" of deep sleep, and the being-experienced of these states themselves. This book asserts that deep sleep (low C, low $\mathscr{B}$, collapse of co-evolution) corresponds to the disappearance of given-ness, and that wakefulness (all three dimensions high, co-evolution maintained) corresponds to the fullness of given-ness. But the "fullness/disappearance of given-ness" is asserted here as a fact correlated with measurable states; its "why it is experienced" does not belong to this chapter. Given a state whose three dimensions of Ψ and co-evolutionary configuration are fully determined, this book can predict the presence, absence, or strength of its given-ness (a matter of the measurable plane), but it cannot, and does not pretend to, explain the being-experienced of that given-ness (the final cell, Chapter 15). This chapter brings the dynamics of conscious states into a measurable geometry, while the being-experienced of each point in that geometry always remains outside the geometry. This is the honest boundary of this chapter, and it is also the C3 prohibition of the whole book applied to the dynamics of states.

24.9. Conclusion of This Chapter

This chapter has unified conscious states such as sleep, anesthesia, split brain, dissociation, and disorders of consciousness as different points and trajectories within one coordinate system—a coordinate system jointly spanned by the three dimensions of Ψ ($\tilde{A}$, C, and $\mathscr{B}$, Chapter 22), the configuration of internal–external co-evolution (whether co-evolution is maintained and whether the exogenous channel is connected, Chapter 14), and the form of the coherent boundary (single / split / ruptured, Chapter 16). The localization of each state is as follows: wakefulness (all three dimensions high, co-evolution maintained, exogenous channel connected, boundary single); deep sleep (global depression of C and $\mathscr{B}$, collapse of co-evolution); dreaming (recovery of C and $\mathscr{B}$, co-evolution maintained, exogenous channel disconnected, endogenous drive); anesthesia (collapse of C and $\mathscr{B}$, co-evolution unsustainable); split brain ($\tilde{A}$ and C largely normal, boundary changes from one into two); conjoined cases (intermediate form of boundary partially crossing two brains); generalized epileptic seizure (extremely high $\tilde{A}$ but failed C and $\mathscr{B}$—a decisive counterexample to "high $\tilde{A}$ with consciousness"); dissociation ($\tilde{A}$ and C largely normal, local boundary rupture); the spectrum of disorders of consciousness (coma: all three dimensions low; vegetative state: relatively high $\tilde{A}$ but low C and $\mathscr{B}$; minimally conscious state: partial recovery of C and $\mathscr{B}$). State transitions are trajectories in this space (falling asleep, anesthesia, emergence, and recovery each form a trajectory). The unified explanatory power lies in "one coordinate system accommodating many states". Three core falsifiable implications follow: consciousness depends on C and $\mathscr{B}$, rather than primarily on $\tilde{A}$ (with epilepsy and vegetative state as examples); given-ness depends on co-evolution rather than behavioral output (with covert consciousness as an example); and boundary form can change independently of global state quantities (with split brain and dissociation as examples). This chapter explains measurable state transitions and does not touch the being-experienced of the states themselves (the final cell, Chapter 15).

The failure conditions of this chapter are as follows (summarized in Appendix E).

- F24.1 States cannot be consistently localized. If a conscious state is shown to be impossible to localize consistently within the three coordinate groups (the three dimensions of Ψ, the internal–external configuration, and boundary form), meaning that its observable features cannot be characterized by this coordinate system, then the explanatory power of the unified geometry is weakened, and the coordinate system must be revised.

- F24.2 Consciousness depends primarily on $\tilde{A}$. If the presence or absence of consciousness is shown to vary primarily with $\tilde{A}$ (metabolism/fuel), rather than with C and $\mathscr{B}$ (that is, if there are no states with high $\tilde{A}$ but no consciousness, and epilepsy and vegetative state are reinterpreted as states in which $\tilde{A}$ is in fact low), then the implication that "consciousness depends on C and $\mathscr{B}$" is falsified.

- F24.3 Given-ness depends on behavioral output. If given-ness is shown to depend on behavioral output (that is, if behaviorally unresponsive subjects necessarily lack given-ness, and covert consciousness does not exist), then the implication that "given-ness depends on the maintenance of co-evolution, not on output" is falsified.

- F24.4 Boundary form is not independent of global state quantities. If changes in boundary form (splitting, rupture) are shown not to be independent of changes in global state quantities (that is, if split brain and dissociation necessarily involve global depression of $\tilde{A}$ and C), then the implication of "boundary-form independence" is falsified.

- F24.5 State explanation oversteps into being-experienced. If this chapter uses the measurable localization of states to explain or support the being-experienced of those states, then it violates the boundary of this chapter and the prohibition of the whole book.

Table 24.1.. The Boundary between What This Chapter Establishes and What It Cannot Derive

What this chapter establishes	What this chapter cannot derive
Conscious states are unified as points and trajectories in three coordinate groups: the three dimensions of Ψ, internal–external configuration, and boundary form.	It cannot explain why each state is experienced as it is (being-experienced, the final cell, Chapter 15).
The localization of individual states and trajectories of state transitions (sleep / anesthesia / split brain / dissociation / disorders of consciousness).	It cannot provide in this chapter the complete list of falsifiable predictions and the testing roadmap (this belongs to Chapter 25).
Three core falsifiable implications: consciousness depends on C and $\mathscr{B}$; given-ness depends on co-evolution rather than output; boundary form is independent.	It cannot decide in this chapter the mechanistic details of specific anesthetic agents (this belongs to empirical neuropharmacology).
The implication of covert consciousness: behaviorally unresponsive subjects with maintained C and $\mathscr{B}$ should be judged to have given-ness.	It cannot use behavioral output as the criterion of given-ness (the criterion lies in the maintenance of co-evolution).

25. Falsifiable Predictions and Testing Roadmap

25.1. The Task of This Chapter, and an Objection That Must Be Answered Directly

By this point, Part IV has placed Ψ as a measurable projection (Chapter 21), characterized the physics and proxies of its three components (Chapter 22), provided the interface from ρ_f to concrete measurement methods (Chapter 23), and used internal–external co-evolution to give a unified explanation of transitions among conscious states (Chapter 24). This chapter is the point of convergence for Part IV, and for the measurable side of the whole book: it gathers the foregoing into a set of falsifiable predictions, while honestly distinguishing what can already be tested from what still belongs to a roadmap. The requirement that the meta-norm M (Chapter 2) imposes on the measurable side is fulfilled in this chapter.

But this chapter must first answer directly an objection that runs through Part IV and bears on the success or failure of the whole book. If it is avoided, the predictions of this chapter will have no foundation. The objection is this: the observable proxies of this book—$\tilde{A}$ close to cerebral metabolism, C close to complexity indices (such as PCI), and $\mathscr{B}$ close to cross-regional coherence and effective connectivity—substantially overlap with the indices already used in existing consciousness science. If the predictions of this book merely restate these existing indices, then the ρ_f level seems to do no work. It would only rename what others are already measuring, without producing any unique and testable empirical content of its own. A framework that cannot make unique predictions, however internally coherent its ontology may be, is empirically idle in the sense of empirical science.

The answer of this book to this objection is clear: the ρ_f framework can indeed make at least one class of unique predictions that existing theories cannot make. These predictions come from the deepest and most distinctive resources of this book: the frequency dimension (Chapter 7) and the transient timescale spectrum (Chapter 10). It must be emphasized in particular that these two unique predictions are not merely qualitative in direction: the second one (the frequency–timescale correspondence) already has a quantitative form through the scaling law $\tau_{\text{retain}} \sim \tau_0(f)/f$, cited in Chapter 10 from *Frozen Cosmology: A Natural Philosophy of Cosmic Evolution in the Perspective of Energy Quantum Theory*. It is therefore a prediction with a concrete inverse tendency and can be tested quantitatively. This chapter accordingly organizes the predictions into three groups: first, ρ_f-specific predictions (25.2, 25.3)—these are the empirical spine of this book and the key answer to the objection above; if they are falsified, the book is seriously damaged at the level of empirical science; second, predictions shared with existing theories or usable for discrimination (25.4)—on these points the book either overlaps with, or diverges from, IIT, GNW, and related theories, and the differences can be experimentally tested; third, the testing roadmap (25.5)—parts that cannot yet be tested immediately but indicate the direction of future work. The chapter finally implements the two-tier standard and preserves the boundary between the measurable plane and the final cell (25.6).

The boundary of this chapter must be restated at the outset: all predictions in this chapter lie on the measurable plane. They concern Ψ, the observable features of internal–external co-evolution, and the projection of frequency structure in neural measurement. This chapter makes no prediction concerning the final cell (the irreducible residual point of being-experienced, Chapter 15), because by its nature that point cannot be falsified through observables. Separating measurable predictions from the final cell is the precondition for the discipline of this chapter.

25.2. ρ_f-Specific Prediction I: Frequency–Quality Correspondence

The most distinctive resource of this book is that it links differences among qualitative kinds to frequency families (Chapter 7). This claim makes a prediction that existing theories cannot make. This book calls it the frequency–quality correspondence prediction.

Existing indices for measuring consciousness—metabolism, PCI, effective connectivity—characterize the overall state of consciousness (its level, complexity, and integration). By design, they are "content-neutral": a high PCI tells us that this is a state with high conscious capacity, but it does not tell us whether what is experienced at this moment is red or pain, visual or auditory. They measure the "presence and strength" of consciousness, not the "kind of quality" of consciousness. This is their nature rather than their defect: they are intended to characterize state, not content.

The frequency dimension of this book makes a content-specific prediction: different qualitative kinds correspond to different and distinguishable frequency structures in neural activity. In particular, experiences belonging to different sensory modalities (vision, hearing, touch, and so on) and different broad qualitative classes should correspond in neural measurement to systematically different features of frequency families, while changes within the same qualitative kind (Chapter 6) should correspond to changes in gradient configuration within that family. More strongly, this book predicts that the discontinuous structure among qualitative kinds (Chapter 7: there are no continuous intermediate transitions among kinds) should correspond to interval structures among frequency families—the major qualitative modalities that are experientially discontinuous from one another should correspond to several separated family regions along the neural frequency axis.

This is a prediction that existing "state indices" cannot in principle make, because it concerns the frequency structure of content rather than an overall state quantity. Its falsifiability is clear: if fine frequency-resolved measurement shows that there is no systematic correspondence between different qualitative kinds and neural frequency structures (for example, if visual and auditory experiences have no distinguishable frequency-structural differences, or if qualitative discontinuities do not correspond to frequency intervals), then the frequency–quality correspondence prediction is falsified, and the core resource by which this book carries qualitative kinds (Chapter 7) is damaged at the empirical level. This book deliberately states the claim in this falsifiable form. This is precisely where the ρ_f level "does work": it makes a content-specific prediction that state indices cannot make and that can be overturned by experiment.

The present status of this prediction must be marked honestly: its direction is consistent with several known findings (different sensory processes are indeed associated with different frequency bands, such as vision with gamma and certain forms of integration with lower frequencies), but the precise "frequency-family–qualitative-kind" map required by the prediction has not yet been systematically measured. Thus it is partly testable now (directional correspondence) and partly a roadmap item (measurement of the precise map; see 25.5).

25.3. ρ_f-Specific Prediction II: Frequency–Timescale Correspondence (Quantitative Form)

The second unique resource of this book is that it links the temporal thickness of experience and the hierarchy of memory to the transient timescale spectrum—high-frequency transients are short, and low-frequency transients are long (Chapter 10). This yields the second unique prediction, which this book calls the frequency–timescale correspondence prediction. Unlike the first prediction, this one is not merely qualitative in direction; it has an explicit scaling law from EQT freezing dynamics as its backbone, and is therefore a quantitatively formulated unique prediction.

According to the scaling law cited in Chapter 10, its content can be stated precisely as follows: the duration for which experiential content is retained, τ_{retain}, and the frequency f of the neural activity that carries it obey the scaling relation $\tau_{\text{retain}} \sim \tau_0(f)/f$. That is, retention duration is approximately inversely proportional to frequency (dominated by $1/f$), where $\tau_0(f)$ is a slowly varying effective background factor (absorbing corrections from dissipation, coupling, and related factors inside the brain). It follows that shorter retention (immediate retention, working memory) corresponds to higher-frequency activity, whereas longer retention (long-term memory) corresponds to lower-frequency, and even structural, activity (extremely low-frequency long-duration transients, Chapters 3 and 10). Moreover, this correspondence is continuous: from the thickness of the present (below the second) to the hierarchy of memory (from seconds to lifelong retention), the same frequency–timescale spectrum unfolds continuously along $\tau_{\text{retain}} \sim \tau_0/f$.

This prediction is likewise not available to state indices: metabolism, PCI, and connectivity characterize the state at a given moment, while frequency–timescale correspondence characterizes a cross-temporal quantitative structural relation between "retention duration and frequency". Its falsifiability is clear, and it is stronger than a qualitative prediction because it has a scaling law: if measurements show that there is no systematic inverse correspondence between the duration of memory/retention and the neural frequency that carries it (for example, if extremely fleeting retention is carried by extremely low-frequency activity, or if long-term memory is unrelated to low-frequency structures), or if an inverse correspondence exists but systematically violates the $1/f$-dominant tendency and this deviation cannot be accommodated by the slow variation of τ_0, then the frequency–timescale correspondence is falsified, and the carrying relation of Chapter 10 is damaged at the empirical level.

The present status of this prediction must be marked honestly, and the word "quantitative" must be used with proper restraint. First, its direction and $1/f$ tendency are consistent with neuroscientific findings that link "high frequency" with "rapid and easily vanishing" processes and "low frequency" with "maintenance and integration" (as already cited in Chapter 10). Its direction, and even a rough inverse tendency, can therefore already be tested preliminarily. Second, it must be emphasized that the slow variation of $\tau_0(f)$ means that $\tau_{\text{retain}} \sim \tau_0/f$ is not an unconditionally exact f^{-1} law. It is a "$1/f$-dominant tendency + slowly varying τ_0 correction", and testing should therefore target this dominant tendency and its calibratable corrections, rather than demand exact pointwise inverse proportionality. Third, a precise segment-by-segment characterization of the entire frequency–timescale spectrum, especially the fine correspondence at the end of present thickness (retention–protention), remains a roadmap item (25.5). This way of stating the prediction—with a quantitative backbone, but also with an honest account of its slowly varying correction and its unmeasured parts—is the proper balance by which this book provides unique quantitative content from ρ_f, without claiming a zero-free-parameter exact law.

Taken together, 25.2 and 25.3 show where the empirical spine of the ρ_f framework lies. Through the frequency dimension and the timescale spectrum, it makes two kinds of content-specific and structural predictions (the frequency correspondence of qualitative kinds, and the frequency–timescale scaling law of retention duration). These two predictions cannot in principle be made by any "content-neutral state index" (metabolism, PCI, connectivity, or even an IIT centered on a single Φ). The second of the two also has the quantitative form $\tau_{\text{retain}} \sim \tau_0/f$. This is the substantive answer to the objection in 25.1: although the proxies of Ψ overlap with existing indices, the unique empirical content of this book does not lie in those state quantities of Ψ, but in the content-specific predictions made by the frequency dimension and the timescale spectrum. The ρ_f level does its work precisely here, and in the second prediction it does so to the extent of a quantitative scaling law.

25.4. Predictions Shared with Existing Theories or Usable for Discrimination

In addition to its unique predictions, this book also makes a series of predictions that are either shared with existing theories or can be used to distinguish this framework from them. Although these predictions are not unique to this book, they remain tests of the framework. The discriminating predictions are especially important because they are the points at which this book confronts other theories directly.

First, given-ness depends on the maintenance of internal–external co-evolution, not on the presence or absence of exogenous input (discriminating). Chapters 14 and 24 established that the survival or disappearance of given-ness depends on the maintenance of the two poles of internal–external co-evolution (when co-evolution collapses under anesthesia, given-ness disappears), rather than on whether the exogenous channel is open or closed (in sensory deprivation and dreaming, exogenous input is closed while given-ness remains). From this follows a set of falsifiable predictions: (a) if exogenous input is cut off while brain-internal co-evolution is preserved (as in early sensory deprivation or REM dreaming), given-ness should be maintained; (b) if brain-internal co-evolution collapses (as in certain forms of anesthesia or deep sleep), then given-ness should disappear even if exogenous channels are preserved. These predictions distinguish the framework from the naive view that "consciousness depends on sensory input", and also from some theories that emphasize peripheral input. Their falsifiability is clear: if the survival or disappearance of given-ness varies primarily with the presence or absence of exogenous input, rather than with the maintenance of co-evolution, then this prediction is falsified.

Second, the subject boundary depends on the extension of coherence (discriminating, with both similarities to and differences from GNW and IIT). Chapter 16 established that the boundary of the "I" is the boundary of dominant coherent co-evolution, and that physically manipulating coherence is manipulating the subject boundary. From this follows the prediction: (a) if the key connection maintaining cross-regional coherence is severed (as in split-brain surgery severing the corpus callosum), the subject should split accordingly into two (Chapter 16); (b) the subject boundary (the observable proxy of $\mathscr{B}$) should vary with the extension and closure of coherence, not with mere information availability (the global broadcasting of GNW) or a mere quantity of integration (the Φ of IIT). Under experimental conditions in which coherent extension and information availability / integration quantity dissociate, this book predicts that the subject boundary follows the former, which is discriminable from GNW and IIT.

Third, transitions among conscious states depend on the joint three dimensions of Ψ, not on any single dimension (discriminating, especially against single-scalar theories). Chapters 21 and 24 established that conscious states are jointly characterized by $\tilde{A}$, C, and $\mathscr{B}$, and that no single dimension is sufficient (in epilepsy, $\tilde{A}$ is high while C and $\mathscr{B}$ are abnormal, and consciousness is lost). This yields the prediction that there are states in which one-dimensional quantities dissociate from consciousness, such as high-$\tilde{A}$ but low-consciousness states (high metabolism with no consciousness, as in generalized epileptic seizure), and states in which C is low while $\tilde{A}$ is high. These dissociations should be explainable consistently by the joint three dimensions, not by any single dimension. This discriminates the book from theories that identify consciousness with one scalar quantity (single-scalar identity theories, including those centered on a single Φ): if conscious states can be completely predicted by one scalar quantity, and the separation among the three dimensions has no empirical consequence, then the three-dimensional claim of this book is weakened.

Fourth, predictions shared with IIT, GNW, and predictive processing. Where the above discriminating points are not involved, this book shares many predictions with the three theories (as Chapter 4 already stated, they are often complementary in division of labor): for example, complexity measures (PCI-like measures) distinguish conscious states (shared with IIT, and explained by this book through the projection C); neural signatures of global availability and reportability (shared with GNW, and explained by this book through the governance of dominant coherence); and prediction–error dynamics (shared with predictive processing, and explained by this book through internally and externally driven perturbative evolution). This book does not claim priority over these shared predictions. Rather, it shows that its framework can naturally accommodate and reinterpret these already supported findings. This is itself support for the compatibility of the framework, although it is not its unique contribution.

25.5. Testing Roadmap: What Cannot Yet Be Tested Immediately

Honesty requires this book to distinguish clearly between "currently testable" and "still belonging to the roadmap", rather than disguising the latter as the former. This section lists the parts of the book's predictions that still require further development in measurement methods or methodology and therefore cannot yet be strictly tested immediately, while indicating their direction.

First, the precise frequency–quality map. The directional form of the frequency–quality correspondence in 25.2 can already be tested, but a complete map of "which qualitative kind corresponds to which precise frequency-family structure" requires measurement tools with sufficient frequency resolution and content-decoding capacity (combining fine spectral analysis with content decoding), and is still under development. Roadmap: use high-frequency-resolution MEG/EEG combined with content decoding to systematically map the correspondence between qualitative kinds and frequency structures, and test the correspondence between kind discontinuities and frequency intervals.

Second, the fine frequency–timescale correspondence at the end of present thickness. For the frequency–timescale correspondence in 25.3, the memory-hierarchy end (above the second) is easier to test, and its $1/f$-dominant tendency can be roughly verified. But the frequency correspondence at the end of present thickness (retention–protention, the fine temporal structure below the second), together with the precise calibration of the slowly varying correction $\tau_0(f)$, requires measurement paradigms with sufficient temporal resolution targeted specifically at the structure of the "extended present". Such paradigms are still lacking. Roadmap: develop time-resolved paradigms targeting the retention–protention structure, test its correspondence with the decay of high-frequency transients, and calibrate the slowly varying form of $\tau_0(f)$.

Third, direct measurement of $\mathscr{B}$ as "subject boundary". Observable proxies of $\mathscr{B}$ (the extension and closure of cross-regional coherence) can already be measured, but the "subject boundary" itself—especially the fine form of coherent boundary in rare states such as dissociation and conjoined cases—requires measurement methods aimed at the topology of coherent closure, and is still under development. Roadmap: develop methods for characterizing the range and topology of coherent closure, and test the correspondence between subject boundary and coherent boundary in split-brain, dissociation, and conjoined cases.

Fourth, cross-subject comparison of qualities. The objectivity of "repeatable co-perturbation" in Chapter 13 in principle allows comparison of experiential types across subjects (different subjects under similar ρ_f structures have comparable experiential types). But strict cross-subject qualitative comparison must solve the methodological difficulty of "how to align qualitative structures across different subjects", which remains exploratory. Roadmap: use the cross-subject alignment of qualitative-space geometry (Chapter 6) as an intermediary, and test whether similar frequency structures correspond to comparable qualitative types.

It must be emphasized that listing these items as a roadmap rather than as already available predictions is not a retreat, but a mark of honesty. A theory that disguises what cannot yet be tested as already testable violates the spirit of M. By marking clearly what is currently testable (the directions in 25.2 and 25.3, the $1/f$ tendency, and the discriminating predictions in 25.4) and what must await methodological development (the four items in this section), this book implements the two-tier standard further within the measurable plane itself: within the measurable plane, there is also a distinction between "currently falsifiable" and "falsifiable in principle but awaiting methods".

25.6. Implementation of the Two-Tier Standard, and the Boundary of the Measurable Plane

All the predictions in this chapter implement the requirement that the meta-norm M imposes on the measurable plane, and they also mark the boundary of that plane. The two-tier standard and this boundary must be made clear at the end, as the closure of Part IV.

Within the measurable plane, falsifiability is strictly preserved. Everything listed in this chapter—the ρ_f-specific predictions (frequency–quality and the quantitative frequency–timescale relation $\tau_{\text{retain}} \sim \tau_0/f$), the discriminating predictions (co-evolution vs. exogenous input, coherence vs. availability/integration, three dimensions vs. one dimension), the shared predictions, and the roadmap items—is falsifiable (either currently falsifiable, or falsifiable in principle once methods are developed). For each item, this book has stated the conditions under which it would be overturned. This is the requirement of M for the measurable plane, and the book exempts none of its claims from it: every genuinely measurable module of this book retains a prediction that is falsifiable in the near or medium term.

Outside the measurable plane lies the final cell, and it is not included among the predictions of this chapter. This chapter must state again (consistent with Chapters 2 and 15): the irreducible residual point of being-experienced (the final cell) cannot, by its nature, be falsified by any observable; therefore this chapter makes no prediction concerning it and does not use the success or failure of any prediction in this chapter to support or refute it. To misread "the frequency–

quality correspondence holds" as "therefore the being-experienced of red has been explained" is an overstep: the former is a content-specific prediction on the measurable plane (red and pain correspond to different frequency structures), whereas the latter is the final cell (why red is experienced as such). However successful the predictions of this chapter may be, they fulfill only the measurable plane. The final cell always belongs to the philosophical boundary (Chapter 27), and not to observable adjudication.

This boundary makes the predictions of this chapter more credible rather than more suspicious: a theory that clearly knows "what my predictions can test and what they cannot test" is more scientifically robust than a theory that claims "my predictions also solve being-experienced". The measurable predictions of this book (especially the two ρ_f-specific predictions) are where its substantive contribution and risk in empirical science lie. They can be overturned, and are therefore genuine predictions. Their success or failure concerns only the truth or falsity of the measurable side of the book, not the final cell, which by its nature cannot be adjudicated by observables.

Part IV is thereby complete. Ψ has been placed as a measurable projection (Chapter 21), its three quantities characterized (Chapter 22) and given measurement interfaces (Chapter 23), transitions among conscious states unified and explained (Chapter 24), and this chapter has gathered all of this into falsifiable predictions. By means of the two ρ_f-specific predictions (frequency–quality and the quantitative frequency–timescale relation), it has answered the objection concerning whether ρ_f does any work, implemented the two-tier standard, and marked the boundary between the measurable plane and the final cell. Part V will then turn to the philosophical closure: the fulfillment of mind–matter unity (Chapter 26), the direct analysis of the final cell (Chapter 27), and the conclusion of the book (Chapter 28).

25.7. Chapter Summary

This chapter is the point of convergence of Part IV. It gathers the measurable plane into falsifiable predictions and directly answers the objection concerning whether ρ_f does any work. First, it states the objection clearly: the proxies of Ψ ($\tilde{A} \approx$ metabolism, $C \approx$ complexity, $\mathscr{B} \approx$ coherent connectivity) overlap with existing indices; without unique predictions, the ρ_f level would be idle. Second, it gives two ρ_f-specific predictions as the empirical spine of the book: frequency–quality correspondence (different qualitative kinds correspond to different neural frequency structures, and discontinuities among kinds correspond to frequency intervals—a prediction that content-neutral state indices cannot in principle make); and frequency–timescale correspondence (retention duration and neural frequency obey the scaling law $\tau_{\text{retain}} \sim \tau_0/f$, with $1/f$ dominant and τ_0 slowly varying, forming one spectrum from present thickness to long-term memory—a unique prediction with quantitative form). Both are content-specific and structural predictions; their directions and tendencies are currently testable, while precise maps and calibration of τ_0 belong to the roadmap. Third, it gives discriminating and shared predictions: givenness depends on co-evolutionary maintenance rather than exogenous input (discriminating against naive views), the subject boundary depends on coherent extension (discriminating from GNW/IIT), and states depend on the joint three dimensions of Ψ rather than one dimension (discriminating against single-scalar theories); it also acknowledges predictions shared with IIT/GNW/predictive processing (which this book naturally accommodates and reinterprets). Fourth, it honestly lists four roadmap items (the precise frequency–quality map, the fine frequency–timescale correspondence of present thickness and the calibration of τ_0, direct topological measurement of $\mathscr{B}$ as subject boundary, and cross-subject qualitative comparison), distinguishing "currently testable" from "awaiting methods". Fifth, it implements the two-tier standard: within the measurable plane, falsifiability is strictly preserved without exception; outside the measurable plane lies the final cell, which is not included among the predictions and must not be supported or refuted by the success or failure of those predictions.

The failure conditions of this chapter are as follows (summarized in Appendix E).

- F25.1 Frequency–quality correspondence is falsified. If fine frequency-resolved measurement shows that qualitative kinds have no systematic correspondence with neural frequency structures (including no correspondence between kind discontinuities and frequency intervals), then the first unique prediction of this book is falsified, and the carrying relation of Chapter 7 is seriously damaged at the empirical level.

- F25.2 Frequency–timescale correspondence is falsified. If measurements show that retention duration has no systematic inverse correspondence with neural frequency, or if an inverse correspondence exists but systematically violates the $1/f$-dominant tendency of $\tau_{\text{retain}} \sim \tau_0/f$ and this deviation cannot be accommodated by the slow variation of τ_0, then the second unique prediction of this book (and its quantitative scaling law) is falsified, and the carrying relation of Chapter 10 is seriously damaged at the empirical level (linked to F10.5 in Chapter 10).

- F25.3 Discriminating predictions reverse. If the survival or disappearance of given-ness varies primarily with exogenous input (rather than co-evolutionary maintenance), if the subject boundary follows information availability / integration quantity (rather than coherent extension), or if conscious states can be completely predicted by a single quantity (so that the separation of the three dimensions has no empirical consequence), then the corresponding discriminating predictions reverse, and this book loses to the competing theory at that point of discrimination.

- F25.4 Roadmap items are shown to be untestable in principle. If any listed roadmap item is shown not merely to await methods but to be untestable in principle by any observable, then it should not be listed as a measurable-plane prediction and must be removed from the measurable plane (or assigned to the final cell, if it indeed belongs there).

- F25.5 Measurable predictions are used to support the final cell. If this book uses the success or failure of any measurable prediction to support or refute the final cell (being-experienced), it violates the boundary between the measurable plane and the final cell and constitutes an overstep.

Table 25.1.. The Boundary between What This Chapter Establishes and What It Cannot Derive

What this chapter establishes	What this chapter cannot derive
Two ρ_f-specific predictions (frequency–quality and the quantitative frequency–timescale relation $\tau_{\text{retain}} \sim \tau_0/f$) form the empirical spine of this book and answer whether ρ_f does work.	These predictions cannot explain or support being-experienced (the final cell, not included among the predictions).
Frequency–timescale correspondence has the form of a quantitative scaling law ($1/f$ dominant, with τ_0 slowly varying), and its test concerns the dominant tendency rather than exact pointwise inverse proportionality.	This chapter cannot provide the exact form of $\tau_0(f)$ (its calibration belongs to the roadmap).
Discriminating predictions (co-evolution vs. exogenous input, coherence vs. availability/integration, three dimensions vs. one dimension) are the points where this book confronts other theories.	This chapter cannot decide the overall success or failure of other theories (only the discriminating points can be adjudicated).
Four testing-roadmap items honestly distinguish "currently testable" from "awaiting methodological development".	Roadmap items cannot be treated as already testable predictions (they require the corresponding methods).

What this chapter establishes	What this chapter cannot derive
The implementation of the two-tier standard: nothing on the measurable plane is exempt from falsifiability; and the boundary between the measurable plane and the final cell is preserved.	The success or failure of measurable predictions cannot be used to support or refute the final cell (overstep, F25.5).

Part IV.

Philosophical Boundary and the Fulfillment of Mind–Matter Unity

26. The Metaphysical Fulfillment of Mind–Matter Unity

26.1. The Moment of Fulfillment: From Parts to Whole

By this point in the book, the first four parts have assembled all the required components. Part I established the general framework, ontology, and positioning; Part II fulfilled the isomorphism between the structure of experience and the gradient configurations of ρ_f (the identity level); Part III characterized the given-ness of experience as internal–external co-evolutionary freezing of a bounded interior; Part IV projected all of this into the measurable Ψ and into falsifiable predictions. Yet the central thesis promised by the title of the book—"mind–matter unity"—has not yet been stated directly and completely: in what sense, exactly, are mind and matter "one"? Is this "unity" a serious metaphysical claim, or merely an attractive title? The task of this chapter is to fulfill that promise directly—to synthesize the components distributed across the first four parts into a complete statement about "unity".

The nature of this chapter must first be stated, because it differs from the first four parts. Each chapter in the first four parts was supported by something testable of its own—relation-by-relation correspondence in isomorphism, the mechanism of given-ness, or measurable predictions. This chapter is a philosophical synthesis and distillation: it introduces no new isomorphism, no new mechanism, and no new prediction, but gathers what the first four parts have already established into a unified claim about the structure of reality. Its persuasiveness does not lie in a new test at any single point, but in whether it can bring everything established in the whole book under the single claim of "unity", without contradiction and with the least remaining residue.

At the same time, a boundary that this chapter must not cross has to be drawn, because this is where it is most easily breached. What this chapter fulfills is **"unity"—the claim that mind and matter are two sides of the same reality; it does not** fulfill, nor pretend to fulfill, the irreducible final cell (the residual point of being-experienced, Chapter 15). The fulfillment of "unity" and the preservation of the final cell are fully compatible: this book can seriously argue that mind and matter are one, while honestly acknowledging that the being-experienced of this unity remains the final cell. To mistake "fulfilling unity" for "also fulfilling the final cell" is the overstep this chapter must guard against most carefully. It would make the book repeat the error of concealing the explanatory gap behind an attractive thesis, which is precisely what Chapter 1 rejected.

26.2. The Complete Statement of First-Level Identity

The exact meaning of "unity" depends on the "identity level" selected in Chapter 5. This section gives the complete statement of that level. Earlier it was introduced gradually and in scattered form as the argument required; here it must be stated once and for all as a complete thesis.

Complete identity thesis: first-person experience and a specific ρ_f process are one and the same reality. Seen from the third-person, external direction, this reality appears as the gradient configuration of ρ_f and its internal–external co-evolution (that is, what physical science can describe); seen from the first-person, internal direction, it is the given experience itself. Mind and matter are not a relation between two realities (correlation, correspondence, or modeling), but two directions of access to the same reality—external and internal.

Each part of this thesis has already been established separately, and is here brought together as one. "A specific ρ_f process"—not ρ_f in general, but the coherent co-evolution of a bounded interior (Chapters 3, 9, and 14); "seen from the outside as gradient configuration and co-evolution"—this is the side accessible to physical science, characterized in Part II (structure) and Part III (the mechanism of given-ness); "seen from the inside as experience itself"—this is the first-person side of given-ness. The essence of the identity level is that the external side and the internal side are not two things, but two directions of access to the same reality, just as the "morning star" and the "evening star" are two names for the same Venus, accessed from two directions (Chapters 4 and 5).

The identity level must again be contrasted with the two weaker levels it rejects, in order to show the weight of "unity" (Chapter 5 already argued this; here it is restated from the standpoint of "unity"). If one adopts the structural-isomorphism level (mind and matter are structurally parallel but remain two kinds of things), then "unity" does not hold—that is "two things in parallel", not "one"; and one must additionally explain why these two kinds of things are so perfectly parallel (the difficulty of pre-established harmony). If one adopts the modeling level (ρ_f is a model of experience), then "unity" holds even less—a model and what is modeled are two different things, so there is no real unity. Only at the identity level does "unity" become a claim about reality: mind and matter are one reality, not two realities that are closely correlated or well modeled. What this book fulfills under the title *Mind–Matter Unity* is precisely this strongest level—not that mind and matter are tightly related, but that they are fundamentally the same reality.

26.3. The Two Sides of "Unity": The Structural Side and the Given Side

"Mind and matter are two sides of the same reality"—these "two sides" must be specified, otherwise "two sides" remains only a metaphor. This section concretizes the two sides: what constitutes the external side (accessible to physical science), what constitutes the internal side (given-ness), and why the two are two sides of the same reality rather than two realities.

The external side: structure from the five sources, and mechanism from co-evolution. Seen from the third-person, external direction, that reality appears as the gradient configuration of ρ_f and its dynamics. This side has two levels, both already characterized in the preceding parts. The first is structure: the fivefold rich structure of gradient configurations (single-frequency configurations, the frequency dimension, master-equation evolution, resonance modes, and transient timescales; Part II), which isomorphically carries the entire structure of experience (quality, kind, flow, unity, and temporal thickness). The second is the mechanism of given-ness: internal–external co-evolutionary freezing of a bounded interior (Part III), which is the physical counterpart of given-ness. Thus the external side is "fivefold-rich gradient structure + the internal–external co-evolutionary mechanism of given-ness"—that is, everything that physical science, and Parts II, III, and IV of this book, can describe, project, and test.

The internal side: the given experience itself. Seen from the first-person, internal direction, that same reality is the given experience itself. It has all the structures characterized by the external side (because it is the same reality, their structures must be the same; this is exactly what the structural sufficiency of Chapter 5 secures), and its given-ness is what that internal–external co-evolution is from within. The internal side is not another reality added to the external side, but the same reality appearing from within.

Why are they two sides of the same reality rather than two realities? This is the key to "unity" and must be answered directly. The book's answer has two parts. First, ontologically: according to Chapter 2, ρ_f is the unique fundamental ontology, and there is no second entity alongside it. Therefore the internal side cannot be another reality; it can only be the internal side of the same ρ_f reality. Second, structurally: Part II has shown relation by relation that the entire structure of the internal side (experience) is isomorphic to the entire structure of the external side (gradient configuration), and complete structural isomorphism is the mark of "the same reality" (two genuinely different realities have no reason to have the same structure everywhere; the most natural explanation for structures that are the same everywhere is that they are one). Ontological uniqueness and structural isomorphism together support the claim that "the two sides are the same reality": the unique ontology rules out the possibility of "two realities", while structural isomorphism positively shows that the two sides share the same structure and are therefore two sides of the same reality.

Thus "unity" receives its concrete fulfillment: the "unity" of mind and matter means that the ρ_f coherent co-evolutionary process of a bounded interior, seen from the outside, is the gradient structure and co-evolutionary mechanism described by physical science (structure from the five sources, mechanism from internal–external co-evolution), and, seen from the inside, is the given experience. The two are two sides of the same process, guaranteed by ontological uniqueness and structural isomorphism. "Unity" is no longer a metaphor or slogan, but this specific thesis.

26.4. "Open Process": The Dynamic Nature of Unity

This section must supply one feature of "unity" that is easily overlooked but crucial to its exact form: the "same reality" is not a static configuration, but an open process. This distinguishes the unity proposed in this book from certain static identity theories, and directly connects it with the co-evolution of Part III and the detection ontology of Chapter 13.

What is an "open process"? First, it is process rather than static configuration: that reality is the continuous evolution of ρ_f (the master equation, Chapter 8), the continuous maintenance of coherence (Chapter 9), and the continuous formation and decay of transients (Chapter 10). It flows, maintains itself, and renews itself in time; it is not a frozen snapshot. Experience as stream and as temporally thick corresponds precisely to this reality as process. Second, it is open: according to Chapters 13 and 14, the coherence of that bounded interior is not closed and self-sufficient, but continuously co-evolves with its exterior (the rest of the brain background, and through the body, the world). Given-ness is precisely the freezing of this open internal–external co-evolution. Therefore the "same reality" is an open, internally and externally co-evolving process, not an isolated and closed configuration.

That "unity" is "two sides of an open process" has an important consequence that must be stated (it echoes a subtle treatment of the identity level in Chapter 14). If that reality were imagined as an isolated and closed configuration, its "internal side" would fall back into the "darkness of the isolated interior" (Chapter 12): something closed within itself has no given-ness. It is precisely because that reality is open and continuously undergoing internal–external co-evolution that it has a side of given-ness. Therefore, among the two sides of "unity", the internal side (given-ness) depends precisely on the openness of that reality: given-ness is not the intrinsic property of an isolated interior, but the freezing of open co-evolution. "Mind and matter are two sides of the same open process"—the word "open" is not a decorative modifier, but carries the key condition under which the given side can obtain. Only an open process can have an internal side of given-ness. The "unity" proposed in this book is therefore a dual-aspect monism of open process: one and the same open process has, as its external side, the physical co-evolutionary structure, and, as its internal side, given experience.

26.5. Advantages and Costs Relative to Dualism and Physicalism

As a metaphysical thesis, "unity" shows its weight through comparison with its main competitors. This section weighs the advantages and costs of the book's position relative to dualism and physicalism. The advantages must be real, and the costs must be acknowledged; any comparison that states only advantages and denies costs is not credible.

Relative to dualism. The strength of dualism lies in its respect for the irreducibility of given-ness (Chapter 4); its weakness lies in positing a second entity, thereby facing problems such as how mind and matter interact and how a second entity is to be placed within the physical world (the problem of psychophysical interaction). The advantage of this book is that, with a monistic ontology, it both respects the irreducibility of given-ness (placing it as the internal side of the same reality, and ultimately as the final cell) and avoids the interaction problem of dualism. Since mind and matter are two sides of the same reality, there is no question of "how two realities interact" (one reality does not need to interact with itself). The cost of this book is that it must accept a thesis that is not intuitively easy: what, seen from the outside, appears only as gradient fluctuation is, seen from the inside, vivid experience (Chapter 5 already acknowledged this cost). Dualism trades "two realities" for intuitive naturalness (mind and matter do indeed seem very different); this book trades "one reality" for ontological unity and the elimination of the interaction problem, at the cost of bearing the counterintuitive character of "the difference between two sides of the same reality". This book regards the exchange as worthwhile: the interaction problem is a hard wound for dualism, whereas "two sides of the same reality look different" is counterintuitive but not logically contradictory (the morning star and the evening star also look different).

Relative to physicalism. The strength of physicalism lies in ontological economy and unity (Chapter 4); its weakness lies in its treatment of given-ness. Reductive physicalism must reduce given-ness to function (falling into topic substitution), while eliminative physicalism must deny given-ness (sawing off the branch on which it sits). The advantage of this book is that it shares monism and economy with physicalism (the same ρ_f ontology), but does not need to reduce or eliminate given-ness. It preserves, within a monistic ontology, an irreducible but non-mysterious and non-substantialized place for given-ness (the internal side and the final cell). The cost of this book is that it must acknowledge one side that physical language cannot exhaust (the being-experienced of the internal side, the final cell). For strict physicalism, this is a concession, and physicalism may regard it as "incomplete" or "mystical". This book regards that cost as honesty rather than a defect: the irreducibility of given-ness is real (as argued in Chapter 15), and a physicalism that denies it buys surface completeness at the cost of truth. This book prefers the honesty of "acknowledging one irreducible cell" to the false completeness of "reduction or elimination".

The place of unity between the two competing positions. It follows from the above that the "unity" proposed in this book stands between dualism and physicalism, taking the strengths of both while avoiding their weaknesses: it is ontologically monistic and economical like physicalism (avoiding dualism's interaction problem), while respecting the irreducibility of given-ness like dualism (avoiding physicalism's reduction or elimination). It can do both because it adopts a dual-aspect monism centered on the identity level—monistic (satisfying physicalism's economy) and dual-aspect (satisfying dualism's respect for given-ness). This position of "standing between and taking both strengths" is the weight of "unity" as a metaphysical thesis: it is not a vague third position outside dualism and physicalism, but a definite thesis centered on the identity level and capable of being compared with both line by line.

26.6. The Limit of Fulfillment: Fulfilling Unity Is Not Fulfilling the Final Cell

This chapter must preserve, with maximum clarity, the boundary drawn in 26.1. This is also the final guarantee of the chapter's honesty.

This chapter has fulfilled "unity": mind and matter are two sides of the same open ρ_f co-evolutionary process; the external side is composed of the structure from the five sources and the mechanism of internal–external co-evolution, while the internal side is given experience; and the two are guaranteed to be one by ontological uniqueness and structural isomorphism. This is a complete and serious metaphysical thesis. The book uses the first four parts as its components and this chapter as its synthesis in order to fulfill it directly.

But the fulfillment of "unity" is not the fulfillment of the final cell. Even if mind and matter are indeed two sides of the same open process, one can still ask the irreducible final question: why is the internal side of this same process experienced, rather than having the same structure and the same co-evolution without experience? In other words, "unity" tells us that mind and matter are two sides of the same reality, but it does not tell us why this internal side is experienced. The latter is the final cell identified in Chapter 15 and not entailed by structure or co-evolution. The fulfillment of "unity" and the preservation of the final cell are therefore two different matters: this book seriously argues for unity (this chapter), while honestly preserving the final cell (Chapters 15 and 27).

The relation between the two must be stated accurately to avoid misunderstanding. "Unity" does not eliminate the final cell, but it places it. Precisely because mind and matter are two sides of the same reality, the irreducible final cell is no longer "a gap between two realities" (as in dualism), but the single point of "the being-experienced of the internal side of one reality". The thesis of unity relocates the residual point from "a gap between mind and matter" to "the final cell inside a monistic reality". Unity does not cross this cell, but it makes it a precisely located single point within a monistic framework, rather than a dualistic abyss. This is the metaphysical manifestation of what Chapters 1 and 15 called "compressing the hard problem to the final cell": the fulfillment of unity compresses the residue to the smallest size and locates it with the greatest precision; that smallest residue is still honestly left to Chapter 27.

This chapter can therefore close its boundary in one sentence: the book has fulfilled the "unity" of mind and matter, and by means of this unity has placed the residual of the hard problem as a final cell inside one monistic reality; the book has not, and does not pretend to, cross that final cell. The fulfillment of unity is the fulfillment that compresses the residue to its minimum and locates it most precisely, not the fulfillment that eliminates the residue as well. If this sentence is preserved, "mind–matter unity" is an honest and defensible metaphysical thesis, not an attractive slogan concealing the explanatory gap.

26.7. Conclusion of This Chapter

This chapter has directly fulfilled the "mind–matter unity" promised by the title of the book. Its main points are as follows. First, it gives the complete statement of first-level identity: first-person experience and a specific ρ_f process (the coherent co-evolution of a bounded interior) are the same reality; seen from the outside, it appears as gradient configuration and co-evolution (accessible to physical science), while seen from the inside, it is given experience. Mind and matter are not a relation between two realities, but two directions of access to the same reality. Second, it concretizes the "two sides": the external side consists of the structure from the five sources + the internal–external co-evolutionary mechanism of given-ness, while the internal side is given experience; that the two are two sides of the same reality is jointly guaranteed by ontological uniqueness (ρ_f as the unique ontology, excluding two realities) and structural isomorphism (the structure of experience and the structure of gradients are isomorphic everywhere, marking the same reality). Third, it clarifies the dynamic nature of "unity": that reality is an open process undergoing internal–external co-evolution, not an isolated closed configuration. Only because it is open can it have a side of given-ness (otherwise it would fall back into the darkness of the isolated interior). The unity proposed in this book is therefore a "dual-aspect monism of open process". Fourth, it weighs the advantages and costs relative to dualism (advantage: monism avoids the interaction problem while still respecting given-ness; cost: it must accept the counterintuitive difference between two sides of the same reality) and physicalism (advantage: economical monism without reducing or eliminating given-ness; cost: it must acknowledge one irreducible cell), and positions unity as "standing between the two, taking both strengths (monistic economy and respect for given-ness) while avoiding both weaknesses (interaction problem and reduction/elimination)". Fifth, it preserves the boundary: the fulfillment of unity is not the fulfillment of the final cell. Unity does not eliminate the final cell, but places it (resetting the residue from "a gap between mind and matter" to "a single point inside one monistic reality"), while the residue remains for Chapter 27.

The failure conditions of this chapter are as follows (summarized in Appendix E; this chapter is a philosophical synthesis chapter, and its failures mostly concern the coherence of the thesis and the faithfulness of its synthesis of the preceding parts).

- F26.1 The synthesis at the identity level is incoherent. If the synthesis that "mind and matter are two sides of the same reality" is shown to be internally inconsistent with what the first four parts have already established (structural isomorphism, the mechanism of given-ness, and the projection Ψ), then the thesis of unity is incoherent and must be restated.

- F26.2 The two sides are not the same reality. If the argument that "the two sides are the same reality" (ontological uniqueness + structural isomorphism) is shown not to hold—for example, if structural isomorphism does not in fact hold everywhere (a failure of some carrying relation in Part II), or if the unique ontology of ρ_f is overturned (Chapter 2, F2.1)—then "unity" is downgraded to dualistic parallelism (the structural-isomorphism level) or to the modeling level, and the promise of the title fails.

- F26.3 Openness is not necessary for given-ness. If given-ness is shown to be able to obtain in an isolated closed configuration (not requiring open internal–external co-evolution), then the characterization of "open process" fails, and the argument of 26.4 is damaged (linked to F14.1 in Chapter 14).
- F26.4 Unity is mistaken for the fulfillment of the final cell. If this book uses the thesis of "mind–matter unity" as an answer to "why the internal side is experienced" (the final cell), then it violates the boundary of this chapter and the prohibition of the whole book, and constitutes the concealment of the explanatory gap behind an attractive thesis (this is the failure this chapter must guard against most carefully).

Table 26.1.. The Boundary between What This Chapter Establishes and What It Cannot Derive

What this chapter establishes	What this chapter cannot derive
The complete statement of first-level identity: mind and matter are two directions of access to the same ρ_f process.	It cannot infer from "unity" why the internal side is experienced (the final cell, Chapter 27).
The concreteness of the "two sides" (external: five-source structure + co-evolutionary mechanism; internal: given-ness), and the guarantee of "same reality" (ontological uniqueness + structural isomorphism).	It cannot re-prove structural isomorphism itself in this chapter (its argument belongs to Part II).
The dynamic nature of unity: dual-aspect monism of open process (only openness allows a side of given-ness).	It cannot provide the details of the mechanism of given-ness in this chapter (these belong to Part III).

What this chapter establishes	What this chapter cannot derive
The advantages and costs of unity relative to dualism and physicalism, and its position of "standing between and taking both strengths".	It cannot decide the overall success or failure of dualism/physicalism in this chapter (it only weighs them point by point).
Boundary preservation: the fulfillment of unity places (but does not eliminate) the final cell, compressing the residue into a single point inside one monistic reality.	It must not use unity to masquerade as the fulfillment of the final cell (the prohibition of the whole book, and the point this chapter must guard against most carefully).

27. The Irreducible Residual Point and the Boundary of Explanation

27.1. Returning to the Final Cell: After the Fulfillment of Unity

Chapter 15 first identified the irreducible final cell in the context of given-ness. It exposed the fissure that, even when all the structures and relations required for being-experienced are present, "being-experienced" is still not entailed, and it established the C3 prohibition: no mechanistic statement may be made to masquerade as an answer to it. Chapter 20 then compressed this cell for a second time: through the restatement of the interface lineage, and through the removal of the components of cognization and $\mathscr{B}$-demarcation, it further compressed and located the residual point from the "general given-ness residual point" to the subject-attribution residual point—why an activity that has been integrated, cognized, and demarcated by $\mathscr{B}$ as "mine" is experienced. This chapter does not repeat those two identifications and compressions. It does something that the previous two chapters could not yet do: after Chapter 26 has fulfilled "unity", it returns to the final cell and gives it a philosophical determination and closure. It clarifies its exact nature and position within a monistic framework, explains why acknowledging it strengthens rather than weakens this book, shows its convergence with phenomenology, and finally draws the boundary of what the theses of "identity", "interface", and "demarcation" can and cannot do.

It must first be stated that what this chapter faces is precisely the residual point handed over by Chapter 20 after two compressions to its most precise location—the subject-attribution residual point. It is smaller and more precise than the given-ness residual point in Chapter 15, because it has already stripped away addressable components such as cognization and demarcation. Yet its irreducibility is of the same kind as that diagnosed in Chapter 15: being-experienced is not entailed by structure and relation, which here take the more specific forms of integration, cognization, and demarcation mechanisms. The philosophical treatment of this chapter therefore applies both to this most precise subject-attribution residual point and to the broader class of irreducible residual points to which it belongs.

The reason it must be revisited "after the fulfillment of unity" is that unity changes the status of this final cell (as already anticipated in Section 26.6). Before unity is fulfilled, the "residue of being-experienced" can be misread as "a gap between mind and matter", as if it were an abyss lying between two kinds of things. After unity is fulfilled, this misreading is excluded: since mind and matter are two sides of the same reality, the residue cannot be a gap "between" the two (within the same reality, there is no such "between"). It can only be a single point inside monistic reality—why the internal side of the same reality, that which has been integrated, cognized, and demarcated as "mine", is experienced. The task of this chapter is precisely to give this single point internal to monism the philosophical treatment it deserves after it has been thus relocated.

It must also be reiterated that this chapter remains governed by the discipline of the whole book. Everything this chapter says about the final cell concerns its nature and position—what kind of problem it is, where it is located, and why it is irreducible—not an answer to it. It has no answer, as Chapters 15 and 20 have already established. To discuss the nature of a problem and to answer that problem are two different matters; this chapter does only the former.

27.2. The Nature of the Final Cell: It Is a Boundary, Not an Object

To give the final cell a philosophical determination, one must first see what kind of thing it is. The core determination in this book is that it is a boundary, not an object. This distinction determines the entire posture toward it.

An object, or a problem awaiting solution, is something that can be studied, answered, and incorporated into knowledge; it lies within our explanatory capacity, even if it has not yet been solved. A boundary, by contrast, marks where a certain explanatory capacity comes to an end. It is not inside that capacity, but is the edge of that capacity itself. The final cell is the latter. It is not a difficult problem within the capacity of physical explanation that remains to be solved, but the edge of physical explanatory capacity. It marks precisely the endpoint of what the language of "structure and relation" can reach. Chapter 15 already argued this point: being-experienced is not entailed by structure and relation, and therefore lies outside structural–relational language. The second compression in Chapter 20 further showed that even after more refined structural–relational characterizations such as cognization and demarcation are added, the residual point still remains outside the boundary, although it is pushed deeper and made smaller. This chapter therefore determines it as follows: the final cell is a boundary marking where physical, structural–relational explanation stops, not an object within physical explanation that remains to be solved.

This determination as "boundary rather than object" clarifies the proper attitude toward the final cell. Toward an object, or a problem awaiting solution, the proper attitude is continued investigation and solution; setting it aside would be a defeat. Toward a boundary, however, the proper attitude is to mark it precisely and to understand why we stop there. To try to "cross" a boundary in order to "answer" it is not progress, but confusion, because what lies beyond the boundary is, by definition, outside the range within which that explanatory capacity can provide answers. This book does not answer the final cell;

this is therefore not an admission of defeat. Defeat would mean giving up on an object that could in principle be solved. It is instead the proper acknowledgment and precise marking of a boundary. Treating the final cell as an "object" (believing that more effort will solve it) and treating it as a "boundary" (understanding that it is already outside physical explanation) are two fundamentally different attitudes. This book adopts the latter. The dissatisfaction caused by earlier drafts of this project partly arose because they oscillated between "treating it as a solvable object and forcing an answer" (the ambitious draft) and "because it is unsolvable, setting aside the solvable parts together with it" (the cautious draft), without clearly determining it as a boundary that should be marked rather than forcibly crossed, while everything on its inner side that can be solved must be fully solved. One further point must be added: the second compression in Chapter 20 is precisely the more thorough realization of the demand that "everything solvable inside the boundary must be fully solved". It stripped away cognization, demarcation, and other addressable components that might previously have been mistakenly included in the residual point, thereby marking the boundary more precisely than Chapter 15 did.

27.3. The Position of the Final Cell: A Single Point Inside Monistic Reality

After its nature has been determined, its position must be specified. Only a precisely located boundary can properly be said to be "marked"; otherwise it becomes vague agnosticism.

The position of the final cell is forced into place by the work of the book part by part, and can be stated precisely as follows. It is not in the structure of experience: structure has been isomorphically carried by the five sources (Part II), so with respect to structure there is no residue. It is not in the mechanism of given-ness: the physical counterpart of given-ness, internal–external co-evolutionary freezing of a bounded interior, has been characterized in Part III, so with respect to mechanism there is no residue. It is not in cognization or demarcation: how experience is compared with and organized by the knowledge base, and how it is demarcated by $\mathscr{B}$ as "mine", have been characterized in the later chapters of Part III (Chapters 17 to 19), so with respect to the structure of cognization and attribution there is no residue. It is not in measurable projection: Ψ and falsifiable predictions have been given in Part IV, so with respect to measurement there is no residue. It is not "between mind and matter": unity has been fulfilled (Chapter 26), and within one and the same reality there is no "between". After all these "not there" locations are excluded one by one, the position of the final cell is forced into a single point: why the internal side of the same reality, that which has been integrated, cognized, and demarcated as "mine", is experienced, rather than operating with the same structure, the same co-evolution, the same cognization, and the same demarcation in the absence of experience.

The precision of this position is the key difference between this book and any vague "mysticism of consciousness". Vague mysticism says that consciousness is mysterious and unfathomable, but cannot specify where the mystery lies; it pushes the entire hard problem wholesale into darkness. This book can say precisely: everything solvable—structure, mechanism, cognization, demarcation, and measurement—has been solved; what remains unsolved is exactly this one point, and it can explain why it is exactly there: being-experienced is not entailed by structural–relational facts (Section 27.2). A precisely located boundary, together with an explanation of why the inquiry stops there, is different from a diffuse atmosphere of mystery. The former is precise demarcation, an epistemological achievement; the latter is a relinquishment of knowledge. The more precise the position of the final cell becomes—and through the two compressions it has now been located as the subject-attribution residual point—the less this book is a mysticism, and the more it is a precise theory that knows its own boundary.

27.4. Why Acknowledging It Strengthens Rather than Weakens This Book

A work titled with "unity" and aiming to unify mind and matter ends by acknowledging one cell that cannot be crossed. Does this amount to an admission of failure? This section argues directly that acknowledging it strengthens, rather than weakens, this book. Chapter 15 already addressed this from the standpoint of credibility; this chapter, after the fulfillment of unity, gives three deeper reasons.

First, acknowledging the boundary defines the achievement. The achievement of a theory becomes determinate only when its boundary has been drawn. What this book has fulfilled—structural isomorphism, the mechanism of given-ness, the characterization of cognization and demarcation, measurable predictions, and mind–matter unity—has its weight precisely because "the final cell lies beyond this". The book knows clearly what it explains (everything inside the boundary) and what it does not explain (the boundary itself). A theory that draws no boundary and claims that "everything has been solved" makes its own alleged achievements suspect, because it has obviously counted the unsolvable as solved. By drawing the boundary, this book makes its achievements inside the boundary more credible. Acknowledging the final cell does not weaken the achievement; it defines the achievement and thereby makes it solid.

Second, the minimal residue is the result of compression. The final cell in this book is not a large and untreated hard problem. It is a single point compressed to the limit: all the solvable components of the hard problem—structure, mechanism, cognization, demarcation, and measurement—have been stripped away, leaving only the cell of why the internal side, that which has been demarcated as "mine", is experienced (Section 27.3). To compress a hard problem that once pervaded the entire study of consciousness and seemed impossible to approach into a single precisely located point through two compressions—first to the given-ness residual point in Chapter 15, and then to the subject-attribution residual point in Chapter 20—is itself a major epistemological advance. Acknowledging this cell, then, is not an admission that "the hard problem remains unsolved" in its original diffuse form, but the declaration that "the hard problem has been compressed to its irreducible core". This compression is the achievement of all the parts of the book, including the two compressions in the later chapters of Part III; it is not their failure.

Third, monistic placement fills the abyss. Section 26.6 already argued that unity relocates the final cell from "a dualistic abyss between mind and matter" to "a single point inside monistic reality". This placement is itself a substantive transformation of the hard problem. In the dualist picture, an abyss separates mind and matter, and the problem is how two kinds of things can be connected. This book fills that abyss through unity. What remains is no longer an abyss "between the two", but only the internal single point of "the being-experienced of the internal side of the same reality". Acknowledging this internal monistic point is very different from acknowledging a dualistic abyss. The former marks one final irreducible location inside an already unified framework; the latter admits that two worlds cannot be connected. This book acknowledges the former, and the former is precisely the result of having filled the latter through unity.

Taken together, these three points show that acknowledging the final cell defines the achievement, declares the result of compression, and fills the dualistic abyss by monistic placement. It strengthens this book, just as a precise map strengthens itself by marking "the edge of the known world". Marking the edge is not the failure of the map, but the mark of its precision. A map that claims to have no edge and to have drawn everything would be less credible.

27.5. Convergence with Phenomenology: Different Paths toward the Irreducibility of Given-ness

This book reaches the conclusion that given-ness is irreducible through a completely independent, bottom-up physical path: from ρ_f, to structure, to given-ness, to cognization and demarcation, and then to the final cell. It is worth noting that a very different philosophical tradition—phenomenology—has long reached a highly convergent insight through a top-down path of reflection on experience itself. This section points out that convergence, because the convergence of two independent paths is itself a kind of support for the conclusion they reach.

A core insight of phenomenology, represented by Husserl and the tradition that follows him, is that given-ness is the starting point of everything and cannot be derived from anything more basic. The being-given of experience—the world's manifestation to me, the fact that experience is experienced—is the beginning point of phenomenology, the irreducible datum that remains after suspension. All knowledge of the world presupposes this given-ness, and therefore given-ness itself cannot be further reduced to something non-given in order to be explained; that would be circular or regressive. Phenomenology therefore places given-ness in a foundational and irreducible position.

The final cell of this book strongly converges with this phenomenological insight. Through the physical path, this book reaches the conclusion that given-ness, the being-experienced of the internal side, is not entailed by structure and relation; it is the final cell and is irreducible. Through the reflective path, phenomenology reaches the conclusion that given-ness is an irreducible beginning point. Both point to the same irreducibility: given-ness cannot be reduced to something non-given. This book and phenomenology therefore arrive at the same point by different paths: one proceeds bottom-up from ρ_f, the other top-down from reflection on experience, and yet both converge on the irreducibility of given-ness.

The significance of this convergence must be assessed accurately and not exaggerated. It does not mean that this book "proves" phenomenology, or that phenomenology "proves" this book. The two paths have different premises and methods, and cannot deductively prove one another. Its significance is this: two completely independent paths converge on the same irreducible point, and such independent convergence is a form of support for the claim that there is indeed an irreducible point here. A convergence of independent evidence is more credible than an assertion along a single path. The final cell of this book is therefore not an isolated posit forced only from within this book's framework; it has an independent counterpart in the phenomenological tradition. This gives the claim that "the final cell truly exists and is irreducible" independent support from another tradition. The boundary this book reaches from below is precisely the beginning point phenomenology confirms from above—the same boundary is marked from two directions.

A note on intentionality is also needed, because it is another core theme in phenomenology and resonates with this book's account of the relationality of given-ness and with Chapter 19's $\mathscr{B}$-demarcation of world-orientation. Phenomenology maintains that experience is essentially intentional: consciousness is always consciousness "of" something, always directed toward an object. This book's characterization of given-ness—internal–external co-evolution, and the claim that given-ness always appears "to an interior" (Chapters 12 and 14)—together with the "world-oriented demarcation" of $\mathscr{B}$ in Chapter 19 (experience is always demarcated as "the world I face"), strongly accords with the directed structure of intentionality. The relational and directed character of given-ness corresponds precisely to phenomenology's claim that consciousness is always directed toward an object. This book characterizes the physical counterpart of that directed structure through internal–external co-evolution and $\mathscr{B}$-demarcation, while leaving the being-experienced of that directedness as the final cell. The book and phenomenology therefore converge on the "directed structure of given-ness" (for which this book gives a physical counterpart), and they also converge on the irreducibility of the "being-experienced of this structure". This is another convergence by different paths.

27.6. Do Interface and Demarcation Dissolve the Hard Problem? Second Compression, Not Dissolution

The later chapters of Part III used the interface lineage (Chapter 20) and $\mathscr{B}$-demarcation (Chapter 19) to compress the residual point a second time into the subject-attribution residual point. This advance is powerful, and precisely for that reason it introduces a new and highly tempting overstep that must be addressed directly here. It is a variant, in the context of the four newly added chapters, of the older temptation addressed in Section 27.7—whether identity softens the hard problem. This book calls it the temptation of "interface/demarcation dissolution".

Let this temptation first be stated in its strongest form, without attacking a straw man. One might argue as follows: Chapter 20 has restated the hard problem as "how interfaces are upgraded along a lineage", and it has clarified the conditions of upgrading from the non-living external interaction interface, to the membranous life interface, to the cellular information-processing interface, to the human-brain subjective-experience interface. Chapter 19 has also clarified how experience is demarcated by $\mathscr{B}$ as "mine". If the question "why an inside arises" has been rewritten as "how interfaces are upgraded" and the conditions have been specified, and if the question "why experience belongs to me" has been explained by $\mathscr{B}$-demarcation, then has the hard problem not been dissolved by interface and demarcation? What irreducible remainder is left?

This book firmly rejects that dissolution. The reason is structurally the same as the argument in Section 27.7 against the temptation of "identity softening", but it must be stated separately for interface and demarcation. What the interface lineage compresses is the addressable part of "why an interior arises"—the conditions and upgrading of interfaces: which level of interface, and under what physical–organizational conditions (enclosure, pooling networks, knowledge base, demarcation), becomes the bearer of experience. This is addressable, and it has been addressed. But the interface lineage answers "what kind of interface"; it does not answer "why that interface is experienced". Even if all the conditions of how interfaces are upgraded along the lineage are fully specified, it still does not follow from them why the highest-level subjective interface is experienced from within, rather than completing the same upgrading and information processing without experience (Chapter 20 established this already, and F20.4 strictly prohibits using this to claim that the hard problem has "completely ceased to exist"). Similarly, what $\mathscr{B}$-demarcation compresses is "how given-ness becomes mine"—the structure of attribution: which activities lock into the dominant coherence and are demarcated as mine. This is addressable, and it has been addressed (Chapter 19). But $\mathscr{B}$-demarcation answers "how attribution as mine occurs"; it does not answer "why this activity demarcated as 'mine' is experienced". Chapter 19 already made clear that $\mathscr{B}$ does not generate given-ness and does not explain being-experienced. Depersonalization is evidence: demarcation fails while given-ness remains, showing that given-ness does not depend on $\mathscr{B}$. Therefore, however clearly the structure of demarcation is explained, it cannot fill the fissure of why that which has been demarcated is experienced.

Viewed together, interface and demarcation can be described generally as follows: like structural identity and co-evolutionary characterization discussed in Chapter 15, they remain on the objective side of the fissure—the structural–relational, third-person side. The interface lineage compresses the addressable part of "why an inside arises", and $\mathscr{B}$-demarcation compresses "how attribution is formed". Both make the residual point smaller and locate it more precisely, from the given-ness residual point to the subject-attribution residual point, but neither has crossed, nor can cross, to the other side of the fissure—the side of being-experienced. The second compression is compression, not dissolution: it strips away another layer of what can be addressed (cognization, interface upgrading, attributional demarcation), and forces the residual point into its most precise location, subject attribution. Yet that cell—why this activity, upgraded and demarcated as "mine", is experienced—remains irreducible and still lies beyond the boundary, only now with greater precision of location.

Thus this book explicitly rejects the temptation that interface/demarcation dissolves the hard problem. The interface lineage and $\mathscr{B}$-demarcation fulfill the second compression: they compress the residual point to its smallest size and locate it with greatest precision. But they have not, and cannot, dissolve the final cell. To use interface upgrading or attributional demarcation to declare that "the hard problem has been dissolved" is to illegitimately extend what this book has established—interface conditions and attributional structure, both on the objective side of the fissure—to what it has not established: being-experienced, on the other side of the fissure. This is the same kind of overstep that Section 27.7 will reject under the form of "dissolving the hard problem by identity". Both are prohibited by C3 at the philosophical closure of the book (F27.5, and linked to Chapter 20, F20.4). This book preserves this guardrail in order to ensure that the second compression remains, from beginning to end, an honest compression rather than a dissolution disguised as compression.

27.7. What the Identity Thesis Can and Cannot Do: On Whether Identity Softens the Hard Problem

This section must draw the boundary of what the identity thesis, or unity, can and cannot do. This is both the final delimitation of the central thesis of the book and a direct response to another real temptation, similar to but more fundamental than the "interface/demarcation dissolution" temptation in Section 27.6: since mind and matter are the same reality, is the question "why the internal side is experienced" itself softened or even dissolved by identity? This temptation must be addressed directly, because many monists cross C3 precisely at this point.

Again, the temptation should first be stated in its strongest form, without attacking a straw man. One might say: to ask "why the internal side of the same reality is experienced" is roughly equivalent to asking "why this reality is what it is". But the latter is not really a question that awaits an answer; a thing is what it is and needs no further explanation. Since the internal side simply is what that reality is from within, "why it is experienced" is like asking "why water is H_2O". Once the identity is established, the "why" question dissolves by itself. Therefore, once identity is established, the hard problem—being-experienced—is softened or even eliminated. This argument has force, and it is indeed taken up by some forms of Russellian monism.

This book does not take that route, and the reason must be made clear, because this is the final guardrail of C3 in the philosophical closure. The flaw in the argument is that it makes an unjustified expansion of the range over which "identity" has been established. The identity level in this book establishes an identity of structure: the structure of experience and the structure of the gradient configuration are the same structure (Chapters 5 and 26). This identity is supported by relation-by-relation structural isomorphism. But "being-experienced" is precisely not included in the structure established as identical (Chapter 15: being-experienced is not entailed by structure). Therefore, although identity has been established, it has been established as identity of structure, while being-experienced lies outside the scope covered by that structural identity. Thus "identity" cannot automatically extend to being-experienced, and one cannot infer from "structural identity" that "being-experienced is thereby dissolved". To replace "identity of structure" with "being-experienced has also been included in the identity" is exactly the unjustified expansion in that argument. "Water is H_2O" dissolves the question "why is water H_2O" because that identity fully covers what is being asked about: all of water is all of H_2O. But the identity in this book covers structure only; it does not cover being-experienced. Therefore it cannot dissolve the question of why being-experienced obtains. That question concerns what lies outside the scope covered by the identity.

From this, what the identity thesis can and cannot do becomes clear. What it can do: the identity thesis establishes that mind and matter are the same reality with respect to structure and co-evolution, thereby filling the dualistic abyss, placing the final cell as a single point inside monistic reality, and securing the structural isomorphism between experience and gradient structure everywhere. This is the full achievement of unity (Chapter 26). What it cannot do: the identity thesis cannot dissolve the residue of being-experienced, because the identity it establishes is an identity of structure, while being-experienced is not contained within structure. To use identity to dissolve being-experienced is to illegitimately extend structural identity into a region it does not cover. Thus this book explicitly rejects the temptation that identity softens the hard problem: identity fulfills unity and places the final cell, but it does not and cannot dissolve the final cell.

Taken together, Section 27.6 and this section reveal a common structure. Whether the overstep takes the form of "dissolution through interface/demarcation" (Section 27.6) or "softening through identity" (this section), it arises in the same way: what the book has established—interface conditions, attributional structure, structural identity, all on the objective side of the fissure—is illegitimately extended to what it has not established: being-experienced, on the other side of the fissure. This rejection is the final and most delicate preservation of the C3 prohibition. In the earlier chapters, C3 prohibited "using mechanistic statements to masquerade as explanations of being-experienced". At this philosophical closure, C3 prohibits more subtle oversteps: using interface, demarcation, or identity to masquerade as a dissolution of being-experienced. These oversteps are more hidden because they do not appeal to a crude mechanism, but to fundamental relations that sound reasonable—"interface upgrading", "subject attribution", and "ontological identity". Yet they belong to the same class as the earlier oversteps. This book preserves this guardrail in the final chapter so that "mind–matter unity" remains, from beginning to end, an honest thesis. It fulfills what unity can do, and what the second compression can do, without pretending that unity can do what it cannot, or that the second compression can do what it cannot.

27.8. Conclusion of This Chapter

After the fulfillment of unity, this chapter has given a philosophical determination and closure to the final cell—the subject-attribution residual point after two compressions. First, it determines the final cell as a boundary (where structural–relational explanation stops), rather than an object (a difficult problem awaiting solution within explanatory capacity). The proper attitude toward it is therefore precise marking and understanding why one stops there, not forcing a crossing. Second, it locates the final cell: after excluding, one by one, structure, mechanism, cognization and demarcation, measurement, and the supposed "between mind and matter", the final cell is forced into a single point—why the internal side of the same reality, that which has been demarcated as "mine", is experienced. The precision of this location, achieved through two compressions into the subject-attribution residual point, distinguishes this book from vague mysticism. Third, it explains why acknowledging it strengthens the book: acknowledging the boundary defines the achievement, the minimal residue is the result of compression, and monistic placement fills the dualistic abyss. Fourth, it shows the convergence with phenomenology: this book reaches the irreducibility of givenness from below, starting from ρ_f, while phenomenology confirms given-ness as an irreducible beginning point from above; the two independent paths converge and support one another. The directed structure of given-ness in this book (internal–external co-evolution and $\mathscr{B}$'s world-oriented demarcation) also accords with phenomenological intentionality. Fifth, it directly addresses the new temptation that "interface/demarcation dissolves the hard problem": the interface lineage compresses the addressable part of "why an interior arises", and $\mathscr{B}$-demarcation compresses "how attribution is formed", but both remain on the objective side of the fissure and cannot cross it. The second compression is compression, not dissolution (preserving C3, F27.5 and F20.4). Sixth, it draws the boundary of what the identity thesis can do (establish unity, fill the abyss, place the residue) and what it cannot do (dissolve being-experienced), and rejects the temptation that "identity softens the hard problem". Taken together, interface, demarcation, and identity all share the same potential overstep: extending what lies on the objective side of the fissure illegitimately to being-experienced. Guarding against this is the preservation of the C3 prohibition at the most delicate point of the whole book.

The failure conditions of this chapter are as follows (summarized in Appendix E).

- F27.1 The final cell is in fact a solvable object. If the final cell is shown to be a difficult problem inside explanatory capacity, rather than a boundary—that is, if there exists a valid path, without appealing to any first-person premise and starting purely from structure and relation, that answers "why the internal side is experienced"—then it is not a boundary but an object, and this book's determination of it fails. This is a failure the book would welcome, because it would mean that the hard problem has truly been solved (linked to Chapter 15, F15.1, and Chapter 20, F20.2).

- F27.2 The position is imprecise. If the final cell still contains components that can be explained physically (that is, if this book has mistakenly assigned something solvable to the boundary), then its location is imprecise and that component must be returned to the solvable side. The second compression in Chapter 20, which stripped away solvable components such as cognization and demarcation, is precisely the practice of this condition.

- F27.3 The convergence with phenomenology is a misreading. If the claimed convergence between this book and phenomenology (the irreducibility of given-ness and intentionality) is shown to be a misreading of phenomenology—if the two do not in fact refer to the same irreducibility—then the support of "convergence by different paths" fails.

- F27.4 Identity is used to dissolve the final cell. If this book, or anyone following its logic, uses "mind and matter are identical" to dissolve the residue of being-experienced (illegitimately extending structural identity to being-experienced), then it violates the C3 prohibition at the philosophical closure and constitutes the use of the identity thesis to masquerade as a dissolution of being-experienced.

- F27.5 Interface or demarcation is used to dissolve the final cell. If this book, or anyone following its logic, uses the upgrading of the interface lineage or the attributional demarcation of $\mathscr{B}$ to claim that the hard problem has been dissolved (illegitimately extending interface conditions and attributional structure on the objective side of the fissure to being-experienced), then it violates the C3 prohibition and constitutes the use of interface/demarcation to masquerade as a dissolution of being-experienced (linked to Chapter 20, F20.4; this is what this chapter and Chapter 20 must guard against most carefully).

Table 27.1.. The Boundary between What This Chapter Establishes and What It Cannot Derive

What this chapter establishes	What this chapter cannot derive
The determination of the final cell: a boundary (where explanation stops), not an object (a difficult problem awaiting solution).	It cannot, and does not attempt to, answer the final cell (marking a boundary, not solving an object).
The precise location of the final cell: a single point inside monistic reality, compressed twice into the subject-attribution residual point.	It cannot thereby assert that it is necessarily unsolvable in philosophy as well (it asserts only that it lies outside physical structural–relational explanation).
Three reasons why acknowledging it strengthens the book: it defines the achievement, is the result of compression (twice), and fills the dualistic abyss.	It cannot thereby deny the achievements of the previous parts (everything inside the boundary still holds).

What this chapter establishes	What this chapter cannot derive
Convergence with phenomenology (the irreducibility of given-ness and intentionality) as independent support reached by different paths.	It cannot derive one path deductively from the other (the two paths have different premises; their convergence is support, not proof).
What interface, demarcation, and identity can do (compress and place) and cannot do (dissolve being-experienced).	Interface, demarcation, or identity must not be used to masquerade as a dissolution of being-experienced (C3 at the closure, the point this chapter must guard against most carefully).

28. Conclusion: The Culmination of the Trilogy and the Unfinished Path

28.1. A Main Spine: The Overall Form of the Book

At the final chapter, it is appropriate first to step back and look at the overall form of the book—not as a list of twenty-three chapters, but as the main spine that runs through them. The value of a theoretical work does not lie in the number of its chapters, but in whether it has a main spine that can be restated in one sentence and can still withstand chapter-by-chapter scrutiny.

The main spine of this book is as follows: the hard problem is not an inaccessible block of stone, but something that can be decomposed. Once decomposed, most of it is addressable, leaving only the final cell as irreducible. More specifically: the entire structure of experience is isomorphic to the fivefold-rich structure of ρ_f gradient configurations, and this isomorphism is identity in the strongest sense (Part II); the given-ness of experience is internal–external co-evolutionary freezing of a bounded interior, the precondition of given-ness (interiority) is produced by enclosure (the boundary of life), and its mechanism is provided by co-evolution in EQT detection (Part III); all of this can be projected into the measurable Ψ and falsifiable predictions, especially the two distinctive predictions of the frequency dimension, which form the empirical spine (Part IV); and mind and matter are thereby fulfilled as two sides of the same open ρ_f process—the structural side and the given side (Chapter 26). Finally, the question "why the internal side is experienced", which is not entailed by structure and co-evolution, is the irreducible final cell. This book marks it precisely, acknowledges that it cannot be crossed, and never uses any mechanism or identity thesis to masquerade as its solution (Chapters 15 and 27).

This main spine can be compressed further into one sentence: through the isomorphism of ρ_f, this book fulfills the unity of mind and matter, compresses and places the hard problem as the final cell inside monistic reality, fully develops what can be developed, and precisely marks what cannot be developed. The twenty-three chapters of the book are the layered unfolding and local testing of this sentence. If the reader remembers only one thing after closing the book, it should be this main spine rather than the detail of any individual chapter. The details serve the main spine; the main spine is what this book is.

28.2. First Account: What This Book Establishes

An honest conclusion must provide a clear account: what the book establishes, and what it honestly does not establish. Let us first list what has been established.

First, a decomposition. This book decomposes the hard problem into addressable parts (structure, the mechanism of given-ness, and measurable projection) and a non-addressable part (the final cell), and argues that this decomposition is complete. The sense of self, the sense of agency, and affective tone, once regarded as possible counterexamples, are placed as types of endogenous perturbation (the perturbation axis), rather than as omitted structures (Chapter 11). This decomposition itself is the first establishment of the book: it transforms a whole problem into the structure of "mostly addressable + one non-addressable cell".

Second, an isomorphism. This book establishes, relation by relation, the isomorphism between experiential structure and ρ_f gradient structure (the five chapters of Part II), and argues that only this fivefold-rich structure, rather than any single scalar, can satisfy the structural sufficiency required by the identity level (Chapter 5). This is the fundamental distinction between this book and all single-scalar identity theories, such as those centered on a single Φ.

Third, a characterization of given-ness. This book characterizes given-ness as internal–external co-evolutionary freezing of a bounded interior (Part III), anchors its precondition (interiority) in the boundary produced by enclosure (Chapter 3, the gate against panpsychism), and derives its mechanism from the co-evolution of EQT detection (Chapter 13). It thereby makes the questions "what physical process given-ness corresponds to" and "when it appears or disappears" directly answerable and falsifiable.

Fourth, a measurable plane and a set of predictions. This book projects the preceding framework into the three-dimensional state space $\Psi = (\tilde{A}, C, \mathscr{B})$ (Part IV), offers a unified explanation of transitions among conscious states (Chapter 24), and gives falsifiable predictions. Among them, the two ρ_f-specific predictions—frequency–quality and frequency–timescale—form the empirical spine and answer the objection concerning whether ρ_f does any empirical work (Chapter 25).

Fifth, a metaphysical fulfillment. This book directly fulfills "mind–matter unity": mind and matter are two sides of the same open ρ_f process, guaranteed by ontological uniqueness and structural isomorphism. It stands between dualism and physicalism, taking the strengths of both—monistic economy and respect for given-ness—while avoiding their weaknesses: the interaction problem, and reduction or elimination. It is a dual-aspect monism of open process (Chapter 26).

28.3. Second Account: What This Book Honestly Does Not Establish

The other half of the account is to state honestly what has not been established. This half is as important as the first, because a work that refuses to state what it has not achieved cannot be fully trusted in what it claims to have achieved.

First, this book does not solve the final cell. "Why is the internal side of the same reality experienced, rather than operating with the same structure and the same co-evolution in the absence of experience?" This cell is not answered by this book, and is explicitly said to be unanswerable by it (Chapters 15 and 27). It is not entailed by structure and relation. It is the boundary of physical, structural–relational explanation, not an object awaiting solution within it. What this book does is to mark it precisely, compress it to its smallest form, and place it through unity as a single point inside monistic reality. But marking is not answering, and placing is not eliminating. This book states solemnly: the innermost core of the hard problem is not solved here, and this book does not pretend that it is solved.

Second, this book does not provide a complete quantitative form. The isomorphism, the characterization of given-ness, Ψ, and the predictions are mostly stated in terms of verbal and structural relations. Their complete quantitative forms—such as the precise scaling law of frequency–timescale correspondence, the exact algorithms for the three quantities of Ψ, and the concrete parameters of the resonance kernel in consciousness—are largely left to appendices or future work. The book establishes a framework and structural relations; their full mathematical fulfillment remains unfinished.

Third, the credibility of this book partly depends on the foundational volume. This book is an application volume of EQT, and its work is conditional: if the ontology of ρ_f in the foundational volume does not hold, then the ground on which this book relies fails (Chapter 2). What this book can independently secure is the rigor of the step "from ρ_f to consciousness"; the justification of the ρ_f

ontology itself lies in the foundational volume, not in this book. This conditionality has been acknowledged from the beginning, and is reiterated here.

Fourth, the book's distinctive predictions still await rigorous testing. The two distinctive predictions, frequency–quality and frequency–timescale, are consistent in direction with existing findings, but their precise maps await the development of measurement methods (the roadmap in Chapter 25). What this book gives is falsifiable predictions and directions for testing, not already confirmed conclusions. The testing of these predictions remains unfinished, and its results may overturn the book. This is precisely what marks them as genuine predictions.

Taken together, the two accounts make the form of this book clear: within the addressable range, it aims to go as far as possible (decomposition, isomorphism, given-ness, measurement, and unity); at the point that cannot be addressed, it stops honestly (the final cell); and it acknowledges the unfinished status of its quantification, its foundation, and the testing of its predictions. This account of "what has been established and what has not been established" is itself the fulfillment of the book's methodology—the two-tier standard and the system of failure conditions introduced in Chapter 2. It does not seek completeness through vagueness, but seeks truth through clear boundaries.

28.4. The Culmination of the Trilogy: The Incorporation of Pre-EQT Phenomenological Observations by ρ_f

This book is the culminating work of a series, and it must therefore explain how it culminates the series: how the ρ_f mechanism incorporates the phenomenological observations made in the three pre-EQT works, and thereby unifies dispersed insights under one mechanism.

The three earlier works were all written before EQT and belong to the level of phenomenological observation. *The Information World of the Cell: Pooling Networks, Continuous-State Computation, and the Architecture of Life Information Processing* observed the complex information processing inside cells (pooling networks). *The Second Axis of Cosmic Aggregation: Membranes, Life, and Consciousness* observed the specificity of life relative to non-life (enclosure). *The Origin of Cellular Proto-Consciousness: Biological Foundations and Operational Criteria* observed that elementary consciousness has the cell as its minimal unit. They saw the right phenomena, but did not reach the underlying mechanism. They answered "what it is", but not "why the mechanism is so" (Chapter 3).

This book incorporates all three through the ρ_f mechanism, giving each phenomenon its mechanism. Enclosure, the insight of *The Second Axis of Cosmic Aggregation: Membranes, Life, and Consciousness*, has as its mechanism the closed organization of ρ_f material transients. It produces an interior, provides the precondition for given-ness, and serves as the gate against panpsychism (Chapter 3). The information processing inside the cell, the insight of *The Information World of the Cell: Pooling Networks, Continuous-State Computation, and the Architecture of Life Information Processing*, has as its mechanism the evolution and resonance of ρ_f gradient configurations within that interior (the isomorphism in Chapters 6 to 10, and the coherence of Chapter 9). Elementary consciousness, the insight of *The Origin of Cellular Proto-Consciousness: Biological Foundations and Operational Criteria*, is incorporated here into the framework that every bounded living system has a minimal degree of given-ness. The cell, as the minimal bounded living unit, has its minimal internal–external co-evolution and its minimal given-ness (Chapters 14 and 16).

This shows where the culmination of the trilogy lies. The earlier works marked, at the phenomenological level and in scattered form, that life has a distinctive interior, that the interior has information processing, and that this constitutes elementary consciousness. This book resettles these phenomena through the unified mechanism of ρ_f: the mechanism of enclosure, the mechanism of information processing, and the mechanism of elementary consciousness are all grounded in the isomorphism of ρ_f gradients and internal–external co-evolution. It then advances beyond what the earlier works reached: the structural isomorphism of experience, the mechanism of givenness, the unity of mind and matter, and the marking of the final cell. This is why the book is a "culmination": it is not a simple continuation of the earlier works, but the use of one unified mechanism to gather, complete, and advance their dispersed phenomenological insights into the problems of the "I" and of "unity". This also confirms again the main line of the book: the main line is the ρ_f gradient, not the fifth mechanism. The fifth mechanism is one part of the "boundary gate" under the main line of ρ_f, and the earlier works are phenomenological observations incorporated by the ρ_f mechanism (Chapter 3).

28.5. The Unfinished Path: Directions for Future Work

An honest conclusion should not only provide an account of what has been done, but also point out the unfinished path—the directions this book has opened but leaves to future work.

First, quantification. The framework and structural relations of this book must be developed into a complete quantitative theory: the precise scaling law for frequency–timescale correspondence, exact algorithms and measurement protocols for the three quantities of Ψ, the concrete parameters of the resonance kernel and thresholds in consciousness, and the dynamical equations governing trajectories in state space. This quantification must be closely connected with the master equation of the foundational volume, and is the most urgent unfinished task.

Second, the testing of distinctive predictions. The two distinctive predictions, frequency–quality and frequency–timescale, together with the topology of coherent closure in $\mathscr{B}$, must be rigorously tested through developing methods for high frequency resolution, content decoding, and coherent-topology measurement (the roadmap in Chapter 25). This testing is the key step by which the book moves from a framework to a theory that is supported or overturned. Its results may confirm, revise, or overturn this book.

Third, inquiry into boundary cases. This book treats boundary as the gate against panpsychism (Chapter 3), and its sharp implications—such as the claim that unenclosed artificial systems lack the precondition for given-ness (Chapter 4)—must be further investigated in boundary cases. Artificial systems, synthetic life, collective systems, and related cases raise questions about how their "boundary" should be judged, whether they possess the precondition for given-ness, and in what sense. These are questions opened by the framework, but not yet fully developed.

Fourth, philosophical inquiry into the final cell. Although the final cell cannot be explained physically, it can still be further investigated philosophically: its relation to phenomenology (given-ness and intentionality, Chapter 27), its exact status within a monistic framework, and whether, and in what non-physical sense, it may be further understood. These remain unfinished philosophical tasks. This book marks and places the final cell, but its philosophical investigation is only beginning here.

These four unfinished paths—quantification, testing of distinctive predictions, inquiry into boundary cases, and philosophical inquiry into the final cell—are the openings this book consciously leaves. The maturity of a theory does not lie in claiming that it has closed everything, but in knowing clearly where it remains open and indicating the direction in which further work must proceed. This book marks these four paths as its account to future work.

28.6. Closing Remarks

This book began from a perplexity: how can physical processes be accompanied by given experience? It has not covered this perplexity with an attractive word, nor has it abandoned the addressable parts because the perplexity is difficult. It has done a third thing: it has decomposed the hard problem, developed what can be developed to the fullest extent—structural isomorphism, the mechanism of given-ness, measurable predictions, and mind–matter unity—precisely marked what cannot be developed—the final cell—and drawn a clear boundary between the two.

As the title of the book, "mind–matter unity" now has its exact meaning: mind and matter are two sides of the same open ρ_f process, fulfilled by ontological uniqueness and structural isomorphism. This is a serious metaphysical thesis that can be weighed point by point against other positions, not a slogan. Yet within this unity there remains an irreducible cell of "being-experienced". This book has not crossed it. It has only relocated it from a dualistic abyss into a precisely located single point inside monistic reality.

Perhaps this is the most honest form a theory of consciousness can reach: to explain to the end what can be explained, to point out precisely what cannot be explained, and to know clearly where the boundary between the two lies. What this book seeks is not to claim that it has solved the entire mystery of consciousness, but to clarify that mystery to the point where what can be done has been done, what cannot be done has been made clear, and the final cell is left to philosophy beyond physics and to a deeper understanding that has not yet arrived.

28.7. Chapter Summary

This chapter is the conclusion of the whole book. First, it extracts the main spine of the book: through the isomorphism of ρ_f, the book fulfills the unity of mind and matter, compresses and places the hard problem as the final cell inside monistic reality, fully develops what can be developed, and precisely marks what cannot be developed. Second, it gives the positive account of what the book establishes: a decomposition (the threefold decomposition of the hard problem, including completeness), an isomorphism (between experiential structure and gradient structure, at the identity level, and as distinct from single-scalar theories), a characterization of given-ness (internal–external co-evolutionary freezing, anchored in boundary), a measurable plane and predictions (Ψ and the two distinctive frequency predictions), and a fulfillment of unity (dual-aspect monism of open process). Third, it gives the negative account of what the book does not establish: it does not solve the final cell, it does not provide a complete quantitative form, its credibility partly depends on the foundational volume, and its distinctive predictions still await rigorous testing. Fourth, it explains the culmination of the trilogy: the ρ_f mechanism incorporates the phenomenological observations of the three earlier works (enclosure, cellular information processing, and elementary consciousness), gives each its mechanism, and confirms that the main line is the ρ_f gradient rather than the fifth mechanism. Fifth, it points to four unfinished paths: quantification, testing of distinctive predictions, inquiry into boundary cases, and philosophical inquiry into the final cell. Sixth, it closes by stating that the aim of the book is not to solve the whole mystery, but to develop what can be developed to the end, to mark precisely what cannot be developed, and to know where the boundary between the two lies.

Table 28.1.. A General Account of What This Book Establishes and What It Honestly Does Not Establish

What this book establishes (positive account)	What this book honestly does not establish (negative account)
A complete decomposition of the hard problem; identity-level isomorphism between experiential structure and ρ_f (as distinct from single-scalar theories).	The final cell (why the internal side is experienced) remains unsolved and cannot be solved here (a boundary, not an object).

What this book establishes (positive account)	What this book honestly does not establish (negative account)
The characterization of given-ness (internal–external co-evolutionary freezing of a bounded interior, anchored in the boundary produced by enclosure).	The complete quantitative form (scaling laws, algorithms, parameters) is largely left to future work and appendices.
The measurable projection Ψ and falsifiable predictions, especially the two ρ_f-specific predictions of frequency–quality and frequency–timescale.	The distinctive predictions still await rigorous testing (their results may revise or overturn this book).
The metaphysical fulfillment of mind–matter unity (dual-aspect monism of open process, standing between dualism and physicalism).	The book's credibility partly depends on the foundational volume (the conditionality of this book's work).
The culmination of the trilogy: the ρ_f mechanism incorporates the earlier phenomenological observations, with the main line being the ρ_f gradient.	Boundary cases (such as artificial systems) and philosophical inquiry into the final cell are only opened, not completed.

Closing Remarks

When a book reaches its end, the author must eventually step out from behind the scaffolding of argument and say a few words that are no longer arguments, but an account of intention.

This book began from a dissatisfaction that I could not set aside. The hard problem of consciousness has stood there for more than a century, and those who try to solve it seem always to fall into one of two outcomes: either they use a resonant word—"emergence", "identity", "information", "unity"—to cover the fissure between physical language and the first person, making it appear that the fissure has been sealed, when in fact it has only been covered by a cloth; or, because the fissure is so deep, they set aside even those things on both sides of it that could have been clarified, retreating to the claim that "consciousness is ultimately mysterious". My own earlier attempts oscillated between these two outcomes: when ambitious, they looked like cheating; when cautious and disciplined, they looked like surrender. I always felt that there had to be a third path.

This book is my account of that third path. Its entire effort can be reduced to one plain task: to clarify to the end what can be clarified, to point out precisely what cannot be clarified, and to know honestly where the boundary between the two lies. What can be clarified is the structure of experience—why it has such richness; the mechanism of given-ness—what it corresponds to physically, and when it appears or disappears; how sensation is illuminated by knowledge and becomes experience; how it is demarcated and becomes "my" experience; how all of this is projected into measurable and fallible predictions; and how mind and matter become "one" as two sides of the same ρ_f process. These I have tried to carry as far as I can. What cannot be clarified is the final cell: why the enclosed, co-evolving interior, illuminated by knowledge and belonging to an "I", is not running silently in darkness, but is lit from within and felt. I have not clarified this cell, and I have not pretended to clarify it. What I have done is to draw it out, step by step, from the fog that once spread across the whole problem, to compress it and compress it again, until it is forced into a single point whose position can scarcely be made smaller, and then to say honestly: here, physical language reaches its edge.

The later I came in this work, the more I believed that acknowledging this edge is not the weakness of the book, but the only ground on which it can stand. An honest map marks the line where "the known world ends"; a map that claims to have drawn everything and to have no edge is less credible. If the truth of consciousness indeed contains a core not entailed by structure and mechanism, then any theory claiming to have solved that core as well has almost certainly crossed a boundary somewhere without admitting it. I would rather draw that boundary in the open and allow it to face the strictest questions: What do you claim to have explained? What do you claim not to have explained? On what grounds is the line drawn here? A book that can withstand such questions, even while leaving one irreducible cell, is closer to the truth than a book in which everything appears connected but everything remains ambiguous.

The cell that remains does not lead to emptiness. On one side, it opens toward philosophy—toward the more than century-long phenomenological reflection on the irreducibility of "given-ness". The edge this book reaches from below, beginning with the energy quantum density field, meets at the same place the beginning point that phenomenology confirms from above, beginning with experience itself. That two unfamiliar paths meet here is, for me, a consolation: it at least suggests that this edge is not a dead corner accidentally produced within my own framework, but a real limit concerning consciousness, one that can be reached from different directions. On the other side, it opens toward the future—toward measurement methods that have not yet arrived, which may confirm or overturn the claims made in this book; toward questions that I have only begun to formulate: whether artificial systems have an "interior", whether a collective can share one "I", and whether the final cell can still be illuminated, beyond physics, by some deeper understanding. These are not questions this book can answer, but they are what the book honestly hands to those who come after it.

This is the culmination of a trilogy. The earliest book saw that complex information flows within the interior of life; the later two saw the boundary that separates life from non-life, and saw that consciousness already has its embryonic form at the level of the cell. At that time, what I saw were phenomena; I had not yet seen the mechanism beneath them. It was ρ_f—the frequency-resolved energy density field and its gradients—that first made me feel that those dispersed phenomena might belong to one deeper principle: the mechanism of force, the mechanism of matter, and the mechanism of given-ness are different frequency bands and different organizations of one and the same process. This book gathers those early observations into that principle, and follows it all the way to the "I", to the "mine", to "unity", and to the edge that cannot be crossed.

Therefore, if there is anything in this book worth preserving, I hope it is not one particular conclusion, but the posture that runs through it from beginning to end: in the face of a genuinely difficult problem, neither claim conquest too easily, nor lay down one's arms too easily; sharpen the knife for the places where it can cut, cut step by step until it can cut no further, and then stop at the place where it can no longer cut, saying truthfully: up to here, what I see is an edge, not a void.

The other side of that edge is left to a deeper understanding, and to those who have not yet arrived.

A. Symbols and Dimensional Table

This appendix provides the unified definitions, dimensions, and proposition types of the symbols used throughout the book. It serves as the citation basis for Appendix B (formalization of isomorphism), Appendix C (formal model of given internal–external co-evolution), and Appendix D (parameters and normalization). Three editorial disciplines must be stated from the outset.

First, citation without re-proof (Chapter 2). Whenever this table concerns the ontology of ρ_f (A1–A4), the master equation 5.4.1, or the Lorentzian resonance kernel, it cites Sections 5.4 and 5.5 of the EQT foundational volume, *EQT Foundational Volume: Frequency, Mass, Range of Action, and the Generation Mechanism of Force*; whenever it concerns the freezing-time scaling law, it cites Section 4.3 of *Frozen Cosmology: A Natural Philosophy of Cosmic Evolution in the Perspective of Energy Quantum Theory*. This appendix does not rederive these formulas. It only summarizes their authoritative definitions and annotates their dimensions and proposition types as used in this consciousness volume.

Second, dimensional distinction. This table clearly marks, for each quantity, whether it is dimensional or dimensionless. For dimensional quantities, it gives the corresponding base-dimensional form in the International System of Units; for dimensionless quantities (including nondimensionalized relative quantities), it explicitly marks them as "dimensionless". All quantities that enter Ψ in this book are dimensionless relative quantities (such as $\tilde{A}$, C, and $\mathscr{B}$); their corresponding dimensional physical quantities are denoted without a tilde (such as A). The two must not be conflated (Chapter 2, F2.4).

Third, restraint concerning interface coefficients and slowly varying factors. Whenever this table concerns interface coefficients (such as χ_g) and slowly varying background factors (such as τ_0), it explicitly marks their status as non-universal constants: they are effective quantities that vary with scale, frequency, and background; they are calibratable and testable, but should not be treated as exact zero-free-parameter constants. This implements the book's discipline of honesty in "not claiming zero free parameters".

A.1. Basic Objects and Spectral Relations

Table A.1.. Basic Symbols, Definitions, Dimensions, and Sources for the Energy Quantum Density Field

Symbol	Definition	Dimension	Source and proposition type
$\rho_f(x,t)$	Energy quantum density field partitioned by frequency families (frequency-resolved energy-density distribution and its gradient structure); the unique fundamental object of EQT	Energy·frequency^{-1} ·volume^{-1} (J·Hz^{-1}·m^{-3})	Foundational volume A1 (postulate)

Continued on next page...

Continued table

Symbol	Definition	Dimension	Source and proposition type
f, ν	Frequency (marking the frequency families of ρ_f; ν is mostly used for specific oscillation frequencies)	s^{-1}(Hz)	Foundational volume A2
h	Planck constant	J·s	Physical constant
$E = h\nu$	Frequency–energy spectral relation (frequency as energetic spectral quantity)	E: J	Foundational volume A2 (postulate)
x, t	Position and time coordinates	m; s	Basic
$\nabla\rho_f, \nabla^2\rho_f$	Spatial gradient and Laplacian of ρ_f	Derived from the dimension of ρ_f	Foundational volume

A.2. The Six Terms of the EQT Master Equation 5.4.1

$$\frac{\partial\rho_f}{\partial t} = k(f)\rho_f^m + D(f)\nabla^2\rho_f - \nabla\cdot(\rho_f v_f) - \Gamma(f)\rho_f + S_f + N(f)$$

This is equation 5.4.1 of the foundational volume, the unified evolution equation of ρ_f (foundational volume A3 postulate). The definitions, dimensions, and uses of its terms in this consciousness volume are as follows.

Table A.2.. Key Symbols, Definitions, Dimensions, and Uses in This Book for the Master Equation

Symbol	Definition	Dimension (derived by equation balance)	Use in this book
$\partial \rho_f / \partial t$	Time-evolution term	$[\rho_f] \cdot s^{-1}$	Evolution law of the stream of consciousness (Chapter 8)
$k(f)\rho_f^m$	Nonlinear self-sustaining term ($k(f)$ is the nonlinear self-sustaining coefficient, and m is the degree of nonlinearity)	$k(f)$: such that the dimension of this term balances with $\partial \rho_f / \partial t$	Resonance enhancement and coherence maintenance (Chapter 9)
$D(f)\nabla^2 \rho_f$	Diffusion term ($D(f)$ is the frequency-dependent diffusion coefficient)	$D(f)$: $\mathrm{m}^2 \cdot s^{-1}$	Spatial unfolding of configurations
$-\nabla \cdot (\rho_f v_f)$	Convection term (v_f is the frequency-dependent effective flow velocity)	v_f: $\mathrm{m} \cdot \mathrm{s}^{-1}$	Directionality of evolution (Chapter 8, arrow of time)
$-\Gamma(f)\rho_f$	Dissipative term ($\Gamma(f)$ is the frequency-dependent dissipation/decay coefficient)	$\Gamma(f)$: s^{-1}	Decay of transients and forgetting (Chapter 10)

Continued on next page...

Continued table

Symbol	Definition	Dimension (derived by equation balance)	Use in this book
S_f	Source term (external input or source)	$[\rho_f] \cdot s^{-1}$	Exogenous perturbation (Chapter 8)
$N(f)$	Fluctuation term (noise, stochastic perturbation)	$[\rho_f] \cdot s^{-1}$	Stochastic component of endogenous perturbation

Terminological translation in the consciousness volume: Chapter 8 separates the law of the master equation from the drive of perturbation ("river channel and water flow")—$k(f)$, $D(f)$, v_f, and $\Gamma(f)$ belong to the side of "law" (describing how evolution is permitted), while S_f and $N(f)$ belong to the side of "drive" (describing how exogenous/endogenous perturbations are injected). The exogenous and endogenous perturbations established on the perturbation axis in Chapter 11 correspond precisely to the consciousness-level use of S_f and $N(f)$.

A.3. Cross-Frequency Coupling Resonance Kernel

$$A(\nu_B, \nu_F) = \frac{A_0}{(\nu_B - \nu_F)^2 + \gamma^2}$$

This is the Lorentzian cross-frequency coupling resonance kernel of the foundational volume (foundational volume A4 postulate).

Table A.3. Key Symbols, Definitions, Dimensions, and Sources for the Cross-Frequency Coupling Resonance Kernel

Symbol	Definition	Dimension	Source
$A(\nu_B, \nu_F)$	Cross-frequency coupling resonance kernel (coupling strength under frequency matching/detuning)	Dimensionless (relative quantity of coupling strength)	Foundational volume A4
ν_B, ν_F	The two frequencies in resonance (background/foreground, or two frequency families)	s^{-1}(Hz)	Foundational volume
$\nu_B - \nu_F$	Frequency detuning	s^{-1}(Hz)	—
γ	Half-width of the resonance kernel (determining the narrowness/broadness of resonance)	s^{-1}(Hz)	Foundational volume A4
A_0	Peak coefficient of the resonance kernel	Dimensionless (same as A)	Foundational volume A4

Use in the consciousness volume: Chapter 9 uses it for the carrying of unity and binding—same-frequency resonance ($\nu_B \approx \nu_F$) corresponds to the coherence of a single subject, while cross-frequency resonance ($\nu_B \neq \nu_F$, but within γ) corresponds to feature binding. The discussion of counterexamples in Chapter 11 also draws on this resonance kernel (the frequency matching determined by the resonance kernel is not the gate against panpsychism; the latter is drawn by the boundary of the membrane. The two gates must not be conflated, Chapter 3, F3.3).

A.4. Freezing Dynamics: Transient Timescale Spectrum and Scaling Law

Table A.4.. Key Symbols, Definitions, Dimensions, and Notes for Freezing Time and Retention Duration Scaling Laws

Symbol	Definition	Dimension	Source and note
$t_{\text{freeze}}(f)$	Freezing time of a ρ_f structure with frequency f (the characteristic duration for which the structure exits active evolution and is stably maintained)	s	*Frozen Cosmology: A Natural Philosophy of Cosmic Evolution in the Perspective of Energy Quantum Theory*, §4.3
$\tau_{\text{retain}}(f)$	Retention duration of a memory transient in the brain with frequency f (the consciousness-volume symbol for ρ_f freezing under the brain's background)	s	Chapters 10 and 17 of this book; consciousness-level use of the same t_{freeze}
$\tau_0(f)$	Slowly varying effective background factor (absorbing corrections from background, coupling, diffusion, decay, and source terms)	s^2 (so that τ_0/f has dimension s and balances with t_{freeze})	*Frozen Cosmology: A Natural Philosophy of Cosmic Evolution in the Perspective of Energy Quantum Theory*, §4.3; not a free parameter or universal constant, but slowly varying, calibratable, and testable

Continued on next page...

Continued table

Symbol	Definition	Dimension	Source and note
$t_{\text{freeze}}(f) \sim \tau_0(f)/f$	Freezing-time scaling law ($1/f$ dominant, with slowly varying correction by τ_0)	Scaling relation	*Frozen Cosmology: A Natural Philosophy of Cosmic Evolution in the Perspective of Energy Quantum Theory*, §4.3
$\tau_{\text{retain}}(f) \sim \tau_0(f)/f$	Retention-duration scaling law for memory transients in the brain (the consciousness-volume form of the same scaling law, with the background changed from cosmic expansion to brain-internal dissipation)	Scaling relation	Claim of this book: memory transients in the brain are directly ρ_f freezing under the brain's background (Chapters 10 and 17)

Key qualifications (consistent with the discipline of the whole book, already established in Chapters 10, 17, and 25):

- $\tau_0(f)$ retains the restrained status of a "slowly varying correction factor". It is not treated as an exact constant, nor is it tied in this book to any specific neural frequency band or metabolic rate; calibration of its exact form belongs to the roadmap of Chapter 25.

- $\tau_{\text{retain}} \sim \tau_0/f$ is not an unconditionally exact f^{-1} law, but a "$1/f$-dominant tendency + slowly varying correction by τ_0"; its test should target the dominant tendency and calibratable corrections, not demand exact pointwise inverse proportionality (see Chapter 25, F25.2).

- This scaling law is the quantitative backbone of the second ρ_f-specific prediction in Chapter 25 (frequency–timescale correspondence).

A.5. Interface Coefficients

Table A.5.. Key Symbols, Definitions, Dimensions, and Notes for Interface Coefficients

Symbol	Definition	Dimension	Note
χ_g	Interface coefficient of gravity/interface (or a similar coupling interface coefficient)	Depends on the specific interface	Foundational volume; not a universal constant, but varies with scale/frequency/background; calibratable and testable, and should not be treated as a zero-free-parameter quantity (Chapter 2, F2.4)
χ_*(general interface coefficient)	General notation for coupling coefficients wherever this book concerns an "interface"	Depends on the interface	Same qualification as above

Disciplinary restatement: wherever this book uses χ_g or other interface coefficients (such as the enclosure interface in Chapter 3 and the discussion of interfaces with existing theories in Chapter 4), it preserves the restraint that "χ is not a universal constant". This restraint is the bottom line against the pretense of a "zero-free-parameter exact law" (Chapter 2, F2.4).

A.6. Experience-Side and Measurable Projection Quantities

Table A.6.. Key Symbols, Definitions, Dimensions, and Sources for the Consciousness-State Tensor

Symbol	Definition	Dimension	Source
A	Dissipative power required to maintain non-equilibrium internal–external co-evolution (dimensional physical quantity)	$W(J \cdot s^{-1})$	Chapters 18 and 22
$\tilde{A}$	Nondimensionalized relative quantity of A (used for cross-system and cross-state comparison, and entering Ψ)	Dimensionless	Chapters 17 and 22
C	Projection quantity of consciousness complexity (differentiation $\times$ integration of the configuration, criticality index)	Dimensionless	Chapters 17 and 22
$\mathscr{B}$	Projection quantity of consciousness boundary (extension and closure of dominant coherent co-evolution, including coherent closure and causal closure)	Dimensionless (relative measure of boundary)	Chapters 17 and 22
$\Psi = (\tilde{A}, C, \mathscr{B})$	Consciousness-state tensor (projection quantity of the isomorphic structural space onto the observable plane)	Dimensionless triple	Chapter 17
$\mathscr{B}$	Subjectivity-demarcation operator (the attribution mechanism that makes activities belong as "mine"; the attributional aspect of dominant coherence, not a new entity)	Operator (non-numerical quantity)	Chapter 19

Key point of dimensional discipline: A is dimensional, while $\tilde{A}$ is dimensionless; the two must not be conflated (as established in Chapters 2 and 22). Whenever physical dissipative power is discussed (such as the absolute value of brain metabolism), A is used; whenever the quantity enters the state tensor Ψ and is used for cross-system comparison, $\tilde{A}$ is used. Conflating them triggers F2.4 (dimensional conflation).

Restatement of the status of $\mathscr{B}$: $\mathscr{B}$ is not a numerical quantity and not a new entity, but a marker for the attributional aspect of dominant coherence (Chapters 9 and 16)—to lock into the same dominant coherence is to be demarcated by $\mathscr{B}$ as "mine". It is listed in the dimensional table as an "operator" only to mark its status as a mechanism, so as to avoid misusing it as a computable numerical variable (Chapter 19, F19.4).

A.7. Markers of the First-Person Side and Phenomenological Side

For the structure of first-person experience (the object carried in Part II), this book does not introduce new physical-dimensional symbols, but uses descriptive phenomenological terms (retention, protention, quality, binding, unity, attention, and so on), in order to avoid mistakenly treating first-person objects as dimensional physical quantities. Three key groups of descriptive objects must be marked in particular, so that subsequent appendices can cite them:

Table A.7.. Descriptive Objects, Referents, and Physical-Side Correspondences

Descriptive object	Referent	Physical-side correspondence (see corresponding chapters)
Inner order of quality	Continuous gradation within a single quality (such as the color continuum within "red")	Single-frequency gradient configuration (Chapter 6)
Qualitative kinds	Discontinuities among different qualities (the difference between red and pain, sound and taste)	Frequency families (Chapter 7)
Temporal thickness	Retention–protention structure of the present and the hierarchy of memory	Transient timescale spectrum (Chapter 10, $\tau_{\text{retain}} \sim \tau_0/f$)

Discipline: first-person objects are used only as descriptive markers, not as physical-dimensional quantities; the physical-side correspondence is a correspondence in the sense of isomorphism (identity level), and does not mean that first-person objects are reduced to physical quantities (as established in Chapters 5 and 15).

A.8. Classification of the Three Types of Propositions (Implementation of the Methodological Discipline of Chapter 2)

All propositions in this book must be explicitly assigned to one of the following three types (they must not be conflated, Chapter 2, F2.4), and the symbols listed in this appendix are classified accordingly:

Table A.8.. Classification of Postulates, Definitions, and Empirical Claims, and the Corresponding Symbols

Category	Content	Relevant symbols
Postulates (not proved in this book; cited from the foundational volume or phenomenology)	A1 (ρ_f), A2 ($E = h\nu$), A3 (master equation 5.4.1), A4 (resonance kernel)	ρ_f, f, h, E; six terms of the master equation; $A(\nu_B, \nu_F), \gamma, A_0$
Definitions (stipulated in this book or the foundational volume to make concepts usable)	Interface coefficient χ_*, construction of $\Psi = (\tilde{A}, C, \mathscr{B})$, attribution mechanism of $\mathscr{B}$, and descriptive objects to be carried in each chapter	$\chi_g, \tilde{A}, C, \mathscr{B}, \Psi, \mathscr{B}$; descriptive objects such as inner order of quality and temporal thickness
Empirical claims (propositions and predictions that can be adjudicated by observation/measurement)	Frequency–quality correspondence, frequency–timescale correspondence ($\tau_{\text{retain}} \sim \tau_0/f$), neural proxies of the three-dimensional state of Ψ, and failure conditions at the end of chapters (Appendix E)	Empirical calibration of $t_{\text{freeze}}, \tau_{\text{retain}}, \tau_0$; proxy relations between $\tilde{A}/C/\mathscr{B}$ and metabolism/PCI/coherent connectivity (Chapters 22 and 23)

Implementation of discipline: whenever postulates are cited from the foundational volume, this book does not re-prove them (it only cites them); whenever definitions are given within this book, their conceptual boundaries must remain consistent; whenever propositions are empirical, failure conditions must be given (summarized in Appendix E). The three types must not be conflated—treating definitions as postulates, or disguising empirical propositions as definitional statements in order to exempt them from falsification, triggers F2.4.

A.9. Overview: Distribution of Symbols across Chapters

For ease of reference, the following table lists the first/main use of the principal symbols across the chapters.

Table A.9.. First Appearance and Principal Chapters of Use for Key Symbols

Symbol	First appearance	Principal chapters of use
ρ_f	Chapters 1 and 2	Throughout the book
Terms of master equation 5.4.1	Chapter 2 (citation)	Chapters 8, 9, and 11
$A(\nu_B, \nu_F), \gamma$	Chapter 2 (citation)	Chapter 9
$t_{\text{freeze}}, \tau_0$	Chapter 10	Chapters 10, 17, and 25
τ_{retain}	Chapter 10	Chapters 10, 17, and 25
χ_g	Chapter 3	Chapters 3 and 4
A(dimensional dissipative power)	Chapter 18	Chapters 18 and 22
$\tilde{A}$	Chapter 17	Chapters 17, 22, 23, and 24
C	Chapter 17	Chapters 17, 22, 23, and 24
$\mathscr{B}$	Chapter 17	Chapters 17, 22, 23, and 24
Ψ	Chapter 17	Throughout Part IV
$\mathscr{B}$	Previewed in Chapter 16, formally introduced in Chapter 19	Chapters 19, 20, and 27

A.10. Citation Conventions for Appendices B, C, and D

When the subsequent appendices cite this appendix, the following conventions must be observed:

1. Citing the master equation: write "equation 5.4.1" or "master equation", and specify the term being discussed (for example, "the dissipative term of the master equation, $-\Gamma(f)\rho_f$").

2. Citing the resonance kernel: write "Lorentzian resonance kernel $A(\nu_B, \nu_F)$", and, where necessary, indicate its half-width γ.

3. Citing the scaling law: write "$\tau_{\text{retain}}(f) \sim \tau_0(f)/f$" or "$t_{\text{freeze}}(f) \sim \tau_0(f)/f$", and state the status of τ_0 as a slowly varying correction.

4. Dimensional check: any newly introduced quantity in Appendices B, C, and D must have its dimension specified and must state whether it enters Ψ (if it enters, use dimensionless quantities such as $\tilde{A}/C/\mathscr{B}$; if it is a dimensional physical quantity, use symbols without a tilde, such as A).

5. Interface coefficients: whenever interface coefficients such as χ_g are used, their restrained status as non-universal constants must be restated.

Any violation of the above conventions triggers F2.4 (conflation of dimensions/proposition types) and must be corrected.

Conclusion of Appendix A: This appendix gathers, in one place, the definitions, dimensions, proposition types, and distribution of use of the symbols used throughout the book, serving as the citation basis for Appendix B (formalization of isomorphism), Appendix C (formal model of given internal–external co-evolution), and Appendix D (parameters and normalization). All postulate-level symbols (ρ_f, the six terms of the master equation, and the resonance kernel) are cited from Sections 5.4 and 5.5 of *EQT Foundational Volume: Frequency,*

Mass, Range of Action, and the Generation Mechanism of Force; all scaling-law symbols (t_{freeze}, τ_{retain}, and τ_0) are cited from Section 4.3 of *Frozen Cosmology: A Natural Philosophy of Cosmic Evolution in the Perspective of Energy Quantum Theory*; all definitional symbols of the consciousness volume ($\tilde{A}$, C, $\mathscr{B}$, Ψ, and $\mathscr{B}$) are introduced for the first time in the corresponding chapters of this book. The restrained status of $\tau_0(f)$, χ_g, and related quantities as "non-universal constants, slowly varying and calibratable" implements, at the symbolic level, the book's discipline of honesty in "not claiming zero free parameters".

B. Formalization of the Isomorphic Structural Space: The Five Sources and the Transfer of the $\tau(f)$ Scaling Law

B.1. The Task and Limits of This Appendix

This appendix gives a precise formal statement of the isomorphism claim established in Part II (Chapters 5–11)—that experiential structure and ρ_f gradient structure are isomorphic (at the identity level). Its purpose is to allow mathematically oriented readers to see clearly what this claim asserts, what formal conditions it requires, and how it can be falsified. The nature of this appendix must be stated first to avoid misunderstanding.

What this appendix does not do. It does not mathematically prove the isomorphism between experience and gradients (the establishment of the isomorphism is fulfilled by the chapter-by-chapter argument and falsifiable comparisons in Part II; the appendix does not re-prove it). It does not provide a mathematical account of the given-ness of experience (given-ness belongs to Part III and is unrelated to the structural level formalized here). It does not provide a "complete mathematical model" of experience (this book does not claim a zero-free-parameter exact law, Chapter 2). What this appendix does is to write, in formal conditions, the content asserted by the claim that "experiential structure and gradient structure are isomorphic", so that the content of the argument is more precise, its failure conditions are clearer, and its connection with the foundational volume is more explicit.

The discipline of this appendix (fully consistent with Appendix A) is as follows: (1) citation without re-proof (formulas from the foundational volume are cited only and not rederived); (2) dimensional checking (any newly introduced quantity must have its dimension specified, and all quantities that enter Ψ must be dimensionless); (3) restraint concerning interface coefficients and slowly varying factors (τ_0, χ, and the like are explicitly marked as non-universal constants); (4) formalization must not overstep into the first-person side (structural formalization characterizes only the structure of experience, not its being-experienced—the latter belongs to Part III and to the final cell).

B.2. The Experiential Structural Space ε and the Gradient Structural Space G

The first step of formalization is to define the "structure of experience" and the "structure of gradients" as two mathematical objects (structural spaces). Isomorphism is then a certain structure-preserving mapping between these two spaces.

Definition of the experiential structural space ε. Let ε be the structural space of all possible experiences. An element $e \in \varepsilon$ is a structural characterization of an experience—that is, a structural description of the inner order of quality, qualitative kinds, law of flow, relations of unity and binding, temporal thickness, and memory hierarchy of a segment of experience (excluding its being-experienced). The structure of ε is constituted by several families of relations:

- Qualitative order relation $\leq_q$: the continuous order within a quality (such as the ordered variation of "red" within color space);

- Qualitative-kind equivalence relation $\sim_\kappa$: equivalence by belonging to the same qualitative kind (for example, all experiences of "red" belong to the same $\sim_\kappa$ equivalence class);
- Evolution relation $\to_t$: temporal transition of experience (including continuity and transition constraints);
- Binding relation $\otimes$: unity and binding of experiential contents (for example, color and shape being bound as attributes of the same object);
- Retention relation $\preceq_\tau$: the ordering of retention durations of experiential contents.

These five groups of relations correspond precisely to the five types of structural features of experience established in the five chapters of Part II. ε is not a numerical space, but a relational structure (in the sense of mathematical structure theory): its elements are structural characterizations of experiences, and its structure is given by the above families of relations.

Definition of the gradient structural space G. Let G be the gradient-configuration space of ρ_f (each $g \in G$ is the total gradient configuration of ρ_f within a certain spacetime region and across several frequency families). The structure of G is likewise constituted by several families of relations:

- Configuration distance d_g: distance within the single-frequency configuration space (a continuous metric between gradient configurations);
- Frequency-family partition P_f: the frequency-family structure of ρ_f (discontinuities among different frequency bands);
- Evolution relation $\mapsto$: gradient evolution constrained by the master equation 5.4.1;
- Resonance relation $\langle \bullet, \bullet \rangle$: same-frequency / cross-frequency coupling determined by the Lorentzian resonance kernel $A(\nu_B, \nu_F)$;

- Transient-duration relation $\preceq_T$: the ordering of transient stability durations determined by the freezing scaling law $t_{\text{freeze}}(f) \sim \tau_0(f)/f$.

The families of relations in G also correspond precisely to the five sources of fivefold richness (B.4–B.8 will correspond to them one by one).

B.3. The Isomorphism Mapping $\Phi : \varepsilon \to G$ and Its Structure-Preserving Requirements

The isomorphism claim of this book can be formalized as follows: there exists a mapping $\Phi : \varepsilon \to G$ such that each family of relations in ε is correspondingly preserved, under Φ, as the relevant family of relations in G. The identity level (Chapter 5) further claims that this Φ is not merely a structural correspondence, but identity: an experiential structure e and its gradient-side counterpart $\Phi(e)$ are two sides of the same reality (externally seen as $\Phi(e)$, internally seen as e). The formal level characterizes only the structure-preserving conditions of Φ; the ontological strength of "identity" is established at the philosophical level in Chapters 5 and 26.

Structure-preserving conditions of Φ (five conditions, corresponding to the fivefold richness):

Table B.1.. Preservation Forms between Experiential-Side and Gradient-Side Relations

Condition	Experiential-side relation	Gradient-side relation	Preservation form
(Φ1)	Qualitative order $\leq_q$	Single-frequency configuration distance d_g	$e_1 \leq_q e_2 \Leftrightarrow d_g(\Phi(e_1), \Phi(e_2))$ is monotonic

Continued on next page...

Continued table

Condition	Experiential-side relation	Gradient-side relation	Preservation form
(Φ2)	Qualitative kind $\sim_\kappa$	Frequency family P_f	$e_1 \sim_\kappa e_2 \Leftrightarrow \Phi(e_1), \Phi(e_2) \in$ the same frequency family
(Φ3)	Evolution $\rightarrow_t$	Master-equation evolution $\mapsto$	$e_1 \rightarrow_t e_2 \Leftrightarrow \Phi(e_1) \mapsto \Phi(e_2)$ (constrained by 5.4.1)
(Φ4)	Binding $\otimes$	Resonance $\langle \bullet, \bullet \rangle$	$e_1 \otimes e_2 \Leftrightarrow \langle \Phi(e_1), \Phi(e_2) \rangle \neq 0$ (resonance-kernel activation)
(Φ5)	Retention $\preceq_\tau$	Transient duration $\preceq_T$	$e_1 \preceq_\tau e_2 \Leftrightarrow \tau_{\text{retain}}(\Phi(e_1)) \leq \tau_{\text{retain}}(\Phi(e_2))$

The fulfillment of each condition is the argumentative task of the corresponding chapter in Part II (Chapters 6, 7, 8, 9, and 10); the failure of each condition corresponds to the relevant F-items in Appendix E (F6.x, F7.x, F8.x, F9.x, F10.x). The structure-preservation of Φ is the formalized content of the isomorphism claim; its empirical fulfillment is the substantive argumentative content of Part II. Appendix B only writes that content formally; it does not decide the argument on its behalf.

Key qualification (isomorphism is not sameness of type; structure preservation is not a quantitatively exact law). The structure preservation of Φ requires preservation of relations (such as order, equivalence, continuity, and the presence/absence of resonance), not exact numerical equality between quantities on the experiential side and the gradient side. Φ is not a zero-free-parameter exact function; it is a relation-preserving mapping at the structural level. Misreading Φ as an exact equation of the form "experiential quantity = gradient quantity" triggers F2.4 (conflation of dimensions/proposition types): the experiential side has no dimension in the physical sense (its structure is given by relations rather than numerical quantities), whereas quantities on the gradient side have dimensions (see Appendix A). The two cannot be directly equated as a numerical law.

B.4. Formalization of the Richness of Single-Frequency Gradient Configurations (Corresponding to Chapter 6)

Object to be carried (experiential side): the continuous order within quality—continuous gradation within the same quality (such as the gradation of "red" within hue–brightness–saturation space; the gradation of "pain" within intensity–bodily-location space).

Carrier (gradient side): the gradient-configuration space within a single frequency family. Suppose ρ_f is restricted to a family with a certain frequency f_0. Then the gradient configurations $g(x) = \nabla\rho_{f_0}(x)$ within that family constitute a high-dimensional configuration space G_{f_0}. The structure of this space is given by the following:

- Direction: the vector direction of $\nabla\rho_{f_0}$;
- Magnitude: $|\nabla\rho_{f_0}|$;
- Topology: topological features of the configuration field (connectedness, holes, distribution of singularities);
- Singularities and critical points: the types of points where $\nabla\rho_{f_0} = 0$ (extrema, saddle points).

Formal mapping Φ_6: structure of the inner order of quality $\mapsto$ distance d_g in G_{f_0}.

Structure-preserving conditions:

- The similarity ordering $\leq_q$ among qualities corresponds to the order preservation of distance d_g in G_{f_0}—the more similar two experiences are, the closer their configurations are.

- The dimensions of qualitative space correspond to the degrees of freedom in G_{f_0}.

- The topological features of qualitative space correspond to the topology of G_{f_0} (for example, the circular topology of hue space corresponds to closed loops in single-frequency configuration space).

- Singularities in qualitative space (such as a gray point) correspond to critical points in G_{f_0}.

Formal failure conditions (linked back to Appendix E): failure of order preservation, dimensional non-correspondence, and non-correspondence of topology or singularities correspond respectively to F6.1, F6.2, and F6.3.

B.5. Formalization of the Richness of the Frequency Dimension (Corresponding to Chapter 7)

Object to be carried (experiential side): qualitative-kind differences (the discontinuities between red and pain, sound and taste).

Carrier (gradient side): the frequency-family structure P_f of ρ_f. The frequency axis is divided into different families, with intervals between different families; there is continuity within the same family and discontinuity between families.

Formal mapping Φ_7: qualitative-kind equivalence relation $\sim_\kappa \mapsto$ frequency-family partition P_f.

Structure-preserving conditions:

- Experiences within the same qualitative kind (a $\sim_K$ equivalence class) correspond to the same frequency family.

- Different qualitative kinds (different $\sim_K$ equivalence classes) correspond to different frequency families.

- Experiential "kind discontinuities" (the absence of intermediate terms in which quality A continuously transitions into quality B) correspond to "frequency intervals" between families on the frequency axis.

- Formally: there exists a set of frequency families $\{P_\alpha\}$ such that each $\sim_K$ equivalence class corresponds to one P_α; between the P_α on the frequency axis there is a minimum interval $\delta_\alpha > 0$ (this interval is the physical counterpart of "discontinuity").

Coordination with Φ_6 (double-structure stitching, corresponding to Section 7.3): within the same frequency family P_α, Φ_6 gives the continuous structure (inner order of quality); between different families, Φ_7 gives discontinuity (qualitative kind). The two layers of structure are jointly carried by the frequency-family structure of ρ_f—continuity and discontinuity are not two mechanisms, but different manifestations of the same frequency-family structure within and between families.

Failure conditions: violation of "kind does not continuously transform with intensity", violation of "discontinuity-pattern correspondence", and failure of double stitching correspond respectively to F7.1, F7.2, and F7.3.

B.6. Formalization of the Richness of Master-Equation Evolution (Corresponding to Chapter 8)

Object to be carried (experiential side): the structure of the stream of consciousness—continuity, transition constraints, directionality, attentional shift, and the arrow of time.

Carrier (gradient side): gradient evolution constrained by the master equation 5.4.1. Citing Appendix A and the foundational volume:

$$\frac{\partial \rho_f}{\partial t} = k(f)\rho_f^m + D(f)\nabla^2\rho_f - \nabla\cdot(\rho_f v_f) - \Gamma(f)\rho_f + S_f + N(f)$$

Formal mapping Φ_8: experiential evolution $\rightarrow_t\mapsto$ gradient evolution $\mapsto$ (constrained by 5.4.1).

Structure-preserving conditions:

- The continuity of experience (no abrupt break) corresponds to the continuity of master-equation evolution (continuous dependence of solutions);
- The transition constraints of experience (which transitions are possible and which are impossible) correspond to the constraints imposed by the master equation on evolutionary paths;
- The arrow of time in experience (irreversibility) corresponds to the dissipative direction of the dissipative term $-\Gamma(f)\rho_f$ in the master equation;
- The shift of attentional focus corresponds to the local enhancement dynamics of the nonlinear self-sustaining term $k(f)\rho_f^m$.

Separation between law and drive (discipline of Chapter 8): the six terms of the master equation 5.4.1 are divided into two sides—the side of law includes $k(f)$, $D(f)$, v_f, and $\Gamma(f)$ (describing how evolution is permitted); the side of drive includes S_f (exogenous perturbation) and $N(f)$ (endogenous/stochastic perturbation). Formally:

- Φ_8 maps the experiential-side "constraints of law" to k, D, v, and Γ in the master equation;
- Φ_8 maps the experiential-side "injection of drive" to S_f and $N(f)$ in the master equation (that is, the exogenous/endogenous types on the perturbation axis in Chapter 11).

Failure conditions: non-correspondence of continuity, transition constraints, directionality, and attentional shift correspond to F8.1–F8.4; if law and drive are shown to be inseparable, F8.5 is triggered.

B.7. Formalization of the Richness of Resonance Modes (Corresponding to Chapter 9)

Object to be carried (experiential side): unity of experience (belonging to the same "I") and binding (attributes of the same object are bound).

Carrier (gradient side): same-frequency and cross-frequency resonance modes determined by the Lorentzian resonance kernel $A(\nu_B, \nu_F)$. Citing Appendix A:

$$A(\nu_B, \nu_F) = \frac{A_0}{(\nu_B - \nu_F)^2 + \gamma^2}$$

Formal mapping Φ_9: unity/binding of experience $\mapsto$ same-frequency/cross-frequency resonance structures.

Structure-preserving conditions:

- Unity: unity within a single subject corresponds to same-frequency coherence—the structure by which experience belongs to the same "I" corresponds to a set of oscillations entering the same coherent resonance mode (frequencies close, $|\nu_B - \nu_F| \ll \gamma$). Formally: the equivalence relation of experiential "belonging to one I" $\mapsto$ the relation of "belonging to one dominant coherent resonance".
- Binding: feature binding (color and shape belonging to the same object) corresponds to cross-frequency resonance—configurations of different frequencies are coupled into one group through the resonance kernel $A(\nu_B, \nu_F) \neq 0$.

Formal dissolution of the combination problem (Chapter 9): the "I" is not defined as the set-theoretic summation $\bigcup_i S_i$ of cellular micro-subjects, but as a dominant coherent resonance mode R^*. Formally: R^* is a global resonance structure in G (a coherent linkage determined by the resonance kernel A), and its boundary of extension is drawn by the actual extension of coherence—this is precisely the formal counterpart of the claim in Chapter 16 that "the boundary is the boundary of coherence" (see B.11).

Failure conditions: lack of correspondence between unity and same-frequency coherence, lack of correspondence between binding and cross-frequency resonance, invalid dissolution of the combination problem, and failure of the distinction between coherence and connection correspond to F9.1–F9.4.

B.8. Formalization of the Richness of the Transient Timescale Spectrum (Corresponding to Chapter 10)

Object to be carried (experiential side): temporal thickness of experience and the hierarchy of memory—from subsecond retention–protention, to working memory over seconds, to lifelong long-term memory.

Carrier (gradient side): the transient timescale spectrum of ρ_f. Each transient with frequency f has its characteristic retention duration $\tau_{\text{retain}}(f)$, given by the freezing scaling law (see B.9).

Formal mapping Φ_{10}: experiential retention duration $\preceq_\tau \mapsto$ transient retention duration $\preceq_T$.

Structure-preserving conditions:

- Preservation of the inverse order between the retention duration of experiential content and the carrying frequency (high frequency–short duration, low frequency–long duration);

- Correspondence of decay gradients (experiential fading $\leftrightarrow$ decreasing vividness of transients; forgetting $\leftrightarrow$ dissipation of transients);

- Correspondence linking present thickness and memory hierarchy (jointly carried by the same scaling law, B.9).

Failure conditions: failure to preserve the ordering of retention timescales, non-correspondence of decay gradients, failure of the continuous correspondence, and inability to incorporate long-term memory into transients correspond to F10.1–F10.4.

B.9. The Isomorphic Transfer of the $\tau(f)$ Scaling Law: From Cosmic Freezing to Memory Transients in the Brain

This section is the direct point of contact between Appendix B and the foundational volume. It formalizes the freezing-time scaling law from Section 4.3 of *Frozen Cosmology: A Natural Philosophy of Cosmic Evolution in the Perspective of Energy Quantum Theory* into the consciousness-volume form used in Chapters 10, 17, and 25 of this book.

The scaling law of cosmic freezing (cited from *Frozen Cosmology: A Natural Philosophy of Cosmic Evolution in the Perspective of Energy Quantum Theory*, §4.3):

$$t_{\text{freeze}}(f) \sim \tau_0(f)/f$$

where:

- $t_{\text{freeze}}(f)$: the freezing time of a ρ_f structure with frequency f under the cosmic background (the characteristic duration for which the structure exits active evolution); dimension $[s]$.

- $\tau_0(f)$: a slowly varying effective background factor, absorbing corrections from the background (cosmic expansion $H(t)$), coupling, diffusion, decay, and source terms; its dimension is inferred from the scaling relation as $[s^2]$ (so that τ_0/f balances dimensionally with t_{freeze}), but this book interprets it as a "scaling relation" rather than a "strict dimensional equation" (see the note at the end of B.9).

- Core tendency: $1/f$ dominance—high-frequency structures freeze earlier (short τ_{retain}), while low-frequency structures freeze later (long τ_{retain}).

Isomorphic transfer in the consciousness volume (the "direct identity" version adopted by this book): this book maintains that memory transients in the brain are directly freezing structures of ρ_f under the brain's background. They obey the same freezing scaling law, except that the "background" changes from cosmic expansion to the brain-internal dissipative background (neural decay, noise, metabolic constraints, and so on), which is absorbed by $\tau_0(f)$. Formally:

$$\tau_{\text{retain}}(f) \sim \tau_0(f)/f$$

(freezing under the brain's background)

where:

- $\tau_{\text{retain}}(f)$: the retention duration of a memory transient in the brain with frequency f.

- $\tau_0(f)$: the effective factor that absorbs the brain-internal dissipative background (the same name and dimension as τ_0 under the cosmic background, but with different concrete content because the background differs).

- Core tendency: $1/f$ dominance; high frequency–short duration corresponds to immediate retention/working memory, and low frequency–long duration corresponds to long-term memory.

Key disciplines of the transfer (fully consistent with the wording of Chapters 10, 17, and 25):

1. Not analogy, but the same mechanism: this book does not treat this as the "analogical transplantation of a cosmological formula". It claims instead that the same ρ_f freezing mechanism has the same scaling law under two backgrounds (cosmic and brain-internal). This is a formal fulfillment, on the problem of memory, of this book's ambition of mind–matter unity—memory retention and the freezing of cosmic structures are manifestations of the same mechanism under two different backgrounds.

2. Restrained status of $\tau_0(f)$: τ_0 retains the status of a "slowly varying correction factor" (consistent with *Frozen Cosmology: A Natural Philosophy of Cosmic Evolution in the Perspective of Energy Quantum Theory*)—it is a slowly varying, calibratable, and testable effective background quantity, not an exact constant and not tied to a specific neural frequency band or metabolic rate. What this book claims is the form of the scaling relation ($\tau_{\text{retain}} \sim \tau_0/f$, with $1/f$ dominance and slowly varying correction by τ_0), not a zero-free-parameter exact law (preserving the restraint of Chapter 2, F2.4).

3. Not an unconditionally exact f^{-1}: the slow variation of $\tau_0(f)$ means that $\tau_{\text{retain}} \sim \tau_0/f$ is not an unconditionally exact f^{-1} law—it is a "$1/f$-dominant tendency + slowly varying correction by τ_0". Formally, for any frequency ratio f_2/f_1, the retention-duration ratio is

$$\tau_{\text{retain}}(f_1)/\tau_{\text{retain}}(f_2) \approx (\tau_0(f_1)/\tau_0(f_2)) \cdot (f_2/f_1)$$

where $\tau_0(f_1)/\tau_0(f_2)$ is a slowly varying ratio (close to 1, but not strictly equal to 1). Tests should target this dominant tendency and the calibratable τ_0 correction, rather than demand exact pointwise inverse proportionality (see Chapter 25, F25.2).

4. Formal failure condition: if the retention duration of memory transients in the brain systematically violates the $1/f$-dominant tendency of $\tau_{\text{retain}} \sim \tau_0/f$, and this deviation cannot be accommodated by the slowly varying correction of τ_0, then the $\tau(f)$ isomorphic transfer formalized in this appendix fails in the case of conscious memory (linked back to F10.5 and F25.2).

Note on the dimension of τ_0: Appendix A and this section take the dimension of τ_0 as $[s^2]$ under the "strict dimensional equation" reading, so that τ_0/f has dimension $[s]$ and balances with t_{freeze}. Another legitimate interpretation is to read "$\sim$" as a scaling relation (not a strict dimensional equation), treat τ_0 as an effective factor with a "timescale dimension" $[s]$, and absorb an implicit frequency reference scale. This book uniformly adopts the former (strict dimensional equation) in order to preserve dimensional closure in the appendix; its physical meaning is concretely settled in the calibration of $\tau_0(f)$ marked in the roadmap of Chapter 25.

B.10. Connection with Parts III and IV

What Appendix B formalizes stops at the structure of experience (the scope reached by the isomorphism of Part II). The connection between this appendix and Parts III and IV must be clarified, so that the boundary of structural formalization can be seen:

- Connection with Part III (given-ness): given-ness is not included within the family of structural relations in ε (ε characterizes only structure, not being-experienced). The "internal–external co-evolutionary freezing" of Part III is a mechanistic characterization of "what ρ_f relation given-ness corresponds to", and is not covered by Φ in Appendix B. Formally, Φ gives the structural isomorphism $\varepsilon \to G$; "the being-experienced of this structure" belongs to the final cell of Chapters 15 and 20 and is not within the image of Φ (this is precisely the formal expression of the C3 prohibition: structure preservation by Φ does not entail being-experienced).

- Connection with the later part of Part III (cognization and demarcation): the cognization, comparison, and $\mathscr{B}$-demarcation established in Chapters 17–19 are further organizations of ε (especially the portions related to memory and knowledge) within the dominant coherence R^* (B.7). Knowledge is the higher-order organization of low-frequency long-duration transients (connected with B.8 and B.9); comparison is resonance between perturbations and knowledge transients (connected with B.7); $\mathscr{B}$-demarcation is the attributional aspect of the dominant coherence R^* (connected with R^* at the end of B.7). These are concrete developments of Φ within the structure of dominant coherence, not additions of a new family of relations to ε (Chapter 11, F11.4: no sixth type of structure is added).

- Connection with Part IV (Ψ): $\Psi = (\tilde{A}, C, \mathscr{B})$ is the projection of G onto the observable plane (Chapter 17), not an element of ε or G itself. Formally, there exists a projection $\pi : G \to \Psi$-space, such that the complex structure of gradient configurations is summarized into a three-dimensional observable projection. π is necessarily incomplete (projection necessarily involves information loss, Chapter 17)—π cannot distinguish qualitative kinds (this requires direct characterization of frequency families, beyond what $\tilde{A}/C/\mathscr{B}$ can summarize). Therefore, the unique predictions of Chapter 25 (frequency–quality and frequency–timescale) appeal to P_f and the transient timescale spectrum themselves, not to the image Ψ of π.

B.11. Summary of Formal Failure Conditions

Each Φ_i and the $\tau(f)$ transfer formalized in this appendix are linked back to the corresponding failure conditions in Appendix E:

Table B.2.. Correspondence between Formalized Items, Failure Conditions, and Appendix E

Formalized item	Formal failure condition	Linked back to Appendix E
Φ_6 (single-frequency configuration)	Failure to preserve distance order, dimensionality, topology, or singularities	F6.1–F6.3

Continued on next page...

Continued table

Formalized item	Formal failure condition	Linked back to Appendix E
Φ_7 (frequency families)	Failure of kind equivalence, family correspondence, or double stitching	F7.1–F7.3
Φ_8 (master-equation evolution)	Failure to preserve continuity, transition constraints, directionality, or attentional shift; failure of the separation between law and drive	F8.1–F8.5
Φ_9 (resonance modes)	Failure to preserve unity or binding; failure to dissolve the combination problem	F9.1–F9.4
Φ_{10} (transient timescale spectrum)	Failure to preserve retention order, decay gradient, or continuous correspondence; inability to include long-term memory within transients	F10.1–F10.4
$\tau(f)$ transfer	Systematic violation of $\tau_{\text{retain}} \sim \tau_0/f$ that cannot be accommodated by the slow variation of τ_0	F10.5 (linked with F25.2)
Φ as a whole, as isomorphism	Structural sufficiency cannot be satisfied; completeness is overturned by counterexample; isomorphism collapses into loose similarity	F5.3, F5.4, F5.1
Identity-level strength of Φ	The identity level cannot be maintained; the reasons for the argument do not hold	F1.1, F5.2

What formalization does is to make the content of these failure conditions more precise and their identification more explicit; their adjudication remains to be completed by the arguments of Part II and by empirical testing.

Conclusion of Appendix B: This appendix formalizes the claim of Part II that "experiential structure and gradient structure are isomorphic (at the identity level)" as the structure-preserving conditions of a mapping $\Phi : \varepsilon \to G$ (the five Φ_i), and formalizes the freezing scaling law $t_{\text{freeze}}(f) \sim \tau_0(f)/f$ from Section 4.3 of *Frozen Cosmology: A Natural Philosophy of Cosmic Evolution in the Perspective of Energy Quantum Theory* into this book's $\tau_{\text{retain}}(f) \sim \tau_0(f)/f$ (the same scaling law under the brain's background). The discipline of formalization is consistent throughout: citation without re-proof, dimensional checking, restraint concerning τ_0 and interface coefficients (non-universal constants), and no overstepping of formalization into the first-person side (the formal expression of C3 at the structural level: structure preservation by Φ does not entail being-experienced). The content of Appendix B is jointly secured by the arguments of Part II (the structural level) and by citations to the foundational volume and the freezing dynamics spectrum; all its failure conditions are linked back to Appendix E. The subsequent Appendix C will address the formal model of given internal–external co-evolution (Part III), while Appendix D will address the calibration rules for parameters and normalization.

C. Formal Model of Given Internal–External Co-Evolution

C.1. The Task of This Appendix and the Strictest Discipline of Formalization

This appendix gives a precise formal statement, in formal language, of the mechanism of given-ness established in Part III (Chapters 12–20): given-ness depends on the internal–external co-evolutionary freezing of a bounded interior and is anchored in the boundary of life. It stands alongside and complements the structural formalization in Appendix B. The measure of this appendix must be stated first, because it lies precisely at the point where the C3 prohibition of the whole book is most easily breached.

What this appendix formalizes: the objective and reproducible side of the internal–external co-evolutionary mechanism established in Part III—the relation between the two poles, the conditions of transient freezing, the prerequisite role of boundary, the retention scaling of $\tau_{\text{retain}}(f)$ under the brain's background, and the formal distinction among three configurations (exogenous connection / endogenous / pure thinking). All these are third-person, objectively describable mechanistic contents, in principle comparable with neural measurements.

What this appendix absolutely does not formalize: the being-experienced of given-ness itself. Being-experienced is not contained within the relational structure of co-evolution (as established in Chapter 15); it is the final cell (compressed in Chapter 20 to the subject-attribution residual point). The relations, operators, and freezing conditions formalized in this appendix do not entail being-experienced. Any interpretation that treats the formal construction in this appendix as a "formula of being-experienced" triggers the red lines of the C3 prohibition, including F1.5, F14.7, F18.5, F19.4, and F20.4. Formalization stops here and leaves that one cell irreducible. This is precisely the formal-level guard of C3, and it is also the dividing line between this appendix and all topic substitutions that use formal formulas to masquerade as explanations of being-experienced.

Continuation of discipline (fully consistent with Appendices A and B): (1) citation without re-proof (formulas from the foundational volume and the freezing dynamics spectrum are cited only); (2) dimensional checking (every newly introduced quantity must have its dimension specified); (3) restraint concerning interface coefficients and slowly varying factors (τ_0, χ, and related quantities are non-universal constants); (4) formalization must not overstep into the first-person side (the formal-level expression of C3).

C.2. Formal Definition of Internal–External Co-Evolution

The mechanism of given-ness established in Part III (especially Chapters 13 and 14) can be formalized as the following structure. Let ρ_f be the foundational field (Appendix A.1).

Division of the two poles. Within a certain spacetime region, according to the division of the current coherent structure, ρ_f is divided into two parts:

- Interior pole I: the currently given coherent configuration (its extension is the boundary of the "I", Chapter 16);
- Exterior pole E: the remaining configurations that co-evolve with I (first of all the brain background, and through the body extending to the world).

It must be especially emphasized (as established in Chapters 14 and 16): the interior pole I and the exterior pole E are not two separate devices (such as different brain regions or different physical systems), but a functional distinction relative to "what is currently given". The ρ_f structure within the same brain is divided, relative to the current dominant coherence, into two sides: "what is locked into coherence as interior" and "what co-evolves with it as exterior". Formally, the I/E division is dynamic and changes with dominant coherence; it is not a pre-fixed topological partition.

Co-evolution operator K. Let the co-evolutionary relation between the two poles be described by the operator $K(I,E;t)$, formalized as follows:

$K : (I,E) \times t \mapsto (I',E')$, where $(I',E') = (I,E)$ is the joint evolution of (I,E) over time t under the common constraint of the master equation 5.4.1.

K satisfies the following conditions:

1. Common master-equation constraint: the evolution of both I and E is constrained by 5.4.1, and they are mutually coupled through the shared structure of the ρ_f field (the coupling is determined by the Lorentzian resonance kernel $A(\nu_B,\nu_F)$, Appendix A.3).

2. Nonlinear coupling: the nonlinearity of co-evolution is carried by the $k(f)\rho_f^m$ term of the master equation—when coupling crosses the threshold, resonance is enhanced; when it does not reach the threshold, it decays.

3. Dissipation: co-evolution is a far-from-equilibrium process and must be maintained by the dissipation of $-\Gamma(f)\rho_f$ (corresponding to the "dissipative power A required to maintain non-equilibrium co-evolution" in Chapters 18 and 22).

4. Transient result: the result of K is not a permanent steady state, but transient freezing—after a period of co-evolution,

a coherent configuration with a finite stable duration forms within I (that is, the physical counterpart of "given content").

Formalization of transient freezing. Suppose that over the co-evolutionary interval $[0, t_f]$, the coherent configuration formed in I is ρ_I^*. Its stable duration is given by the freezing scaling law (Appendix B.9):

$$\tau_{\text{retain}}(f_I) \sim \tau_0^{\text{brain}}(f_I)/f_I$$

where f_I is the characteristic frequency of this configuration, and $\tau_0^{\text{brain}}(\bullet)$ is the slowly varying background factor under the brain-internal dissipative background (see C.5).

Meaning of the formalization: this construction shows that the physical quantity corresponding to "given content" is the transient freezing ρ_I^* produced under the action of K; its presence/absence, stability/collapse, and retention duration are all constrained by the co-evolutionary conditions of K and by the freezing scaling law, and are in principle observable and falsifiable. This is the objective reproducible side of Part III.

What the formalization does not entail: the formation, stability, and retention of ρ_I^* do not entail that "this ρ_I^* is experienced". The formalization asserts only that "this transient freezing is formed through this co-evolution"; it does not assert that "this transient freezing is experienced from within". Being-experienced belongs to the final cell (see C.7).

C.3. Boundary Condition: The Membrane as the Formal Prerequisite of Interior Existence

The co-evolution operator K in C.2 presupposes the "existence of an interior pole I"; the existence of the interior pole I, in turn, presupposes a boundary (the anti-panpsychist gate of Chapters 3 and 14). This section formalizes that prerequisite.

Boundary condition (B): the meaningful construction of $K(I,E;t)$ requires that there exist a genuine topological boundary ∂ between I and E—that is, a membrane structure that encloses a region of ρ_f, distinguishes it from the remaining region, and allows selective exchange. Formally:

Let ∂ be the topological boundary determined by the membrane, such that a certain region Ω of ρ_f is enclosed: the inside of Ω is the "interior", coupling on ∂ defines "exchange", and the outside of Ω is the "exterior". Condition (B): the meaningful construction of $K(I,E;t)$ requires that ∂ exist.

Formalization of anti-panpsychism: if ∂ does not exist (that is, if the system has no topological boundary and no distinction between interior and exterior), then K has no interior pole on which it can act, and the whole formalization does not apply to that system. This does not mean that the system has a "co-evolution without interiority"; rather, it means that the relational structure of "internal–external co-evolution" has no anchoring point in that system. Formally:

If ∂ does not exist, then there is no division between I and E, K is undefined, and τ_{retain} cannot be said to be the "retention of this content"—all structures in this appendix run empty for that system.

This is precisely the formal-level expression of the anti-panpsychist gate in Chapter 3: the gate does not lie in frequency matching, nor in a threshold (the γ of the resonance kernel A has nothing to do with this gate, Appendix A.3 and F3.3), but in **whether ∂ exists**. Without ∂, all formalization runs empty; with ∂, the formalization becomes applicable, and given-ness has something to which it can be anchored.

Formal position of the fifth mechanism: the fifth mechanism established in Chapter 3 (enclosure / membrane demarcation) formally functions to guarantee the existence of ∂, that is, to generate the topological boundary that makes K meaningful. Formally, the fifth mechanism is not inside K (it is not one term of K), but belongs to the prerequisite of K—it allows the interior pole I to exist, and thereby allows the entire mechanism of given-ness to apply.

C.4. Three Configurations: Exogenous Connection, Endogenous, Pure Thinking

Chapter 14 established an analysis of the internal–external configurations of six states. Its core can be formalized into three types of co-evolutionary configurations, distinguished according to the source and connection status of the exterior pole E.

Configuration (I): exogenous connection (normal wakefulness). E is connected to the real environment through the body's sensorimotor channels, and the exogenous perturbation S_f of the master equation is systematically injected from the external world; $K(I,E;t)$ yields transient freezing through co-evolution with the world. Formal conditions:

- $S_f \neq 0$ (continuous injection of exogenous perturbation);
- The main body of E is constituted by the ρ_f structure of the external environment;
- Transient freezing corresponds to "given experience of the world".

Configuration (II): exogenous closure and endogenous dominance (pure thinking, meditation, dreaming). The exogenous channel is closed or greatly weakened (the exogenous part of $S_f \to 0$), but co-evolution between the endogenous perturbation $N(f)$ and the brain's own ρ_f structure is maintained as usual; $K(I,E^{\text{interior}})$ yields transient freezing through co-evolution between the brain-internal ρ_f structure and dominant coherence, without requiring connection to the environment. Formal conditions:

- $S_f^{\text{exterior}} \to 0$; perturbations from $N(f)$ and the brain-internal ρ_f structure are maintained;

- The main body of E is constituted by the brain's own remaining configurations (not by the external world);
- Transient freezing corresponds to "inward given-ness" (dreaming, imagination, reasoning).

Configuration (III): collapse of co-evolution (deep sleep, anesthesia). The co-evolution of the two poles cannot be maintained—either because the dissipative term $\Gamma(f)\rho_f$ of the master equation is too strong (dissipation overwhelms maintenance), or because the coherence of dominant coherence collapses (low C, failed $\mathscr{B}$, Chapter 22), so that the meaningful action of K disappears. Formal conditions:

- The coherent structure within I collapses;
- K does not produce stable transient freezing;
- The physical counterpart of given-ness does not form—therefore given-ness disappears.

The key point of formalization (consistent with Chapter 14): configurations (I) and (II) jointly show that given-ness depends on the "maintenance of co-evolution", not on the "connection of the exogenous channel"—this is where this book is discriminated from the naive view that "consciousness depends on sensory input" (one of the discriminating predictions in Chapter 25, F25.3). Configuration (III) shows that the disappearance of given-ness corresponds to the collapse of co-evolution, not merely to a decrease in metabolism ($\tilde{A}$)—this is the formal-level fulfillment of the physical characterization of the three-dimensional state space in Chapter 22.

C.5. Formal Expression of $\tau_{\text{retain}}(f)$ under the Brain's Background

Appendix B.9 already gave the isomorphic transfer of $\tau_{\text{retain}}(f) \sim \tau_0(f)/f$. This section further formalizes $\tau_0^{\text{brain}}(f)$ under the brain's background and connects it with Chapter 17's claim that "memory is the higher-order organization of transients".

Characterization of $\tau_0^{\text{brain}}(f)$. Under the brain's background, the slowly varying content of $\tau_0(f)$ is jointly absorbed from the following factors:

- Dissipative background: the decay rate of neural activity $\Gamma^{\text{brain}}(f)$ (the consciousness-volume counterpart of the dissipative term in the master equation);
- Coupling background: coupling density and effective-connectivity structure in the neural network (connected with the resonance structure in Appendix B.7);
- Diffusion and convection background: diffusion and convection of ρ_f in the neural medium (the consciousness-volume counterparts of the $D(f)$ and v_f terms of the master equation);
- Metabolic constraint: the dissipative power A required to maintain non-equilibrium co-evolution (Appendix A.6, the dimensional counterpart of consciousness intensity).

Formally:

$$\tau_0^{\text{brain}}(f) = F(\Gamma^{\text{brain}}(f), D^{\text{brain}}(f), v_f^{\text{brain}}, A; f)$$

where $F(\bullet)$ is a slowly varying function (its variation over f is much slower than the dominant $1/f$ term). Its concrete form awaits the experimental calibration marked in the roadmap of Chapter 25. This book does not claim the concrete form of $F(\bullet)$, but asserts only that:

1. $F(\bullet)$ is slowly varying—its relative variation within the frequency bands of interest is much smaller than the relative variation of $1/f$;
2. $F(\bullet)$ is calibratable—in principle, it can be inferred from neural measurements (dissipation rate in EEG, connection structure, metabolic rate);

3. $F(\bullet)$ is testable—its calibrated result can be independently verified or overturned by observation (linked back to Chapter 25, F25.2).

Connection with Chapter 17's "knowledge as the higher-order organization of memory transients". Chapter 17 maintains that knowledge is the higher-order organization of low-frequency long-duration memory transients, whose stability is supported by the low-frequency end of $\tau_{\text{retain}}(f) \sim \tau_0^{\text{brain}}(f)/f$ (small f, large τ_{retain}); learning corresponds to the refreezing of new transients into low-frequency long-duration bands. Formally, the knowledge-transient organization K^{know} established in Chapter 17 can be characterized as a set of transient configurations in low-frequency bands ($f \in$ low-frequency band). Each transient is granted long-term stability by the τ_{retain} scaling law and is connected with the others by the resonance kernel $A(\nu_B, \nu_F)$ (Appendix B.7) into a higher-order organization ("knowledge as a structure of resonance–coherence linkages among transients", Section 17.3).

Restatement of restraint: the above formalization does not claim a specific frequency range for knowledge transients (it is not tied to any specific brain-wave band), nor does it claim an exact numerical law for τ_0^{brain}. The formalization establishes only structural relations (low frequency and long duration carry long-term memory; resonance linkages constitute higher-order organization). Empirical calibration belongs to the roadmap of Chapter 25.

C.6. The Formal Interface of the Later Part of Part III: Cognization, Comparison, and $\mathscr{B}$-Demarcation

The three mechanisms established in the newly added four chapters (Chapters 17–20)—cognization (Chapter 17), comparison (Chapter 18), and $\mathscr{B}$-demarcation (Chapter 19)—can all be treated, at the formal level, as further structuring of the co-evolution operator K in C.2, rather than as newly added independent mechanisms. This section gives their formal interface, consistent with the wording of the newly added four chapters.

Cognization mechanism K^{know}. The cognization established in Chapters 17 and 18 can be formalized as follows: when the resulting transient ρ_I^* of $K(I,E;t)$ undergoes resonant coupling (triggered by the resonance kernel A) with an element of the existing knowledge base K^{know} (the higher-order organization of low-frequency long-duration transients, C.5), ρ_I^* is cognized as "recognized experience". Formally:

If $\langle \rho_I, \rho_K \rangle \neq 0$ (resonance-kernel activation, $\rho_K \in K^{\text{know}}$), then ρ_I is cognized as "recognition".

When there is no match (the resonance kernel is not activated), the result is "strangeness"—driving heightened attention, search, reorganization, and refreezing (the cycle of comparison outcomes in Section 18.5), formally corresponding to the dynamic reconstruction of K^{know} and the refreezing of a new low-frequency long-duration transient.

Comparison mechanism (Chapter 18): the resonance relation $\langle \bullet, \bullet \rangle$ in the above expression is itself the formalization of "comparison"—the resonance relation between the ρ_I^* excited by perturbation (exogenous S_f or endogenous $N(f)$) and existing knowledge transients. The four types of comparison outcome (matching / conflict / reorganization / refreezing) correspond to four dynamics of this resonance relation (resonance enhancement, destructive interference, organizational reconstruction, and stabilization of a new transient).

$\mathscr{B}$-demarcation (Chapter 19): within the framework of C.2, $\mathscr{B}$ can be formalized as an attribution marker—it marks the ρ_f structures locked into the dominant coherence R^* (Appendix B.7) as "belonging to the I supported by that coherence". Formally:

$$\mathscr{B}(\rho) = 1 \quad \text{iff} \quad \rho \text{locks into} R^*$$

(that is, the coherent coupling between ρ and R^* exceeds the coherence threshold θ_{coh}).

It must be restated (Chapter 19, F19.4): $\mathscr{B}$ does not produce ρ_I^* (given content is produced by K); it does not explain the being-experienced of ρ_I^* (the final cell lies beyond C.7); and it does not introduce an independent criterion (the demarcation by $\mathscr{B}$ is the attributional marker of the physical fact of coherence locking, avoiding the homunculus problem). The failure modes of $\mathscr{B}$ (depersonalization / thought insertion / somatoparaphrenia / made experiences, Section 19.4) are formalized as local failures of the coherence threshold θ_{coh}: certain configurations fail to lock into R^* (resonant coupling below θ_{coh}) while remaining given, and are therefore experienced as "non-mine contents".

Formal interface of the second compression (Chapter 20): after cognization (participation of K^{know}) and $\mathscr{B}$-demarcation, the givenness residual point (the fissure marked in C.7) is further compressed and located as the subject-attribution residual point—that is, "why the ρ_I^* cognized by K^{know} and demarcated by $\mathscr{B}$ as 'mine' is experienced". Formally, the residual point of the first compression lies in the being-experienced of (K, ρ_I^*); the residual point of the second compression lies in the being-experienced when $(K, K^{\text{know}}, \rho_I^*, \mathscr{B})$ are all in place. The latter has stripped away more formalizable content than the former, and is therefore more precisely located and more compressed, but it still cannot be reduced by formalization (C.7).

C.7. The Boundary of Formalization: It Does Not Reach the Final Cell

Everything formalized in this appendix—the co-evolution operator K, the boundary condition ∂, the three configurations, the τ_{retain} scaling law, and the formal interfaces of cognization and demarcation—lies on the objective side of the fissure (Chapter 15). This section explicitly states the boundary of formalization as the final formal-level guard of the C3 prohibition.

The range reached by formalization:

- K and transient freezing: what kind of ρ_f relation given-ness corresponds to (mechanistic characterization);
- ∂ and boundary condition: given-ness presupposes the existence of an interior (the anti-panpsychist gate);
- Three configurations: the presence or absence of given-ness varies with the configuration of co-evolution (observable and falsifiable);
- τ_{retain} scaling law: the structural relation of the retention duration of given content (calibratable and testable);
- Formal interfaces of cognization, comparison, and $\mathscr{B}$-demarcation: how given content is organized and attributed (further characterization at the structural level).

What formalization does not reach (the final cell): when all of the above are in place, "why this ρ_I^* (produced by K, cognized by K^{know}, and demarcated by $\mathscr{B}$ as 'mine') is experienced from the first-person interior"—this cell is not entailed by any construction in this appendix. Formalization stops here and leaves that cell open.

Why formalization cannot cross it (consistent with Chapters 15, 20, and 27): all constructions in this appendix (operators, relations, conditions, and scaling laws) are third-person, objectively reproducible descriptions (objectivity in the detection volume means repeatable co-perturbation). Being-experienced is a first-person fact; it does not appear in any term of third-person description. The entire vocabulary of formalization concerns "what operates and how"; being-experienced asks "why this operation is experienced from within". To treat any formal construction (such as $\mathscr{B}(\rho) = 1$, activation of the resonance kernel, or formation of transient freezing) as an answer to "why it is experienced" is to extend what belongs to the objective side illegally into the subjective side, triggering all the C3 red lines (F14.7, F18.5, F19.4, F20.4, F27.5).

Honesty of formalization: this appendix pushes formalization to everything it can reach (mechanism, boundary, configurations, scaling, cognization, and demarcation are all formalized), while precisely identifying what cannot be formalized (being-experienced, the final cell). This posture of "formalizing to the end, and honestly leaving what cannot be formalized" is fully consistent with the honest posture of the whole book, and with the structural formalization in Appendix B (Φ does not entail being-experienced).

C.8. Summary of Formal Failure Conditions

Each formalized item in this appendix is linked back to the corresponding failure conditions in Appendix E:

Table C.1.. Formalized Items, Formal Failure Conditions, and Correspondence with Appendix E (Continued)

Formalized item	Formal failure condition	Linked back to Appendix E
Relational structure of K	Co-evolutionary relation cannot be transferred; the functional two poles cannot be separated; the relational approach fails	F13.2, F13.3, F14.1
Boundary condition ∂	Interiority is not a necessary precondition of given-ness; an unbounded system has given-ness; the two gates are conflated	F3.2, F1.4, F14.5, F3.3
Three configurations	The states are overturned by counterexamples; characterization of the functional exterior fails	F14.3, F14.6

Continued on next page...

Continued table

Formalized item	Formal failure condition	Linked back to Appendix E
Brain-background expression of the τ_{retain} scaling law	The scaling law systematically fails; $F(\bullet)$ cannot be calibrated	F10.5, F25.2
Cognization mechanism K^{know}	The cognization proposition does not hold; the knowledge base is in fact a substantive region; comparison has no correspondence	F18.1, F18.2, F18.3
Attribution marker of $\mathscr{B}$	$\mathscr{B}$ has no correspondence with attributional phenomena; failure modes do not correspond; $\mathscr{B}$ bears too much	F19.2, F19.3, F19.4
Formal location of the second compression	The second compression does not hold; the subject-attribution residual point is in fact reducible	F20.1, F20.2
Formalization oversteps into being-experienced	Mechanism / Ψ / interface / demarcation / identity is made to masquerade as an answer to being-experienced	F14.7, F18.5, F19.4, F20.4, F27.5

The adjudication of the final category (formalization overstepping) is not empirical observation (these are not empirical propositions), but the identification of a violation of the discipline of formalization: if this book, or anyone following its logic, uses any formal construction in this appendix as an answer to "why it is experienced", the corresponding red line is triggered. This appendix establishes the formal-level guard of C3: formalization stops on the objective side and does not reach the final cell.

Conclusion of Appendix C: This appendix formalizes the mechanism of given-ness established in Part III (including the newly added four chapters) as the internal–external co-evolution operator K, the boundary condition ∂, the three configurations, the brain-background expression of $\tau_{\text{retain}}(f) \sim \tau_0^{\text{brain}}(f)/f$, and the formal interfaces of cognization, comparison, and $\mathscr{B}$-demarcation. The discipline of formalization is consistent throughout: citation without re-proof, dimensional checking, restraint concerning the slow variation of τ_0^{brain} and $F(\bullet)$ (non-universal constants, not tied to specific neural frequency bands), and no overstepping into the first-person side (the formal-level guard of C3). All formalized items are linked back to the corresponding failure conditions in Appendix E; the C3 guard items (F14.7, F18.5, F19.4, F20.4, and F27.5) are concentrated here as a line of defense: formalization stops on the objective side of the fissure, while the final cell is left to the philosophical treatment of Chapters 15, 20, and 27. The subsequent Appendix D will address the calibration rules for parameters and normalization (involving the calibration discipline of $\tilde{A}$, χ, τ_0, and related quantities).

D. Parameters and Normalization

D.1. The Task and Discipline of This Appendix

This appendix provides the calibration rules and normalization discipline for the dimensional quantities, interface coefficients, and slowly varying background factors used throughout the book in actual measurement and cross-system comparison. It is an operational supplement to Appendix A (symbols and dimensions), Appendix B (formalization of isomorphism), and Appendix C (formal model of co-evolution). Its purpose is to allow empirical scientists and later researchers, when using the quantities introduced in this book—such as $\tilde{A}$, C, $\mathscr{B}$, χ, and τ_{retain}—to know which operations are legitimate and which operations overstep, so that the book's "structural relations and slowly varying tendencies" are not misused as "zero-free-parameter exact laws".

What this appendix does: it establishes rules and discipline for calibration and normalization—the principle of nondimensionalizing A into $\tilde{A}$, the conventions for cross-system comparison of the three quantities of Ψ, the restrained calibration of interface coefficients such as χ_g, the testable form of the slowly varying constraint on $\tau_0(f)$, and the point at which these operations must stop.

What this appendix does not do: it does not provide any concrete numerical calibration for any quantity. Concrete numerical values must be determined by real experimental data; this book has no right to fabricate them. What this book establishes is the rule of "how to calibrate", not the conclusion of "what value the calibration has". Misreading a rule as a numerical value would itself trigger F2.4 (conflation of dimensions / proposition types): rules are definitional propositions (belonging to "definitions"), whereas numerical values are empirical propositions (belonging to "empirical claims"); the two must not be conflated (the discipline of Chapter 2).

Continuation of discipline (fully consistent with Appendices A, B, and C): (1) citation without re-proof; (2) dimensional checking (all normalization operations must pass dimensional checking); (3) restraint concerning interface coefficients and slowly varying factors (χ, τ_0, and the like must never be treated as universal constants); (4) separation of the three types of propositions (postulates / definitions / empirical claims, Chapter 2, F2.4); (5) no formalization or calibration of being-experienced (the final cell lies outside the scope of calibration; the operational-level guard of C3).

D.2. Normalization from Dimensional Quantities to Dimensionless Quantities

Appendix A has already established that "all quantities entering Ψ are dimensionless relative quantities". This section gives the general principle for the operation of normalizing dimensional quantities into dimensionless quantities.

General principle (three requirements for normalization):

Given a dimensional physical quantity X (for example, A, dissipative power, measured in W), its normalization into a relative quantity $\tilde{X}$ (used for cross-system and cross-state comparison) must satisfy the following:

1. The reference scale must be specified explicitly: in the normalization $\tilde{X} = X/X_{\text{ref}}$, the reference scale X_{ref} must be explicitly

specified (for example, the system's baseline value, wakeful-state value, or group mean), and must not remain implicit. Any "dimensionless quantity" whose reference scale is not explicitly specified triggers F2.4.

2. The physical meaning of the reference scale must be calibratable: X_{ref} itself must be an independently measurable or independently stipulated physical quantity (it must not be inferred backward from $\tilde{X}$); otherwise, "normalization" becomes circular.

3. Cross-system comparison must use the same type of reference scale: when comparing $\tilde{X}$ across different systems, the X_{ref} used in each system must be of the same type (for example, all must be baseline metabolic rates, or all must be wakeful-state values). It is not permissible to compare system A using A_{wakeful} with system B using $A_{\text{group mean}}$—that is not cross-system comparison, but misuse of normalization.

Specific normalization from A to $\tilde{A}$ (Chapters 18 and 22):

$$\tilde{A} = A/A_{\text{ref}}$$

where A is the dissipative power required to maintain non-equilibrium internal–external co-evolution (W, Appendix A.6), and A_{ref} is the reference scale. This book does not fix a single choice of A_{ref}, but gives three typical uses and their conventions:

- Within-individual cross-state comparison: A_{ref} is taken as the individual's wakeful baseline value A_{baseline}; $\tilde{A}$ describes that individual's state relative to his or her own baseline. This is the most commonly used normalization, used to characterize transitions among sleep, wakefulness, anesthesia, epilepsy, and related states (Chapter 24).

- Cross-individual comparison: A_{ref} is taken as a group-level reference scale (for example, the mean value of healthy wakefulness in the same age group); $\tilde{A}$ describes an individual's state relative to the group. This must be used cautiously, because group heterogeneity may obscure individual differences.

- Cross-species comparison: the choice of A_{ref} is itself a research question, and there is no ready-made answer. Any cross-species comparison of $\tilde{A}$ must explicitly state the rationale and limits of the chosen reference scale.

Warning against misuse: to read "$\tilde{A}$ is higher than 1" as meaning "this state is absolutely higher than some universal standard" is incorrect. $\tilde{A}$ is a relative quantity; its being "higher than 1" means only "higher than the chosen A_{ref}". Mistaking a relative quantity for an absolute quantity is the most common misuse of normalization.

D.3. Normalization and Cross-System Comparison of the Three Quantities of Ψ

The three quantities of $\Psi = (\tilde{A}, C, \mathscr{B})$ serve as the three-dimensional coordinates of conscious states in Ψ-space (Chapters 17 and 22), and their use must obey the following normalization discipline.

Normalization of each of the three quantities:

- $\tilde{A}$: normalized from A to $\tilde{A}$ according to D.2.
- C: projection of consciousness complexity. Its dimensional counterpart, if any (such as the raw value of PCI), depends on the measurement method (PCI is a dimensionless index within the interval 0–1 and is itself already a kind of normalization). The C used in this book takes a dimensionless form, but its reference normalization (for example, taking the wakeful-state value as 1) must be explicitly specified.
- $\mathscr{B}$: projection of consciousness boundary. Its dimensional counterpart (the extension and closure of cross-regional coherence) is usually given as a relative measure (such as a coherence fraction). The $\mathscr{B}$ used in this book likewise takes a dimensionless form, and its reference normalization must be explicitly specified in the same way.

Trajectories and state points in Ψ-space:

Chapters 22 and 24 have already established that conscious states correspond to points or regions in Ψ-space, and that state transitions correspond to trajectories. Normalization operations must ensure the following:

1. Continuity of trajectory: after normalization, the Ψ trajectory should preserve its continuity (except for genuine state discontinuities, normalization should not introduce artificial breaks).

2. Independence of the three dimensions: after normalization, $\tilde{A}$, C, and $\mathscr{B}$ should preserve their physical independence (the three-dimensional orthogonality of Chapter 22), and should not lose their independence because of a shared normalization operation (for example, dividing all three quantities by one common scale).

3. Uncertainty region: according to Section 23.8, "Ψ must be reported as an uncertainty region rather than an exact point". After normalization, Ψ should therefore be reported as a region in Ψ-space (reflecting the joint uncertainty of measurement and normalization), and the size of this region must be determined by the strength of the data constraints.

Discipline for cross-system comparison (the discriminating prediction that "generalized epilepsy has high $\tilde{A}$ but no consciousness" depends on this discipline):

One of the core claims of Chapters 18, 22, and 24 is that conscious states are jointly characterized by three dimensions, and that one dimension is insufficient. Testing this claim requires cross-state comparison in Ψ. The normalization discipline for such comparison is as follows:

When stating that "during a generalized epileptic seizure, $\tilde{A}$ is high, C is low, and $\mathscr{B}$ fails", the reference scale of each of the three quantities must be explicitly specified, and the three scales should be taken from the same subject's wakeful baseline (or from the same group), so that the joint claim $\tilde{A} > 1$ but $C < 1$ has a comparable meaning. Mixing reference scales deprives the discrimination of empirical content.

Formal failure-condition chain: misuse of normalization causing cross-state comparison in Ψ to fail is linked back to the misuse of F22.4 (the three dimensions are reducible to one dimension) or F24.1 (states cannot be consistently localized).

D.4. Calibration Discipline for Interface Coefficients such as χ_g

Appendix A.5 has already established that "χ_g is not a universal constant, but varies with scale / frequency / background; it is calibratable and testable, and should not be treated as a zero-free-parameter quantity". This section gives the calibration discipline for interface coefficients such as χ_g.

What is an interface coefficient? Coefficients of the χ type appear at the interface between two levels or two mechanisms—for example, the interface between ρ_f and concrete mechanical descriptions in the foundational volume (χ_g), the interface between the ρ_f mechanism in this book and neural measurement proxies, and the coupling strength between the interior pole I and the exterior pole E (formalized as resonance coupling in Appendix C.2). The physical meaning of an interface coefficient is the coupling strength or conversion ratio between two levels at a given interface.

Calibration discipline for interface coefficients (four rules):

1. Its variability must be explicitly stated: when χ is used, the variables on which it depends must be specified

   $$chi = \chi(\text{scale}, \text{frequency}, \text{background}, \ldots).$$

   Any use of χ as a "universal constant" triggers F2.4 and F4.4 (overstrong anti-functionalist inference).

2. Calibration must be scale-specific and frequency-band-specific: χ should be calibrated within a specified scale / frequency band, and a calibrated value at one scale must not be extrapolated to another scale. Formally: if $\chi(\text{scale } s_1, \text{frequency } f_1)$ is measured as a certain value, this does not imply that $\chi(s_2, f_2)$ has the same value—the latter must be calibrated independently.

3. Testability: the calibrated result for χ must be subject to independent observational testing. If the value of χ for the same interface is systematically inconsistent across different experiments or subjects, and this inconsistency cannot be explained by differences in its dependent variables, then the physical characterization of that interface fails.

4. It must not enter the class of "universal physical quantities": χ must not be listed among "universal physical constants" (such as h, c, or G), nor must it be treated as a state quantity entering Ψ in parallel with $\tilde{A}$, C, and $\mathscr{B}$. In this book, χ is always a calibration parameter of an interface, not a state quantity.

Examples of typical uses (for illustrating the discipline only, not for giving concrete numerical values):

- χ_{membrane} at the enclosure interface in Chapter 3 (coupling strength of selective exchange through the membrane): it varies with membrane type, temperature, and ionic environment, and should not be fixed at a single numerical value.

- χ_{proxy} at the interface between ρ_f and consciousness measurement proxies in Chapter 4 (for example, the conversion coefficient between $\tilde{A}$ in ρ_f and the measured value of brain metabolism): it varies with measurement method, individual, and state, and must be calibrated separately.

- The calibration coefficient of the resonance coupling $\langle I, E\rangle$ in Appendix C.2: it varies with the specific frequency bands, coherence structure, and exogenous/endogenous configuration of I/E.

Warning against misuse: to "assign a value to χ and substitute it into the formula" in order to obtain an absolute prediction is incorrect. The calibration of χ is an independent experimental task, and its result carries uncertainty. Wherever calibrated values of χ are used to make predictions, the calibration uncertainty and dependent variables of χ must be reported at the same time.

D.5. Normalization Treatment of the Slowly Varying Constraint on $\tau_0(f)$

Appendices A.4, B.9, and C.5 have already established that $\tau_0(f)$ is a slowly varying effective background factor that absorbs corrections from brain-internal dissipation, coupling, diffusion, decay, and source terms. It is not a universal constant, and must be slowly varying, calibratable, and testable. This section gives the normalization and calibration discipline for $\tau_0(f)$.

Formalization of the slowly varying constraint on $\tau_0(f)$:

The "slow variation" of $\tau_0(f)$ can be formalized as the following condition: within the frequency band of interest $[f_{\min}, f_{\max}]$,

$$\left| \frac{d \ln \tau_0(f)}{d \ln f} \right| \ll 1$$

that is, the logarithmic rate of change of $\tau_0(f)$ is far smaller than the logarithmic rate of change of the dominant $1/f$ term (the latter being 1). This condition ensures that the $1/f$-dominant tendency of $\tau_{\text{retain}}(f) \sim \tau_0(f)/f$ is maintained within that frequency band—the slowly varying correction of τ_0 does not become dominant and does not "twist" the $1/f$ tendency into another scaling form (such as $1/f^2$ or const).

Frequency-band-specific calibration of $\tau_0(f)$:

Because $\tau_0(f)$ absorbs different physical backgrounds in different frequency bands (the high-frequency range is mainly governed by neural decay $\Gamma_{\text{high}}(f)$, while the low-frequency range is mainly governed by structural stability and metabolic constraints), its concrete form may differ across frequency bands. This book therefore advocates frequency-band-specific calibration:

Within the frequency band $[f_a, f_b]$, the calibrated result for $\tau_0(f)$ is $\tau_0^{f_a,f_b}$, which must be reported separately and must not be extrapolated to other frequency bands.

Normalized form of $\tau_0(f)$:

For ease of cross-system comparison, $\tau_0(f)$ may be normalized as follows: choose a reference frequency f_{ref} and a reference value of τ_0, $\tau_{0,\text{ref}} = \tau_0(f_{\text{ref}})$. Then the normalized form is

$$\tilde{\tau}_0(f) = \tau_0(f)/\tau_{0,\text{ref}}$$

Cross-system comparison should compare the shape of the curve $\tilde{\tau}_0(f)$ (its functional dependence on f), rather than the absolute value of $\tau_0(f)$. The advantage of this normalization is that the shape of $\tau_0(f)$ (its mode of slow variation) can be compared independently of the absolute reference scale.

Testability of $\tau_0(f)$ (linked back to Chapter 25, F25.2):

The slowly varying constraint on $\tau_0(f)$ must be subject to empirical testing. The testing procedure is as follows:

1. Measure the relationship between the retention duration and frequency of neural activity within a given frequency band, and obtain the empirical curve of $\tau_{\text{retain}}(f)$;

2. Compare it with the $1/f$-dominant tendency and extract the deviation;

3. Test whether this deviation can be accommodated by a slowly varying $\tau_0(f)$ (that is, whether the deviation satisfies)

$$|d \ln \tau_0(f)/d \ln f| \ll 1;$$

4. If the deviation systematically exceeds the slowly varying constraint ($|d\ln\tau_0(f)/d\ln f|$ approaching or exceeding 1), then the $1/f$-dominant tendency of $\tau_{\text{retain}} \sim \tau_0/f$ is falsified—triggering F10.5 and F25.2.

Warning against misuse: it is dangerous to "fit $\tau_0(f)$ as a specific function (for example, $\tau_0(f) = a \cdot f^\beta$) in order to obtain precise predictions". Any concrete functional fitting of $\tau_0(f)$, if used as a "universal law", becomes a disguised "zero-free-parameter exact law" (F2.4) and violates the restrained status of τ_0 as slowly varying. A fitting result can only serve as the calibration for that experiment and that frequency band, and must not be extrapolated.

D.6. The Boundaries among the Three Types of Propositions in Normalization

Chapter 2 established three types of propositions (postulates / definitions / empirical claims), and Appendix A.8 has classified the corresponding symbols. This section explicitly states the boundaries among the three types in normalization operations, in order to prevent overstepping.

Postulate-level propositions (A1–A4, the master equation 5.4.1, and the Lorentzian resonance kernel): these do not participate in normalization operations (they are postulates cited from the foundational volume, and this book does not reshape their form at the level of normalization). They are the starting point of normalization operations, not their objects.

Definitional propositions (the definitions of $\tilde{A}$, C, $\mathscr{B}$, Ψ, χ_g, the restrained positioning of the slow variation of τ_0, and the attribution mechanism of $\mathscr{B}$): definitions themselves are not tested through calibration—their legitimacy is adjudicated by conceptual clarity and fidelity to the ontology on which they rely (the coherence–philosophical level). However, their conditions of application and calibration methods in concrete empirical contexts (that is, the discipline established in this appendix) must be empirically tested. For example, the definition of $\tilde{A}$ is not overturned by calibration, but whether "a particular calibration method for $\tilde{A}$ is usable" is an empirical question.

Empirical propositions (concrete numerical values of τ_{retain}, calibrated neural proxies for the three quantities of Ψ, scale-specific numerical values of χ, and measured shapes of state trajectories): these are fully exposed to empirical testing. Their failure conditions are summarized in Appendix E (F10.x, F22.x, F23.x, F24.x, F25.x, and so forth).

Core discipline: normalization operations bring definitional propositions (such as $\tilde{A}$) into empirical contexts and generate empirical propositions (such as concrete numerical values and trajectories of $\tilde{A}$); but empirical results must not be made to masquerade backward as revisions of definitions (unless the definition itself is shown, independently, to contain an internal problem). This one-way relation—definitions provide the framework for empirical measurement, while empirical results provide calibration of the definition in concrete contexts—is the core discipline of normalization operations in this book.

Special guard: being-experienced is not within the scope of normalization (the operational-level guard of C3):

All normalization operations in this appendix ($A \to \tilde{A}$, comparison among the three quantities of Ψ, calibration of χ, and testing of the slow variation of τ_0) are located on the objective side of the fissure—they normalize, calibrate, and compare what belongs to the objective side, but they never normalize or calibrate being-experienced. Any attempt to normalize "the degree of being-experienced" into some dimensionless quantity (such as a "Q factor" or a "phenomenal intensity index") triggers the red line of C3 (F1.5, F18.5, F19.4, F20.4, F27.5): being-experienced is not contained within structure and mechanism, and therefore lies outside the scope that normalization and calibration can reach.

Formally: the domains of all normalization mappings ($X \to \tilde{X}$) established in this appendix consist of dimensional physical quantities or observable proxies; they never include being-experienced. Being-experienced appears neither in the domain nor in the codomain of any normalization operation, and remains the uncalibratable item of the final cell.

D.7. Summary of Formal Failure Conditions

Every violation of the calibration discipline in this appendix is linked back to the corresponding failure condition in Appendix E:

Table D.1.. Calibration Items, Calibration Failure Conditions, and Correspondence with Appendix E

Calibration item	Calibration failure condition	Linked back to Appendix E
$A \to \tilde{A}$ normalization	Reference scale not specified; cross-system comparison mixes reference scales	F2.4 (conflation of dimensions / proposition types)
Cross-system comparison of the three quantities of Ψ	Mixing the three scales deprives the discrimination of empirical content; independence of the three dimensions is lost	F22.4, F24.1
χ_g interface coefficient	Treated as a universal constant; extrapolated to uncalibrated scales / frequency bands	F2.4, F4.4

Continued on next page...

Continued table

Calibration item	Calibration failure condition	Linked back to Appendix E
Slowly varying constraint on $\tau_0(f)$	Slowly varying constraint on $\tau_0(f)$ fails (scaling law shown to be violated)	F10.5, F25.2
Extrapolation of concrete functional fitting of $\tau_0(f)$	Fitting result is extrapolated as a universal law	F2.4
Conflation of the three proposition types in normalization	Empirical results are made to masquerade backward as definitional revisions without independent grounds	F2.4
Normalization oversteps into being-experienced	Attempt to normalize or calibrate being-experienced as a quantity	F1.5, F18.5, F19.4, F20.4, F27.5

The adjudication of the last category (normalization overstepping) is not empirical observation, but the identification of a violation of operational discipline. Any attempt to bring being-experienced into the normalization system under names such as “phenomenal intensity index”, “measure of being-experienced”, or similar expressions triggers the corresponding C3 red line. This appendix establishes this guard at the operational level: normalization stops on the objective side, and the final cell is not included within it.

Conclusion of Appendix D: This appendix establishes the calibration and normalization discipline for the dimensional quantities, interface coefficients, and slowly varying background factors used throughout the book in actual measurement and cross-system comparison. Its key points are as follows: the normalization from A to $\tilde{A}$ must explicitly specify the reference scale, and cross-system comparison must use reference scales of the same type (D.2); cross-state comparison of the three quantities of Ψ must use reference scales from the same source for all three quantities, and must be reported as an uncertainty region rather than an exact point (D.3); interface coefficients such as χ_g are always non-universal constants and must be calibrated scale by scale and frequency band by frequency band (D.4); the slowly varying constraint on $\tau_0(f)$ is formalized as $|d \ln \tau_0 / d \ln f| \ll 1$, calibrated by frequency band, and normalized in shape for cross-system comparison (D.5); the three types of propositions preserve a one-way relation in normalization—definitions provide the framework, empirical results provide calibration, and empirical results must not masquerade backward as definitions (D.6); normalization and calibration are wholly located on the objective side, and being-experienced is not included (the operational-level guard of C3). All failure conditions of this appendix are linked back to Appendix E.

Thus Appendices A (symbols and dimensions), B (formalization of isomorphism), C (formal model of co-evolution), and D (parameter normalization) together constitute the complete formal-operational support of this book: A establishes the symbols, B establishes the isomorphism, C establishes the mechanism, and D establishes the operations. Their consistent discipline—citation without re-proof, dimensional checking, restraint concerning interface coefficients and slowly varying factors, separation of the three proposition types, and no formalization or calibration of being-experienced (C3 guarded at both the formal and operational levels)—is the full implementation, at the level of formalization, of the book's honest posture.

E. General Table of Failure Conditions (Two-Tier Classification)

This appendix gathers in one place the failure conditions listed at the end of each chapter of the book, and classifies them into two tiers according to the two-tier standard established in Chapter 2: the falsifiable layer (whose failures can be adjudicated by empirical observation or testable structural correspondence) and the coherence–philosophical layer (whose failures are adjudicated by internal coherence, conceptual consistency, or fidelity to the cited/preceding framework, rather than directly by observables). This classification itself is an implementation of the meta-norm M: the book marks in advance every possible point of failure and explicitly states which tier of adjudication it belongs to.

Three points must be clarified. First, most chapters contain failure conditions at both layers, and therefore the items from the same chapter may be distributed across the two tables. Second, all items concerning the "final cell being made to masquerade as explained / dissolved / overstepped into given-ness" (such as F1.5, F13.4, F14.7, F15.x, F18.5, F19.4, F20.4, F21.1, F22.5, F23.4, F24.5, F25.5, F26.4, F27.4, and F27.5) have a special status: what they guard is the C3 prohibition. Their "failure" is not empirical falsification, but argumentative overstepping, and therefore they belong to the coherence–philosophical layer. Third, this table is a summary; the complete formulation of each item appears at the end of the corresponding chapter.

E.1. Failure Conditions at the Falsifiable Layer

The conditions at this layer can, in principle, be adjudicated by empirical observation, neural measurement, or testable structural correspondence; their failure means that the corresponding empirical or structural claim of this book is overturned.

Table E.1.. Appendix E: Core Falsifiability Propositions, Failure Conditions, and Corresponding Claims in the Book

Number	Failure condition (brief)	Corresponding claim
F1.2	The five sources are incomplete: there exists some structural feature of experience for which none of the five sources has a counterpart	Completeness of the isomorphism program
F1.3	The relational characterization of given-ness fails: given-ness is unaffected when internal–external co-evolution is severed	Given-ness as internal–external co-evolution
F1.4	The boundary gate fails: an unbounded (non-living) system has first-person given-ness	The anti-panpsychist gate
F3.1	Passive aggregation can produce an interior: the four forces (without enclosure) can produce a self-maintaining interior–exterior distinction	The interior must be produced by enclosure
F3.2	Interiority is not a necessary precondition of given-ness: given-ness can obtain in a system without topological interiority	Interiority as the precondition of given-ness

Continued on next page...

Continued table

Number	Failure condition (brief)	Corresponding claim
F5.3	The structural sufficiency criterion cannot be satisfied: gradient structure is in principle unable to carry some type of experiential structure	Structural sufficiency at the identity level
F5.4	The completeness claim is overturned by counterexample: there exists a sixth type of structure that cannot be carried by the five sources	Completeness of the five sources
F6.1	The similarity order is not preserved: the similarity order within qualities is inconsistent with the distance order of single-frequency configurations	Carrying the inner order of quality
F6.2	The dimensions do not correspond: the dimensionality of quality space does not correspond to the degrees of freedom of single-frequency configurations	Carrying the inner order of quality
F6.3	Topology or singularities do not correspond: topological features of quality space have no counterpart in configuration space	Carrying the inner order of quality
F7.1	Kinds transform continuously with intensity: one qualitative kind changes continuously into another as intensity changes	The frequency dimension carries qualitative kinds

Continued on next page...

Continued table

Number	Failure condition (brief)	Corresponding claim
F7.2	Discontinuity patterns do not correspond: discontinuities among qualitative kinds do not correspond to intervals among frequency families	The frequency dimension carries qualitative kinds
F7.3	Double-structure stitching fails: internal continuity and inter-kind discontinuity cannot be carried by the same frequency-family dimension	Stitching of Chapters 6 and 7
F8.1	Continuity does not correspond: experience has genuine instantaneous breaks while evolution is everywhere continuous (or vice versa)	Carrying the stream of consciousness
F8.2	Transition constraints do not correspond: experiential transition constraints do not correspond to evolutionary constraint patterns	Carrying the stream of consciousness
F8.3	Directionality has a different source: the arrow of experiential time has no systematic relation to the direction of dissipation	The arrow of time has a common origin in dissipation
F8.4	Attentional shift does not correspond: attentional dynamics do not correspond to nonlinear enhancement dynamics of configurations	Carrying the stream of consciousness

Continued on next page...

Continued table

Number	Failure condition (brief)	Corresponding claim
F9.1	Isomorphic carrying of unity fails: experiential unity has no systematic relation to same-frequency coherence	Unity is carried by coherence
F9.2	Isomorphic carrying of binding fails: binding (including illusory conjunction) has no correspondence with cross-frequency coupling	Binding is carried by cross-frequency resonance
F10.1	The retention-timescale spectrum does not preserve order: retention duration and transient frequency do not form an order-preserving relation	Carrying temporal thickness and memory
F10.2	Decay gradients do not correspond: experiential decay (forgetting) does not correspond to transient decay	Carrying temporal thickness and memory
F10.3	Continuous correspondence fails: present thickness and memory hierarchy cannot be carried by the same timescale spectrum	Carrying temporal thickness and memory
F10.4	Long-term memory cannot be included within transients: long-term memory is in principle impossible to understand as low-frequency long-duration transient	Matter as stable transient

Continued on next page...

Continued table

Number	Failure condition (brief)	Corresponding claim
F10.5	The scaling law systematically fails: memory retention systematically violates the $1/f$-dominant tendency of $\tau_{\text{retain}} \sim \tau_0/f$ and the slow variation of τ_0 cannot accommodate the deviation	The quantitative backbone of frequency–timescale correspondence
F11.2	The three candidates contain irreducible structural components: self, agency, or affect contains structure that the five sources cannot carry	The three candidates belong to the perturbation axis
F14.3	The characterization of the functional exterior fails: there exists a state of given-ness with neither exogenous input nor brain-internal co-evolutionary background	Characterization of the "exterior" in co-evolution
F14.4	Freezing has no correspondence with given content: the determinacy of given content does not correspond to transient freezing	Transient freezing carries given content
F14.5	The boundary gate fails: an unbounded system has first-person given-ness	The anti-panpsychist gate
F14.6	State explanation is overturned by counterexample: some conscious state cannot be consistently explained by internal–external configuration	Unified explanation of states

Continued on next page...

Continued table

Number	Failure condition (brief)	Corresponding claim
F16.1	Boundary and coherence do not correspond: the subject boundary has no systematic relation to the extension of dominant coherence	The boundary of the I is the boundary of coherence
F16.2	Split brain does not support the splitting prediction: when coherence is severed, the subject boundary does not split	Split brain as a critical test
F17.1	Memory is not gradient transient: memory retention has no systematic inverse relation with carrying frequency	Memory as gradient transient
F17.2	Knowledge is not the higher-order organization of transients: the invocation, association, and combination of knowledge have no systematic correspondence with activation, resonance, and reconstruction of transient organizations	Knowledge as higher-order transient organization
F17.3	The comparison mechanism does not hold: the comparison between knowledge and perturbation has no systematic correspondence with observable recognition, attention, and learning	Comparability of knowledge (leading to conscious activity)
F18.1	The cognization proposition does not hold:	Experience = cognized perception

Continued on next page...

Continued table

Number	Failure condition (brief)	Corresponding claim
	sensation is already complete experience without the participation of knowledge	
F18.2	The experiential knowledge base is in fact a substantive region: the knowledge base must be understood as a specialized and sharply bounded storage region	The knowledge base as distributed transient organization
F18.3	The comparison mechanism has no counterpart: conscious activities (recognition, surprise, search, learning) have no systematic correspondence with comparison between perturbations and knowledge transients	Core mechanism of conscious activity
F18.4	The two kinds of initiation cannot be unified: exogenous initiation (perception) and endogenous initiation (thinking) require two different mechanisms	The two kinds of initiation share the same mechanism
F19.2	$\mathscr{B}$-demarcation has no correspondence with attributional phenomena: fivefold demarcation has no systematic correspondence with observable structures of attribution	The demarcation mechanism of $\mathscr{B}$
F19.3	Failure modes do not correspond: attributional abnormalities (depersonalization, thought insertion, somatoparaphrenia,	$\mathscr{B}$ as a falsifiable mechanism

Continued on next page...

Continued table

Number	Failure condition (brief)	Corresponding claim
	made experiences) do not correspond to failures of the corresponding demarcations	
F20.3	The restatement by the interface lineage is mistaken: the hard problem cannot be consistently restated as a conditional problem of interface upgrading along a lineage	Restatement framework of the interface lineage
F21.2	A component has no observable proxy: one of $\tilde{A}/C/\mathscr{B}$ is in principle unmeasurable	Ψ as measurable projection
F21.3	The necessity of $\tilde{A}$ does not hold: strong experience can occur at zero power	$\tilde{A}$ projects dissipative power
F21.4	The three-dimensional projection omits a core measurable feature: there exists a necessary measurable feature that cannot be summarized by $\tilde{A}/C/\mathscr{B}$	Sufficiency of the three-dimensional Ψ
F22.1	$\tilde{A}$ has no systematic relation with dissipative power/metabolism	Physical characterization of $\tilde{A}$
F22.2	C has no systematic relation with differentiation$\times$integration/criticality	Physical characterization of C
F22.3	$\mathscr{B}$ has no systematic relation with coherent–causal closure	Physical characterization of $\mathscr{B}$

Continued on next page...

Continued table

Number	Failure condition (brief)	Corresponding claim
F22.4	The three dimensions are reducible to one: conscious states can be completely characterized by a single quantity	Three dimensions are required (against single-scalar theories)
F23.1	The proxy relation between window and quantity does not hold: a method does not project the quantity it claims to proxy	Measurement interface
F23.2	Frequency structure cannot be projected by EEG/MEG	Measurement basis of the consciousness frequency-band hypothesis
F23.3	Multimodal projections cannot be assembled: multiple windowed projections are in principle impossible to integrate into a consistent inference	Multimodal complementarity
F24.1	States cannot be consistently localized: some state cannot be located within the three coordinate groups	Unity of the state space
F24.2	Consciousness depends primarily on $\tilde{A}$: the presence or absence of consciousness varies primarily with metabolism rather than C and $\mathscr{B}$	Consciousness depends on C and $\mathscr{B}$
F24.3	Given-ness depends on behavioral output: behaviorally unresponsive subjects necessarily lack given-ness	Given-ness depends on co-evolution (covert consciousness)

Continued on next page...

Continued table

Number	Failure condition (brief)	Corresponding claim
F24.4	Boundary form is not independent of global state quantities: split brain/dissociation necessarily involve depression of $\tilde{A}$ and C	Independence of boundary form
F25.1	Frequency–quality correspondence is falsified: qualitative kinds have no systematic correspondence with neural frequency structures	First ρ_f-specific prediction
F25.2	Frequency–timescale correspondence is falsified: retention duration has no systematic inverse correspondence with neural frequency, or systematically violates the $1/f$ tendency of $\tau_{\text{retain}} \sim \tau_0/f$	Second ρ_f-specific prediction (quantitative)
F25.3	Discriminating predictions reverse: given-ness depends on exogenous input / boundary follows availability / states can be predicted by one dimension	Discrimination from other theories

E.2. Failure Conditions at the Coherence–Philosophical Layer

The conditions at this layer are adjudicated by internal coherence, conceptual consistency, fidelity to the cited ontology / preceding framework, or adherence to the C3 prohibition, rather than directly by observables. Their failure means that the corresponding argument, positioning, or boundary of this book has a problem and must be restated or corrected.

Table E.2.. Appendix E: Core Falsifiability Propositions, Failure Conditions, and What They Guard (Continued)

Number	Failure condition (brief)	What is guarded
F1.1	The identity level cannot be maintained: structural sufficiency is in principle impossible to satisfy, and the theory must retreat to a weaker level	The level of the primary thesis
F1.5	The final cell is made to masquerade as explained: a mechanistic statement, interface, demarcation, or identity alone is used as the answer to "why it is experienced"	The C3 prohibition
F2.1	The cited ontology does not hold: the ρ_f ontology (A1–A4) is falsified	The conditional foundation of this book
F2.2	Citation is not faithful: the citation of A1–A4 or M departs from the original meaning in the foundational volume	Citation without re-proof
F2.3	Observing the two-tier standard becomes exemption from falsification: the "final cell" is used to exempt measurable modules from falsification	Honesty of the two-tier standard
F2.4	Dimensions or proposition types are conflated: $A/\tilde{A}$ are conflated, χ is misused, or the three proposition types are confused	Methodological discipline

Continued on next page...

Continued table

Number	Failure condition (brief)	What is guarded
F3.3	The boundary gate and the resonance-kernel gate are conflated: the resonance kernel is used as the anti-panpsychist gate (or vice versa)	Distinction between the two gates
F3.4	The membrane cannot be included in ρ_f transients: the membrane (matter) is in principle impossible to understand as stable transient	Reconnecting the fifth mechanism to the ontology
F3.5	The positioning of the earlier work is mistaken: the fifth mechanism is in fact an indispensable foundation rather than one link under the ρ_f main line	ρ_f as the main line
F4.1	The positioning is unfair: the presentation of a position distorts its strongest form	Positioning within the intellectual map
F4.2	The distinction from Russellian monism does not hold: the two advances do not provide content it lacks	Independent positioning of this book
F4.3	The "division of labor" with scientific theories is in fact avoidance: division of labor is used as an excuse to avoid direct competition	Relation to GNW and related theories
F4.4	The anti-functionalist inference is too strong or inconsistent: the enclosure criterion cannot be applied consistently to artificial systems	Anti-functionalist inference

Continued on next page...

Continued table

Number	Failure condition (brief)	What is guarded
F5.1	Isomorphism collapses into loose similarity: what is actually achieved is only "similarity", not a strict structure-preserving correspondence	Strictness of isomorphism
F5.2	The reasons for adopting the identity level do not hold: the alleged advantage of removing the explanatory burden of correspondence is false	Reasons for choosing the identity level
F6.4	The structure is merely a coordinate convention: the structure of quality space is a convention rather than a coordinate-independent intrinsic structure	Reality of the object to be carried
F7.4	The correspondence is merely arbitrary labeling: the frequency-band–kind correspondence is unconstrained by isomorphism and can be assigned at will	Carrying is not labeling
F8.5	Law and drive cannot be separated: the law of evolution and the drive of perturbation are shown to be inseparable	Framework of law / drive
F9.3	The dissolution of the combination problem is invalid: the I must be understood as the real summation of micro-subjects	Dissolution of the combination problem

Continued on next page...

Continued table

Number	Failure condition (brief)	What is guarded
F9.4	The distinction between coherence and connection does not hold: coherence does not go beyond "tight connection"	Ontological distinction in the dissolution
F11.1	The distinction between the two axes does not hold: sources of structure and sources of perturbation cannot be separated	Two-axis framework
F11.3	The two layers of internal/external cannot be separated: the internal/external distinction in perturbation sources and the internal/external distinction between the two poles of co-evolution are one and the same	Distinction between the two layers of internal/external
F11.4	There exists an unlisted sixth type of structure (neither one of the five types nor a perturbation type)	Completeness (structural axis)
F12.1	The distinction between structure and given-ness does not hold: the two can be logically derived from one another	Separation of structure / given-ness
F12.2	The intrinsic-property model is not truly excluded: an isolated and yet given interior can be conceived	Exclusion of the intrinsic model
F12.3	The external-reading model is not truly excluded: there exists a reading without regress and without presupposing undisturbed copying	Exclusion of the reading model

Continued on next page...

Continued table

Number	Failure condition (brief)	What is guarded
F12.4	The direction has been misidentified: given-ness is neither intrinsic, nor reading, nor any form of co-evolution	Identification of the co-evolutionary direction
F13.1	The cited mechanism does not hold: the detection-volume ontology of observation is falsified	The cited detection mechanism
F13.2	The relational structure cannot be transferred: the co-evolutionary relation cannot be applied to a system's internal–external co-evolution with itself	Transfer of the mechanism
F13.3	The reinterpretation of the two functional poles fails: the internal/external distinction in consciousness must consist of two separate devices	Internal/external as functional distinction
F13.4	The objective mechanism is misused to explain being-experienced	The C3 prohibition
F14.1	The relational approach does not hold: an interior can be constructed that is disconnected from all exterior relations and yet remains given	Relationality of given-ness
F14.2	The two layers of internal/external cannot be separated (same as F11.3, restated in Chapter 14)	Distinction between the two layers of internal/external

Continued on next page...

Continued table

Number	Failure condition (brief)	What is guarded
F14.7	The final cell is made to masquerade as explained: a mechanistic statement is used as the answer to "why it is experienced"	The C3 prohibition
F15.1	The final cell can in fact be filled: there exists a valid derivation that infers "being-experienced" purely from structure and process	Irreducibility of the final cell (a failure welcomed by this book)
F15.2	The location is imprecise: the final cell still contains components that can be physically explained	Precise location of the residual point
F15.3	The boundary drawing is incoherent: the heterogeneity between structure / the third-person side and being-experienced / the first-person side does not hold	Diagnosis of the fissure
F16.3	The two layers of internal/external cannot converge: the nesting of functional boundary and the material boundary of the membrane is incoherent	Convergence of the two layers of internal/external
F16.4	The claim of non-I micro-subjects violates anti-panpsychism: one must admit that unbounded things have an interior	Consistency with anti-panpsychism

Continued on next page...

Continued table

Number	Failure condition (brief)	What is guarded
F17.4	The transition of language and symbols is mistaken for the source of given-ness: the higher-order character of knowledge organization is used as the source of given-ness / being-experienced	The C3 prohibition Knowledge organization is not the source of given-ness
F18.5	Cognization is made to masquerade as an explanation of being-experienced: "experience is cognized perception / the result of comparison" is used as the answer to "why it is experienced"	The C3 prohibition
F19.1	Given-ness and attribution are inseparable: there is no case in which "content is given but not demarcated as mine"	Distinction between given-ness and attribution ($\mathscr{B}$ is responsible specifically for attribution)
F19.4	$\mathscr{B}$ bears too much: $\mathscr{B}$ is used to generate given-ness, or to explain being-experienced, or $\mathscr{B}$ becomes an autonomous reviewer with a newly introduced standard	The C3 prohibition Restraint of $\mathscr{B}$
F20.1	The second compression does not hold: what is stripped away (integration, cognization, demarcation) itself contains a residual of being-experienced	The second compression
F20.2	The subject-attribution residual point is in fact reducible:	Irreducibility of the subject-attribution residual point

Continued on next page...

Continued table

Number	Failure condition (brief)	What is guarded
	there exists a valid derivation that infers "being-experienced" purely from the structural facts of integration / cognization / demarcation	(a failure welcomed by this book)
F20.4	Further compression is made to masquerade as elimination: the second compression, interface restatement, or $\mathscr{B}$-demarcation is used	The C3 prohibition
	to claim that the hard problem has "completely ceased to exist / been solved"	(the Version B overreach most carefully guarded against)
F21.1	Ψ is mistaken for a definition or generator	The C3 prohibition / the status of Ψ
F22.5	The three quantities overstep into given-ness: the three quantities are used to explain / produce being-experienced	The C3 prohibition
F23.4	Measurement oversteps into given-ness: measurement results are used to support being-experienced	The C3 prohibition
F24.5	State explanation oversteps into being-experienced	The C3 prohibition
F25.4	Roadmap items are shown to be untestable in principle: some roadmap item is in fact in principle unfalsifiable	Honest division within the measurable plane
F25.5	Measurable predictions are used to support the final cell	The C3 prohibition / boundary between the measurable and the residual point

Continued on next page...

Continued table

Number	Failure condition (brief)	What is guarded
F26.1	The synthesis at the identity level is incoherent: unity has an internal contradiction with what the first four parts have established	Coherence of unity
F26.2	The two sides are not the same reality: the argument from ontological uniqueness + structural isomorphism does not hold	The thesis of unity
F26.3	Openness is not necessary for given-ness: given-ness can obtain in an isolated closed configuration	Characterization of open process
F26.4	Unity is mistaken for the fulfillment of the final cell	The C3 prohibition
F27.1	The final cell is in fact a solvable object: there exists a valid path that answers "why the internal side is experienced"	Boundary-status of the final cell (a failure welcomed by this book)
F27.2	The position is imprecise: the final cell still contains components that can be physically explained	Location of the residual point
F27.3	The convergence with phenomenology is a misreading: the two do not in fact refer to the same irreducibility	Support from convergence by different paths

Continued on next page...

Continued table

Number	Failure condition (brief)	What is guarded
F27.4	Identity is used to dissolve the final cell: identity is illegitimately extended to being-experienced in order to dissolve it	The C3 prohibition (guarded at the conclusion)
F27.5	Interface or demarcation is used to dissolve the final cell: interface upgrading or $\mathscr{B}$-demarcation is used to claim that the hard problem has been dissolved	The C3 prohibition (linked with F20.4)

E.3. Special Note on the Two Kinds of "Failures Welcomed by This Book"

Several failure conditions with a special status must be explicitly marked. If they hold, they are welcomed by this book, because they would mean that the hard problem has been genuinely solved, rather than merely that this book has been overturned:

- F15.1 / F20.2 / F27.1 (the final cell can in fact be filled / the subject-attribution residual point is in fact reducible / it is in fact a solvable object). If someone provides a valid derivation that appeals to no first-person premise and proceeds purely from structural and process facts (including integration, cognization, demarcation, and so forth) to infer that "this configuration is experienced", then this book's core claim concerning the irreducibility of the final cell (whether the given-ness residual point of the first compression or the subject-attribution residual point of the second compression) fails. But this failure would mean that the innermost core of the hard problem has truly been answered. This book regards it as a welcomed failure—not a defeat for the book, but a victory for consciousness research.

The book states the final cell in a way that can be overturned in this manner precisely in order to leave a clear place for that victory, should it occur.

E.4. Overall Pattern of the Failure Conditions

Seen together, the two tables show that the overall pattern of the failure conditions of this book is precisely the mirror image of its two-tier standard (Chapter 2) and of the C3 prohibition:
The conditions at the falsifiable layer (E.1) are concentrated in Part II (the carrying relation of structural isomorphism), Part III (the mechanism of given-ness, the mechanisms of cognization and demarcation, and conscious states), and Part IV (Ψ and predictions). They are where the book takes its risks in the sense of empirical science. Among them, F25.1 and F25.2, the two ρ_f-specific predictions, are especially tied to the empirical spine of the book (F25.2 has quantitative form because of the scaling law $\tau_{\text{retain}} \sim \tau_0/f$, and is linked with F10.5). The falsifiable conditions of the four newly added chapters (new Chapters 17 to 20: F17.1–3, F18.1–4, F19.2–3, F20.3) place the mechanisms of "cognization, comparison, attributional demarcation, and interface restatement" under testability as well.

The conditions at the coherence–philosophical layer (E.2) are concentrated in Part I (positioning and method), and in the philosophical portions of Part III and Part V (the characterization of given-ness, the two compressions of the final cell, and unity). What they guard is the argumentative coherence of the book, fidelity to what it cites, and, most importantly, the C3 prohibition. It is worth noting that the C3-guarding items run through all parts of the book and expand with the newly added four chapters and the second compression: C3 no longer only prohibits "using mechanistic statements to masquerade as explanations", but prohibits the use of mechanistic statements (F1.5, F13.4, F14.7), Ψ (F21.1), the three quantities (F22.5), measurement (F23.4), state explanation (F24.5), measurable predictions (F25.5), knowledge organization (F17.4), cognization (F18.5), $\mathscr{B}$-

demarcation (F19.4, F27.5), interface restatement (F20.4, F27.5), and the identity thesis (F26.4, F27.4). These more than ten guardrails are the same prohibition fortified at different levels: nothing on the objective side of the fissure may be used to masquerade as, or to dissolve, the irreducible final cell. That this prohibition is defended in more than ten places is the most concentrated expression of the book's honest posture; F20.4 is especially the item that the whole book must guard against most carefully, because it directly rejects the Version B overreach of "claiming through the second compression that the hard problem has completely ceased to exist".

Finally, the welcomed failures marked in E.3 (F15.1, F20.2, and F27.1) indicate the relation between this book and the genuine solution of the hard problem: this book does not claim to have solved it, but explicitly leaves a place for "if it is solved". That would be a failure welcomed by the book and a victory for consciousness research.

F. Detailed Comparison with Existing Theories and Philosophical Positions (Including References)

F.1. The Task and Reading Method of This Appendix

Chapter 4 has already provided an argumentative positioning of this book in relation to dualism, physicalism, functionalism, panpsychism, and existing theories of consciousness. What this appendix does is different from Chapter 4: it is an expanded reference-style comparison, allowing readers to begin from the name of a given position and quickly locate the key points of its relation to this book—what is shared, where the divergence lies, where the discriminating point is, and what route this book takes. At the same time, this appendix supplements several comparisons that had not yet been addressed when Chapter 4 was drafted—especially those arising from the cognization, comparison, $\mathscr{B}$-demarcation, and second compression established in the newly added four chapters (Chapters 17–20), and from the refusal, established in Chapter 27, to "dissolve the hard problem through identity / interface / demarcation"—including further comparisons with several positions such as higher-order theories, the dissolution route in Russellian monism, self-representational theories, and phenomenology of life.

Reading method: the tables in this appendix are organized according to five columns: "position–what this book shares–where this book diverges–discriminating prediction / argument–failure-condition chain-back". Readers who wish to examine the arguments in detail should return to the corresponding chapters of the main text; this appendix serves only as a quick comparison. This appendix does not restate the arguments of Chapter 4, but only cites its conclusions. The comparisons newly added in this appendix (especially F.3 on HOT, F.4 on the fine comparison with intentionality, and F.5 on the refusal of the dissolution route in Russellian monism) are those not sufficiently developed in Chapter 4.

Discipline (consistent with the whole book): (1) each position is stated in its strongest form (F4.1), without distortion; (2) shared points are acknowledged and divergences are explicitly marked (F4.3 does not use division of labor as an excuse to avoid direct competition); (3) wherever the book refuses to "dissolve the final cell through identity / interface / demarcation", it is explicitly marked as belonging to the C3 guardrail (F20.4, F27.4, F27.5), and is not mixed into the "empirical discrimination" tier.

F.2. The General Landscape of the Mind–Matter Relation

This section compares the major ontological positions on the mind–matter relation. This book takes itself to be an "open-process dual-aspect monism" (Chapter 26), standing between dualism and physicalism, taking the strengths of both (monistic economy and respect for given-ness) while avoiding the weaknesses of both (the interaction problem and reduction / elimination).

Table F.1.. Comparison between This Book and Classical Positions in Philosophy of Mind

Position	Shared points	Divergence	Discrimination / failure conditions
Substance dualism (Cartesian tradition)	Shared point: it respects the irreducibility of given-ness (neither reducing nor eliminating it).	Divergence: this book is monistic (the same ρ_f ontology), whereas dualism requires two substances; this book has no mind–matter interaction problem, while dualism cannot avoid it.	Chapter 26; F26.2
Property dualism (Chalmers and others)	Shared point: it acknowledges that being-experienced is not entailed by structure and function, and that the explanatory gap is real.	Divergence: this book places mind and matter as two sides of the same ρ_f process (structure carried by the five sources, given-ness by co-evolution), and does not posit a "mental property" as an independent fundamental property.	Chapter 26

Continued on next page...

Continued table

Position	Shared points	Divergence	Discrimination / failure conditions
Eliminativism / illusionism (Dennett, Frankish, and others)	Shared point: it is alert to the excessive mystification of being-experienced.	Fundamental divergence: this book does not eliminate given-ness. Givenness is the starting point of the entire discussion in this book; eliminating it is equivalent to sawing off the branch one is sitting on, and does not save effort ("why there is this stubborn illusion" itself becomes another problem).	Section 15.4
Strong reductive physicalism (radical forms of functionalism)	Shared point: monistic economy.	Divergence: this book does not reduce given-ness to function (functionalism falls into topic substitution); this book acknowledges one cell that is not entailed by structure and function.	Chapters 4 and 27; F1.5, F27.4

Continued on next page...

Continued table

Position	Shared points	Divergence	Discrimination / failure conditions
Russellian monism (Russell, Strawson, Goff, and others)	Strong shared point: being-experienced arises from the intrinsic nature of reality rather than from structural relations; mind and matter are two sides of the same reality. This book acknowledges itself as one branch of Russellian monism (Chapter 4).	Divergence (two advances): (1) this book fulfills structural sufficiency through the fivefold richness of ρ_f, while Russellian monism may not explicitly specify where the structural richness of its fundamental physics lies; (2) this book refuses the temptation to "soften the hard problem through identity" (Section 27.7), whereas some Russellian monists cross the boundary through that route.	Chapter 4, Section 27.7; F27.4

Continued on next page...

Continued table

Position	Shared points	Divergence	Discrimination / failure conditions
Panpsychism (Strawson, Goff, Chalmers, and others)	Shared point: it attempts to respect being-experienced within a monistic framework.	Fundamental divergence: this book adopts anti-panpsychism—given-ness is anchored in the boundary of life (the interior produced by the fifth mechanism); without boundary, there is no given-ness. Rocks, atoms, and gravitational-wave detectors have no given-ness.	Chapters 3 and 14; F1.4, F14.5
New mysterianism (McGinn)	Shared point: it acknowledges the hardness of the hard problem.	Divergence: this book does not push the entire hard problem wholesale into darkness; it extracts everything that can be solved (structure, the mechanism of given-ness, measurable predictions, unity), and precisely locates what cannot be solved (the final cell, compressed twice into the subject-attribution residual point).	Chapters 15 and 27

The position of this book (Chapter 26): open-process dual-aspect monism (monistic yet dual-aspect), standing between dualism and physicalism, closest to Russellian monism while refusing its dissolution route.

F.3. Major Theories in Contemporary Consciousness Science

This section compares the major theories in contemporary consciousness science. The relation between this book and those theories is discussed in detail in Chapters 4, 22, 23, 24, and 25; this appendix provides a reference-style summary.

Table F.2.. Comparison between This Book and Contemporary Neural, Cognitive, and Physical Theories of Consciousness

Theory	Shared points	Divergence	Discriminating predictions / failure conditions
Integrated Information Theory (IIT) and related work (Tononi and others)	Shared points: consciousness is related to integration× differentiation in neural activity (the counterpart of C); measures such as PCI distinguish states.	Divergence: (1) IIT centers on a single scalar Φ, whereas this book uses the three-dimensional $\Psi = (\tilde{A}, C, \mathscr{B})$ and argues that structural sufficiency requires the fivefold richness; (2) IIT treats Φ as consciousness itself (definition / generator), whereas this book treats Ψ as a projection quantity (descriptive, not definitional, and not generative; Chapter 21, F21.1).	Chapters 4 and 21; F22.4, F25.3
Global Neuronal Workspace (GNW) and related work (Dehaene, Baars, and others)	Shared points: neural markers of reportability and global availability (corresponding to the governance of dominant coherence in this book).	Divergence: this book holds that the subject boundary depends on the extension of coherence, not on information availability; explanations of boundary abnormalities such as split brain and dissociation are discriminating cases (Chapters 16 and 25).	F25.3 (coherence vs. availability)

Continued on next page...

Continued table

Theory	Shared points	Divergence	Discriminating predictions / failure conditions
Higher-Order Theories (HOT/HOP) and related work (Rosenthal, Lau and LeDoux, and others)	Shared points (related to the newly added four chapters): the cognization of experience and higher-order representation of one's own states have conceptual affinity with cognized perception in Chapters 17 and 18.	Divergence (key point): HOT holds that "when a state is represented by a higher-order state, that state is conscious"; this book rejects that position—(1) given-ness does not depend on being made higher-order (in depersonalization, cognization and demarcation are weakened while given-ness remains; Section 19.4); (2) HOT easily falls into topic substitution (using "being represented" to masquerade as "being experienced"), whereas this book strictly observes C3: cognization explains only the structure and organization of experience, not being-experienced (F18.5, F20.4).	Chapters 17 and 18; F18.5, F19.4

Continued on next page...

Continued table

Theory	Shared points	Divergence	Discriminating predictions / failure conditions
Predictive processing / free-energy principle (Friston, Clark, Hohwy, and others)	Shared points: experience is driven by prediction–error dynamics of internal and external perturbations (consistent with this book's perturbation axis and comparison between knowledge base and perception; Chapters 8 and 18).	Divergence: predictive processing tends to reduce consciousness to the functional properties of inference; this book does not claim that given-ness is reducible to prediction–error function (an extension of F1.5).	Chapters 8, 18, and 22; F4.3
Attention Schema Theory (AST) and related work (Graziano)	Shared points: it focuses on the system's modeling of its own state (with affinity to the self-related role of $\mathscr{B}$ in Chapter 19).	Divergence: AST identifies consciousness with the "system's internal model of its own attention", tending toward an eliminativist reading; this book does not eliminate given-ness, and does not treat the "self-model" as a sufficient condition of consciousness.	Chapter 19; F19.4

Continued on next page...

Continued table

Theory	Shared points	Divergence	Discriminating predictions / failure conditions
Quantum theories of consciousness (Penrose–Hameroff Orch-OR and others)	Shared point: they explore new mechanisms in fundamental physics to approach consciousness.	Divergence: this book appeals to the frequency-resolved gradient structure of ρ_f (a unified physical ontology not based on quantum mechanics), and does not depend on mechanisms such as microtubule wave-function collapse.	Shares with this book the posture that "fundamental physics participates"

The position of this book: at the level of shared indicators of conscious states, it is largely compatible with IIT, GNW, and predictive processing (which this book reinterprets through the ρ_f mechanism); at the level of content-specific and structural predictions, it makes ρ_f-specific predictions (frequency–quality and frequency–timescale correspondence, Chapter 25; F25.1 and F25.2); in relation to HOT and AST, it explicitly refuses to "dissolve being-experienced through higher-order representation / self-model" (F18.5, F19.4, echoing F27.5 in Chapter 27).

F.4. The Phenomenological Tradition: Given-ness and Intentionality

This section compares the phenomenological tradition. This book proceeds from below, starting from ρ_f as fundamental physics, and reaches the "irreducibility of given-ness"; this converges with the same irreducibility that phenomenology confirms from above, beginning with experience itself (as established in Chapters 23 and 27). This appendix provides a more detailed comparison.

Table F.3.. Shared Points, Divergences, and Chapter Correspondences between This Book and Phenomenological Positions

Phenomenological position	Shared points	Divergence / supplementation	Chapter
Husserlian given-ness (Husserl)	Strong shared point: given-ness is an irreducible beginning point, the datum that remains after suspension; it cannot be derived from the non-given. The final cell of this book is of the same kind.	Supplementation: this book gives given-ness a physical counterpart (internal–external co-evolution of a bounded interior), whereas phenomenology does not appeal to physics. The two paths converge on the point of "irreducibility" (convergence by different paths).	Chapters 15 and 27.5
Husserlian intentionality	Strong shared point: consciousness is always consciousness "of" something; it is always directed toward or oriented to an object.	Supplementation: this book provides its physical counterpart through internal–external co-evolution (Chapter 14) and $\mathscr{B}$'s world-oriented demarcation (Chapter 19). The being-experienced of intentionality is left as the final cell (again, convergence by different paths).	Section 27.5

Continued on next page...

Continued table

Phenomenological position	Shared points	Divergence / supplementation	Chapter
Heidegger's "being-in-the-world" (Heidegger)	Shared point: experience precedes the subject–object division and involves the entanglement of internal and external (consistent with this book's internal–external co-evolution and the perturbation axis).	Supplementation / divergence: Heidegger's route is hermeneutic and non-physical; this book gives a physical counterpart to the structure of "being-in-the-world" through the ρ_f mechanism, without eliminating its phenomenological description.	Chapter 14
Merleau-Ponty's embodiment (Merleau-Ponty)	Shared point: experience and body are inseparable (consistent with this book's claims that "the boundary of the I depends on coherence" and that bodily attribution of "mine" is demarcated by $\mathscr{B}$, Chapters 16 and 19).	Supplementation: this book gives a precise treatment of embodiment through the claim "I have a body, but I am not that whole body" (Section 16.3)—the boundary of the I is drawn by coherence, and bodily attribution is demarcated by $\mathscr{B}$; both have physical counterparts.	Chapters 16 and 19

Continued on next page...

Continued table

Phenomenological position	Shared points	Divergence / supplementation	Chapter
Zahavi's "minimal self"	Shared point (related to the newly added four chapters): experience carries a dimension of "mineness", which is a structural feature of experience rather than an added judgment.	Supplementation: this book assigns the physical counterpart of mineness to the subjective-core demarcation of $\mathscr{B}$ (Chapter 19); its failure (depersonalization) corresponds to a failure of $\mathscr{B}$-demarcation—precisely the physical counterpart of the "weakening of mineness" observed by Zahavi and others.	Chapter 19; F19.3

Continued on next page...

Continued table

Phenomenological position	Shared points	Divergence / supplementation	Chapter
Phenomenology of life (Jonas, Thompson, and others)	Shared point (core): consciousness and life are inseparable, and consciousness emerges from the minimal forms of life.	Most closely aligned with this book's route: the anchoring of "enclosure–life–given-ness" in Chapter 3 is precisely the physical fulfillment of this tradition; the elementary consciousness established in the cellular trilogy (such as *The Origin of Cellular Proto-Consciousness: Biological Foundations and Operational Criteria*) is consistent with the direction of phenomenology of life.	Chapters 3 and 14; F1.4

The position of this book: the bottom-up physical path and the top-down reflective path of phenomenology converge at several points on the same edge of "the irreducibility of given-ness" (given-ness, intentionality, mineness, and the binding of life and consciousness). The convergence of these independent paths provides independent support for the final cell of this book; however, this book and phenomenology do not prove one another deductively, and each preserves its own method.

F.5. Routes of Response to the Hard Problem

This section compares the major routes of response to the hard problem, clarifying which route this book takes and which routes it rejects.

Table F.4.. Comparison between Different Dissolution Routes for the Hard Problem and the Route of This Book

Route	Representatives	Relation to this book
Strong reduction (functionalism): consciousness = some function; the hard problem = a collection of easy problems.	Early Dennett, Lewis, and others	Rejected: topic substitution (F1.5); this book acknowledges that being-experienced is not entailed by structure and function.
Eliminativism / illusionism: phenomenal consciousness does not exist; it is a cognitive illusion.	Frankish, and some later versions of Dennett	Rejected: given-ness is the starting point of this book; eliminating it is equivalent to sawing off the branch one is sitting on, and does not save effort (Section 15.4).
New mysterianism: the hard problem is in principle insoluble for human intelligence.	McGinn	Partly rejected: this book agrees that "there is one cell not entailed by structure and mechanism", but rejects the claim that "the whole hard problem is wholesale insoluble"—this book solves everything solvable and leaves only one cell as irreducible (precise location rather than vague mystery; Section 27.3).

Continued on next page...

Continued table

Route	Representatives	Relation to this book
Phenomenal consciousness as fundamental (one branch of property dualism)	Chalmers and others	Partly shared: it acknowledges the explanatory gap; divergence concerns whether independent mental properties are needed (as stated in F.2).
The dissolution route in Russellian monism: once identity is established, the hard problem is softened or even dissolved.	Some Russellian monists	Explicitly rejected: this book acknowledges itself as a branch of Russellian monism, but refuses this dissolution route (Section 27.7). What identity establishes is identity of structure; being-experienced is not inside structure, and therefore identity cannot extend to being-experienced. Using identity to dissolve being-experienced triggers F27.4.
Dissolving the hard problem through higher-order representation / interface / demarcation: higher-order representation, interface upgrading, or demarcation of experience explains being-experienced.	Some HOT theorists; the radical version of the route in *The Final Attempt*	Explicitly rejected: this is the same kind of overstep as "dissolution through identity"—illegitimately extending what lies on the objective side of the fissure to the subjective side (F18.5, F19.4, F20.4, F27.5). The second compression of this book is compression, not dissolution (Chapter 20).

Continued on next page...

Continued table

Route	Representatives	Relation to this book
The route of this book: precise boundary-drawing + two compressions	This book	It extracts everything solvable (structural isomorphism, mechanism of given-ness, measurable predictions, and mind–matter unity), and precisely locates what cannot be solved as the final cell (compressed twice into the subject-attribution residual point); the boundary between the two is clear, and each is placed where it belongs.

Self-positioning of this book (Section 15.4): between the "ambitious draft" (forcing a crossing and concealing the fissure with a single word) and the "disciplined draft" (setting aside the solvable parts because the discipline is strict), this book takes a third route—everything solvable is solved; what cannot be solved is precisely marked; and the boundary between the two is clear. This posture receives external reference in the comparison of routes in F.5: the relation between this book and each competing position is the implementation of that posture in a concrete theoretical confrontation.

F.6. Position Comparisons in Several Key Boundary Cases

The differences among theories of consciousness appear most clearly in boundary cases. This section compares this book with several major positions in the following boundary cases.

F.6.1. Case 1: Do Artificial Systems Have Given-ness?

Table F.5.. Comparison of Positions and Reasons Concerning Consciousness in Artificial Intelligence and Silicon-Based Entities

Position	Judgment	Reason
Functionalism / HOT / GNW	Yes (if function / higher-order status / availability matches)	Consciousness depends on function and is independent of material.
Strong physicalism	Depends on the concrete physical realization	It must be identical with the physical process of a conscious being.
Panpsychism	Yes (with degree depending on the concrete constitution)	Everything has weak experience.
This book	No (without enclosure / membrane, there is no precondition of given-ness)	The boundary gate (Chapters 3 and 14, F1.4): without ∂, there is no interior pole I, and the formalization of K runs empty for that system (Appendix C.3). Discriminating prediction: synthetic life, if it has a lipid membrane and the boundary of life, has the precondition of given-ness, whereas silicon-based computers, however functionally powerful, do not.

F.6.2. Case 2: Can a Collective (Bee Colony, Human Society, the Internet) Share One "I"?

Table F.6.. Comparison of Different Positions on Group / Collective Consciousness

Position	Judgment
Physicalism	Generally no (a collective lacks a unified neural basis)
Panpsychism	A matter of degree
This book	No, unless there is a physical connection among the members of the collective sufficient to support cross-brain phase locking (except for rare conjoined cases, Section 16.5). Societies and bee colonies usually lack a sufficiently coherent physical basis to form a single dominant coherence. Within this book's framework, "collective consciousness" is physical co-evolution without an interior, not given-ness.

F.6.3. Case 3: Animal Consciousness?

Table F.7.. Comparison of Different Positions on the Degree and Scope of Consciousness

Position	Judgment
Functionalism	A matter of degree
Panpsychism	A matter of degree
This book	All beings with a boundary (living beings) have a minimal degree of given-ness (the minimal given-ness of the cell is established by this book); but differences in degree are determined by the complexity of co-evolution, the degree of cognization, and the fineness of $\mathscr{B}$-demarcation. Mammals have more advanced cognization and demarcation, and therefore richer experience; the given-ness of the cell is extremely minimal (only minimal internal–external co-evolution). This is the physical fulfillment of the "elementary consciousness" observed in the cellular trilogy.

F.6.4. Case 4: Conjoined Twins and Split Brain (Intermediate States between One and Two)?

Table F.8.. Comparison of Different Positions and Judgments Concerning the Boundary of Cognitive Subjects

Position	Judgment
Functionalism	Depends on the degree of cognitive integration
This book	The boundary is drawn by the actual extension of coherence: if a cross-brain physical connection is sufficient to support cross-brain coherence, then a coherence spanning two brains is formed, and the subject boundary correspondingly crosses the two brains (Section 16.5). Split brain is the opposite case—severing the corpus callosum severs the cross-hemispheric coherence bridge, and the subject boundary correspondingly splits. This is precisely the critical test of F16.2.

F.6.5. Case 5: "Covert Consciousness" in Vegetative-State Patients?

Table F.9.. Comparison of Different Positions on Behavioral Output and the Determination of Consciousness

Position	Judgment
Behaviorism / strict physicalism	No behavioral response means no consciousness

Continued on next page...

Continued table

Position	Judgment
This book	For behaviorally unresponsive subjects whose C and $\mathscr{B}$ are maintained, this book predicts that they have given-ness (Chapter 24). Given-ness depends on the maintenance of co-evolution, not on behavioral output (the same reasoning as in locked-in syndrome). This is the discrimination between this book and the position that "consciousness depends on behavioral output" (F24.3).

F.7. References

The references are classified by discipline and period. The works listed here are the main references for the development of this book's thought; specific citations appear in the footnotes of each chapter and in the foundational volume. The main internal EQT works cited by this book are the foundational volume and the freezing-dynamics spectrum, as detailed in F.7.1; representative works in consciousness science and philosophy are listed in F.7.2–F.7.4.

F.7.1. EQT Series (The Theoretical Framework to Which This Book Belongs)

- Kaisheng Li and Longji Li (2026). *EQT Foundational Volume: Frequency, Mass, Range of Action, and the Generation Mechanism of Force*: the EQT foundational volume. It establishes the ontology of the ρ_f frequency-resolved energy quantum density field (postulates A1–A4), the unified master equation (5.4.1), and the form of the Lorentzian resonance kernel. Directly cited in Chapter 2 and Appendix A of this book.

- Kaisheng Li and Longji Li (2026). *Frozen Cosmology: A Natural Philosophy of Cosmic Evolution in the Perspective of Energy Quantum Theory*: the EQT volume on freezing dynamics. Section 4.3 establishes the freezing-time scaling law $t_{\text{freeze}}(f) \sim \tau_0(f)/f$, directly cited in Chapters 10, 17, and 25 of this book, as well as Appendices B and C.

- Kaisheng Li and Longji Li (2026). *EQT Detection Theory: From the Ontology of Observation to Frequency Matching, Three-State Response, and Falsifiable Predictions*: the EQT detection volume. It establishes the ontology of observation in EQT and clarifies the idea that observation is participation. Directly cited in Chapters 13 and 14 of this book.

- Kaisheng Li and Longji Li (2026). *The Information World of the Cell: Pooling Networks, Continuous-State Computation, and the Architecture of Life Information Processing*: the first work of the cellular trilogy. It establishes cellular pooling networks, two-state memory, and the PFR loop—the phenomenological foundation of Chapters 17 and 18 of this book.

- Kaisheng Li and Longji Li (2026). *The Second Axis of Cosmic Aggregation: Membranes, Life, and Consciousness*: establishes the second axis of aggregation (enclosure / life threshold). It serves as the phenomenological foundation and mechanistic interface of Chapter 3 of this book.

- Kaisheng Li and Longji Li (2026). *The Origin of Cellular Proto-Consciousness: Biological Foundations and Operational Criteria*: establishes the cell as the minimal candidate system for elementary consciousness. It serves as phenomenological support for Chapters 14 and 16 of this book; its restrained discipline of "functional / candidate / elementary" is inherited by the newly added four chapters.

F.7.2. The Hard Problem and the Mind–Matter Relation (Philosophy)

- Chalmers, D. J. (1995). "Facing Up to the Problem of Consciousness." The classical formulation of the hard problem.

- Chalmers, D. J. (1996). *The Conscious Mind.* A systematic elaboration of property dualism.
- Levine, J. (1983). "Materialism and Qualia: The Explanatory Gap." A classical work on the explanatory gap.
- Jackson, F. (1982). "Epiphenomenal Qualia." The Mary argument.
- Nagel, T. (1974). "What Is It Like to Be a Bat?" A classical work on subjective character.
- Russell, B. (1927). *The Analysis of Matter.* The origin of Russellian monism.
- Strawson, G. (2006). *Consciousness and Its Place in Nature.* Contemporary Russellian monism / panpsychism.
- Goff, P. (2017). *Consciousness and Fundamental Reality.* A systematic elaboration of contemporary Russellian monism.
- McGinn, C. (1989). "Can We Solve the Mind-Body Problem?" New mysterianism.
- Dennett, D. (1991). *Consciousness Explained.* The eliminativist / functionalist route.
- Frankish, K. (2016). "Illusionism as a Theory of Consciousness." Contemporary illusionism.

F.7.3. Major Theories in Consciousness Science

- Tononi, G. et al. (2016). "Integrated Information Theory: From Consciousness to Its Physical Substrate." The latest form of IIT.
- Dehaene, S. (2014). *Consciousness and the Brain.* Global Neuronal Workspace.
- Baars, B. J. (1988). *A Cognitive Theory of Consciousness.* The earliest systematic account of GWT.

- Rosenthal, D. M. (2005). *Consciousness and Mind.* Higher-order thought theory (HOT).
- Lau, H. & Rosenthal, D. (2011). "Empirical Support for Higher-Order Theories of Conscious Awareness." Empirical support for HOT.
- Friston, K. (2010). "The Free-Energy Principle: A Unified Brain Theory?" The free-energy principle.
- Clark, A. (2013). "Whatever Next? Predictive Brains, Situated Agents, and the Future of Cognitive Science." Predictive processing.
- Graziano, M. S. A. (2013). *Consciousness and the Social Brain.* Attention Schema Theory.
- Penrose, R. & Hameroff, S. (2014). "Consciousness in the Universe: A Review of the 'Orch OR' Theory." Quantum consciousness.

F.7.4. The Phenomenological Tradition

- Husserl, E. (1913). *Ideen I.* A classical work on given-ness, intentionality, and epoché.
- Husserl, E. (1928). *Vorlesungen zur Phänomenologie des inneren Zeitbewusstseins.* A classical work on inner time-consciousness—retention / protention (cited in Chapter 10 of this book).
- Heidegger, M. (1927). *Sein und Zeit.* Being-in-the-world.
- Merleau-Ponty, M. (1945). *Phénoménologie de la perception.* Embodiment.
- Zahavi, D. (2005). *Subjectivity and Selfhood.* Minimal self.
- Jonas, H. (1966). *The Phenomenon of Life.* A classical work in phenomenology of life.

- Thompson, E. (2007). *Mind in Life.* A systematic elaboration of contemporary phenomenology of life / embodied cognition.
- Varela, F., Thompson, E., & Rosch, E. (1991). *The Embodied Mind.* A dialogue between embodied cognition and Buddhist thought.

F.7.5. Experimental Research in Neuroscience and Consciousness

- Casali, A. G. et al. (2013). "A Theoretically Based Index of Consciousness Independent of Sensory Processing and Behavior." Foundational work on PCI (related to the proxy for the projection C in this book).
- Massimini, M. & Tononi, G. (2018). *Sizing Up Consciousness.* A systematic overview of PCI and conscious states.
- Owen, A. M. et al. (2006). "Detecting Awareness in the Vegetative State." A key work on covert consciousness (related to F24.3 in this book).
- Sperry, R. W. (1968). "Hemisphere Deconnection and Unity in Conscious Awareness." A classical work on split brain (related to Chapter 16 of this book).
- Gazzaniga, M. S. (2014). *Tales from Both Sides of the Brain.* An overview of split-brain research.

F.7.6. Free Will, Self, and Attribution (Related to the Newly Added Four Chapters)

- Frith, C. D. (2007). *Making Up the Mind.* On the neural basis of the sense of subjecthood and sense of agency (related to the perturbation axis in Chapter 11 and $\mathscr{B}$-demarcation in Chapter 19).
- Gallagher, S. (2000). "Philosophical Conceptions of the Self." An overview of philosophical conceptions of the self.

- Metzinger, T. (2003). *Being No One*. Self-model theory.
- Sass, L. A. & Parnas, J. (2003). "Schizophrenia, Consciousness, and the Self." Self-disturbance in schizophrenia (related to the failure modes in Section 19.4).

Conclusion of Appendix F: This appendix gathers in one place the comparison between the book and existing theories and philosophical positions, and appends the main references. Its core conclusion is consistent with Chapters 4, 22, and 27: this book stands between dualism and physicalism; it acknowledges itself as one branch of Russellian monism, but refuses its dissolution route; it converges with the phenomenological tradition at several points by different paths; it is largely compatible with contemporary consciousness science at the level of state indicators, while making genuine discriminating predictions at the level of ρ_f-specific predictions (frequency–quality and frequency–timescale); and, through the precise boundary-drawing of the two compressions, it avoids the various oversteps of "dissolving the hard problem through identity / interface / demarcation" (F18.5, F19.4, F20.4, F27.4, F27.5).

At this point, the appendices of the book—Appendix A (symbols and dimensions), Appendix B (formalization of isomorphism), Appendix C (formal model of co-evolution), Appendix D (parameter normalization), Appendix E (general table of failure conditions), and Appendix F (comparison with existing theories and references)—are all complete. Together with Chapters 1–28 of the main text and the closing remarks, they constitute the complete form of the book.

——The End of the Book——

www.ingramcontent.com/pod-product-compliance
Lightning Source LLC
LaVergne TN
LVHW052015200926
848481LV00005B/1494